Seeing Through Statistics

SECOND EDITION

Seeing Through Statistics

Jessica M. Utts

University of California, Davis

Duxbury Press

An Imprint of Brooks/Cole Publishing Company

I(T)P® *An International Thomson Publishing Company*

Pacific Grove • Albany • Belmont • Boston • Cincinnati • Detroit • Johannesburg • London
Madrid • Melbourne • Mexico City • New York • Scottsdale • Singapore • Tokyo • Toronto

Sponsoring Editor: *Carolyn Crockett*
Marketing Team: *Laura Hubrich and Deanne Brown*
Editorial Assistant: *Kimberly Raburn*
Production Editor: *Laurel Jackson*
Manuscript Editor: *Carol Dondrea*
Permissions Editor: *Catherine Gingras*
Interior and Cover Design: *Carolyn Deacy*

Cover Photo: *Zefa Germany / The Stock Market*
Art Editor: *Jennifer Mackres*
Interior Illustration: *Suffolk Technical Illustrators and Delgado Design, Inc.*
Typesetting: *TBH Typecast, Inc.*
Printing and Binding: *Malloy Lithographing*

For more information, contact:

BROOKS/COLE PUBLISHING COMPANY
511 Forest Lodge Road
Pacific Grove, CA 93950
USA

International Thomson Publishing Europe
Berkshire House 168-173
High Holborn
London WC1V 7AA
England

Thomas Nelson Australia
102 Dodds Street
South Melbourne 3205
Victoria, Australia

Nelson Canada
1120 Birchmount Road
Scarborough, Ontario
Canada M1K 5G4

International Thomson Editores
Seneca 53
Col. Polanco
11560 México, D. F., México

International Thomson Publishing GmbH
Königswinterer Strasse 418
53227 Bonn
Germany

International Thomson Publishing Asia
60 Albert Street
#15-01 Albert Complex
Singapore, 189969

International Thomson Publishing Japan
Hirakawacho Kyowa Building, 3F
2-2-1 Hirakawacho
Chiyoda-ku, Tokyo 102
Japan

Printed in the United States of America

10 9 8 7 6 5 4 3 2 1

Library of Congress Cataloging-in-Publication Data
Utts, Jessica M., [date]
   Seeing through statistics / Jessica M. Utts. — 2nd ed.
      p.   cm.
   "An International Thomson Publishing Company".
   Includes bibliographical references (p.   ) and index.
   ISBN 0-534-35786-5 (alk. paper)
   1. Statistics. I. Title.
QA276.12.U88   1999
001.4'22—dc21   98-40811
                              CIP

DEDICATED TO

**Patricia and her terrific grandchildren:
Jennifer, Jerald, Charles, Julia,
Patrick, Alex, and Shaun**

Contents

Preface

Statistics deals with complex situations involving uncertainty. We are exposed daily to information from surveys and scientific studies concerning our health, behavior, attitudes, and beliefs, or revealing scientific and technological breakthroughs. This book's first objective is to help you understand this information and to sift the useful and the accurate from the useless and the misleading. My aims are to allow you to rely on your own interpretation of results emerging from surveys and studies and to help you read them with a critical eye so that you can make your own judgments.

A second purpose of this book is to demystify statistical methods. Traditional statistics courses often place emphasis on how to compute rather than on how to understand. This book focuses on statistical ideas and their use in real life.

Finally, the book contains information that can help you make better decisions when faced with uncertainty. You will learn how psychological influences can keep you from making the best decisions, as well as new ways to think about coincidences, gambling, and other circumstances that involve chance events.

Approach

The focus of this book is on the use of statistical methods in the real world. There are dozens of real life, in depth case studies drawn from various media sources as well as scores of additional real-life examples. The emphasis is on understanding rather than computing, but the book also contains examples of how to compute important numbers when necessary.

Although this book is written as a textbook, it is also intended to be readable without the guidance of an instructor. Each concept or method is explained in plain language and is supported with numerous examples.

Organization

There are 25 chapters divided into four parts. Each chapter covers material more or less equivalent to a one-hour college lecture. The final chapters of Part 1 and Part 4 consist solely of case studies and are designed to illustrate the thought process you should follow when you read studies on your own.

By the end of Part 1, "Finding Data in Life," you will have the tools to determine whether or not the results of a study should be taken seriously; you will be able to detect false conclusions and biased results. In Part 2, "Finding Life in Data," you will learn how to turn numbers into useful information and to quantify relationships between such factors as aspirin consumption and heart attack rates or meditation and aging. You will also learn how to detect misleading graphs and figures and to interpret common economic statistics.

Part 3 is called "Understanding Uncertainty in Life" and is designed to help you do exactly that. Every day, we have to make decisions in the face of uncertainty. This part of the book will help you understand what probability and chance are all about and presents techniques that can help you make better decisions. The material on probability will also be useful when you read Part 4, "Making Judgments from Surveys and Experiments."

Part 4 is slightly more technical than the rest of the book, but once you have mastered it, you will truly understand the beauty of statistical methods. Henceforth, when you read the results of a statistical study, you will be able to tell whether the results represent valuable advice or flawed reasoning. Unless things have changed drastically by the time you read this, you will be amazed at the number of news reports that exhibit flawed reasoning.

Thought Questions: Using Your Common Sense

Each chapter, except those that consist solely of case studies, begins with a series of "thought questions" that are designed to be answered before you read the chapter. Most of the answers are based on common sense, perhaps combined with knowledge from previous chapters. Answering them before reading the chapter will reinforce the idea that most information in this book is based on common sense. You will find answers to the thought questions—or to similar questions—later in the chapter.

In the classroom, the thought questions can be used for discussion at the beginning of each class. For relatively small classes, groups of students can be assigned to discuss one question each, then to report back to the class. If you are taking a class in which one of these formats is used, try to answer the questions on your own before class. By doing so, you will build confidence as you learn that the material is not difficult to understand if you give it some thought.

Case Studies and Examples: Collect Your Own

The book is filled with real-life case studies and examples covering a wide range of disciplines. These studies and examples are intended to appeal to a broad audience. In the rare cases where technical subject-matter knowledge is required, it is given with the example. Sometimes, the conclusion presented in the book will be different from the one given in the original news report. This happens because many news reports misinterpret statistical results.

I hope you find the case studies and examples interesting and informative; however, you will learn the most by examining current examples on topics of interest to you. Follow any newspaper, news magazine, or Internet news site for a while, and you are sure to find plenty of illustrations of the use of surveys and studies. If you start collecting them now, you can watch your understanding of them increase as you work your way through this book.

Formulas: It's Your Choice

If you dread mathematical formulas, you should find this book comfortably readable. In most cases where computations are required, they are presented step by step rather than in a formula. The steps are accompanied by worked examples so that you can see exactly how to carry them out.

On the other hand, if you prefer to work with formulas, each relevant chapter ends with a section called "For Those Who Like Formulas." The section includes all the mathematical notation and formulas pertaining to the material in that chapter.

Exercises and Mini-Projects

Numerous exercises appear at the end of each chapter. Many of them are similar to the "thought questions" and require an explanation for which there is no one correct answer. Answers to some of those with concise solutions are provided at the back of the book. *Teaching Seeing Through Statistics: An Instructor's Resource Manual,* which is available to instructors, explains what is expected for each exercise.

In most chapters, the exercises contain many real-life examples. However, with the idea that you learn best by doing, most chapters also contain "mini-projects." Some of these ask you to find examples of studies of interest to you; others ask you to conduct your own small-scale study. If you are reading this book without the benefit of a class or instructor, I encourage you to try some of the projects on your own.

Covering the Book in a Quarter, in a Semester, or on Your Own

I wrote this book for a one-quarter course taught three times a week at the University of California at Davis as part of the general education curriculum. My aim was to allow one lecture for each chapter, and in most cases, that practice works quite well. A few chapters may spill over into two lectures, still allowing for completion of the book (and a midterm or two) in the usual 29- or 30-lecture quarter. When I teach the course, I do not cover every detail from each chapter; I expect students to read some material on their own.

If the book is used for a semester course, it can be covered at a more leisurely pace and in more depth. For instance, two classes a week can be used for covering new material and a third class for discussion, additional examples, or laboratory work. Alternatively, with three regular lectures a week, some chapters can be covered in two sessions instead of one.

Instructors can obtain a copy of *Teaching Seeing Through Statistics: An Instructor's Resource Manual,* which contains additional information on how to cover the material in one quarter or semester. The manual also includes tips on teaching this material, ideas on how to cover each chapter, sample lectures, additional examples, and exercise solutions.

Instructors who want to focus on more in-depth coverage of specific topics may wish to exclude others. Certain chapters can be omitted without serious consequences in later chapters. These include Chapter 9, Chapters 13 and 14 (but Chapter 14 relies on Chapter 13), Chapter 16, and Chapter 24. The chapters in Part 3 (15 through 17) rely on each other, but only some sections of Chapter 15 are needed to understand Part 4 of the book. Thus, Chapters 16 and 17 (as a unit) could be omitted.

If you are reading this book on your own, you may want to concentrate on selected topics only. Parts 1 and 3 can be read alone, as can Chapters 9 and 13. Part 4 relies most heavily on Chapters 8, 12, and 15. Although Part 4 is the most technically challenging part of the book, I strongly recommend reading it because it is there that you will truly learn the beauty as well as the pitfalls of statistical reasoning. If you get stuck, try to step back and reclaim the big picture. Remember that although statistical methods are very powerful and are subject to abuse, they were developed using the collective common sense of researchers whose goal was to figure out how to find and interpret information to understand the world. They have done the hard work; this book is intended to help you make sense of it all.

About the Second Edition

A book like this one is probably only as interesting as the examples and stories it relates, so for the second edition, I have introduced numerous fresh examples and case studies. The book includes over 100 new exercises, many based on news stories. In the short time since the first edition appeared, Internet use has skyrocketed, and this edition contains examples from and references to web sites with some interesting data.

The most substantial structural change to the second edition is in Part 3. Using feedback from instructors, I have combined and altered Chapters 15 and 16 from the first edition to make the material more relevant to daily life and have moved some of that material to the subsequent two chapters (now Chapters 16 and 17). Box plots have been added to Chapter 7, Chapter 13 has been rewritten to reflect changes in the Consumer Price Index, and wording and data have been updated throughout the book as needed.

Web Site for <u>Seeing Through Statistics</u>

The Duxbury Resource Center for Statistical Literacy has been established for users of this book; the URL is **www.duxbury.com/utts_seeingthrustat.**

This site will include links to related sites and data sets, a discussion group, and other information of interest to users of this book.

Acknowledgments

I would like to thank Robert Heckard and William Harkness from the Statistics Department at Penn State University, as well as their students, for providing data collected in their classes. I would also like to thank the following individuals for reviewing the manuscript and for providing valuable insights: Mary L. Baggett, Florida State University; Paul Cantor, formerly of Lehman College, City University of New York; Deborah Delanoy, University of Edinburgh; Hariharan K. Iyer, Colorado State University; Richard G. Krutchkoff, Virginia Polytechnic Institute and State University; Scott Plous, Wesleyan University; Larry Ringer, Texas A&M University; Ralph R. Russo, University of Iowa; and Farroll T. Wright, University of Missouri. In addition, I want to express my appreciation to the following reviewers for their many helpful comments and suggestions: Dale Bowman, University of Mississippi; James Casebolt, Ohio University–Eastern Campus; Richard Krutchkoff, Virginia Polytechnic Institute and State University; Lawrence M. Lesser, University of Northern Colorado; Vivian Lew, University of California, Los Angeles; Lawrence Ries, University of Missouri–Columbia; Barb Rouse, University of Wyoming; Laura J. Simon, Pennsylvania State University; Eric Suess, California State University–Hayward; Larry Wasserman, Carnegie Mellon University; Sheila Weaver, University of Vermont; and Arthur B. Yeh, Bowling Green State University.

Finally, I want to thank my sisters, Claudia Utts-Smith and Melissa Utts, for helping me realize the need for this book; Alex Kugushev, former Publisher of Duxbury Press, for persisting until I agreed to write it (and beyond); Carolyn Crockett, Duxbury Press, for her encouragement and support during the writing of the second edition; and Robert Heckard, Penn State University, for instilling in me many years ago, by example, the enthusiasm for teaching and educating that led to the development of this material.

Jessica Utts

Seeing Through Statistics

Finding Data in Life

By the time you finish reading Part 1 of this book, you will be reading studies reported in the newspaper with a whole new perspective. In these chapters, you will learn how researchers should go about collecting information for surveys and experiments. You will learn to ask questions, such as who funded the research, that could be important in deciding whether the results are accurate and unbiased.

Chapter 1 is designed to give you some appreciation for how statistics helps to answer interesting questions. Chapters 2 to 5 provide an in-depth, behind-the-scenes look at how surveys and experiments are supposed to be done. In Chapter 6, you will learn how to tie together the information from the previous chapters, including seven steps to follow when reading about studies. These steps all lead to the final step, which is the one you should care about the most. You will have learned how to *determine whether the results of a study are meaningful enough to encourage you to change your lifestyle, attitudes, or beliefs.*

CHAPTER ONE

The Benefits and Risks of Using Statistics

THOUGHT QUESTIONS

1. A recent newspaper article concluded that smoking marijuana at least three times a week resulted in lower grades in college. How do you think the researchers came to this conclusion? Do you believe it? Is there a more reasonable conclusion?

2. It is obvious to most people that, on average, men are taller than women, and yet there are some women who are taller than some men. Therefore, if you wanted to "prove" that men were taller, you would need to measure many people of each sex. Here is a theory: On average, men have lower resting pulse rates than women do. How could you go about trying to prove or disprove that? Would it be sufficient to measure the pulse rates of one member of each sex? Two members of each sex? What information about men's and women's pulse rates would help you decide how many people to measure?

3. Suppose you were to learn that the large state university in a particular state graduated more students who eventually went on to become millionaires than any of the small liberal arts colleges in the state. Would that be a fair comparison? How should the numbers be presented in order to make it a fair comparison?

4. In its March 3–5, 1995 issue, *USA Weekend* magazine asked readers to return a survey with a variety of questions about sex and violence on television. Of the 65,142 readers who responded, 97% were "very or somewhat concerned about violence on TV" (*USA Weekend,* 2–4 June 1995, p. 5). Based on this survey, can you conclude that about 97% of U.S. citizens are concerned about violence on TV? Why or why not?

1.1   STATISTICS

When you hear the word *statistics*, you probably either get an attack of math anxiety or think about lifeless numbers, such as the population of the city or town where you live, as measured by the latest census, or the per capita income in Japan. The goal of this book is to open a whole new world of understanding of the term *statistics*. By the time you finish reading this book, you will realize that the invention of statistical methods is one of the most important developments of modern times. These methods influence everything from life-saving medical advances to which television shows remain on the air.

The word **statistics** is actually used to mean two different things. The better known definition is that statistics are numbers measured for some purpose. A more appropriate, complete definition is the following:

> *Statistics is a collection of procedures and principles for gaining and processing information in order to make decisions when faced with uncertainty.*

Using this definition, you have undoubtedly used statistics in your own life. For example, if you were faced with a choice of routes to get to school or work, or to get between one classroom building and the next, how would you decide which one to take? You would probably try each of them a number of times (thus gaining information) and then choose the best one according to some criterion important to you, such as speed, fewer red lights, more interesting scenery, and so on. You might even use different criteria on different days—such as when the weather is pleasant versus when it is not. In any case, by sampling the various routes and comparing them, you would have gained and processed useful information to help you make a decision.

In this book, you will learn ways to intelligently improve your own methods for collecting and processing complex information. You will learn how to interpret information that others have collected and processed and how to make decisions when faced with uncertainty. In Case Study 1.1, we will see how one researcher followed a casual observation to a fascinating conclusion.

CASE STUDY 1.1   ## Heart or Hypothalamus?

SOURCE: Salk (1973), pp. 26–29.

You can learn a lot about nature by observation. You can learn even more by conducting a carefully controlled experiment. This case study has both. It all began when psychologist Lee Salk noticed that, despite his knowledge that the hypothalamus plays an important role in emotion, it was the heart that seemed to occupy the thoughts of poets and songwriters. There were no everyday expressions or song titles such as "I love you from the bottom of my hypothalamus" or "My hypothalamus longs for you." Yet, there was no physiological reason for suspecting that the heart should be the center of such attention. Why had it always been the designated choice?

Salk began wondering about the role of the heart in human relationships. He also noticed that when, on 42 separate occasions, he watched a rhesus monkey at the zoo holding her baby, she held the baby on the left side, close to her heart, on 40 of those occasions. He then observed 287 human mothers within 4 days after giving birth and noticed that 237, or 83%, held their babies on the left. Handedness did not explain it; 83% of the right-handed mothers and 78% of the left-handed mothers exhibited the left-side preference. When asked why they chose the left side, the right-handed mothers said it was so their right hand would be free. The left-handed mothers said it was because they could hold the baby better with their dominant hand. In other words, both groups were able to rationalize holding the baby on the left based on their own preferred hand.

Salk wondered if the left side would be favored when carrying something other than a newborn baby. He found a study in which shoppers were observed leaving a supermarket carrying a single bag; exactly half of the 438 adults carried the bag on the left. But when stress was involved, the results were different. Patients at a dentist's office were asked to hold a 5-inch rubber ball while the dentist worked on their teeth. Substantially more than half held the ball on the left.

Salk speculated that "it is not in the nature of nature to provide living organisms with biological tendencies unless such tendencies have survival value." He surmised that there must indeed be survival value to having a newborn infant placed close to the sound of its mother's heartbeat.

To test this conjecture, Salk designed a study in a baby nursery at a New York City hospital. He arranged for the nursery to have the continuous sound of a human heartbeat played over a loudspeaker. At the end of 4 days, he measured how much weight the babies had gained or lost. Later, with a new group of babies in the nursery, no sound was played. Weight gains were again measured after 4 days.

The results confirmed what Salk suspected. Although they did not eat more than the control group, the infants treated to the sound of the heartbeat gained more weight (or lost less). Further, they spent much less time crying. Salk's conclusion was that "newborn infants are soothed by the sound of the normal adult heartbeat." Somehow, mothers intuitively know that it is important to hold their babies on the left side. What had started as a simple observation of nature led to a further understanding of an important biological response of a mother to her newborn infant. ■

1.2   DETECTING PATTERNS AND RELATIONSHIPS

Some differences are obvious to the naked eye, such as the fact that the average man is taller than the average woman. If we were content to know only about such obvious relationships, we would not need the power of statistical methods. But had you noticed that babies who listen to the sound of a heartbeat gain more weight? Have you ever noticed that taking aspirin helps prevent heart attacks? How about the fact that people are more likely to buy blue jeans in certain months of the year than in others? The fact that men have lower resting pulse rates than women do? The fact that listening to Mozart improves performance on the spatial reasoning questions of an IQ test? All of these are relationships that have been demonstrated in studies using proper statistical methods, yet none of them are obvious to the naked eye.

Let's take the simplest of these examples—one you can test yourself—and see what's needed to properly demonstrate the relationship. Suppose you wanted to verify the claim that, on average, men have lower resting pulse rates than women do. Would it be sufficient to measure only your own pulse rate and that of a friend of the opposite sex? Obviously not. Even if the pair came out in the predicted direction, the singular measurements would certainly not speak for all members of each sex.

It is not easy to conduct a study properly, but it is easy to understand much of how it should be done. We will examine each of the following concepts in great detail in the remainder of this book; here we just introduce them, using the simple example of comparing male and female pulse rates.

To conduct a study properly, one must

1. Get a representative sample.

2. Get a large enough sample.

3. Decide whether the study should be an observational study or an experiment.

---

1. Get a representative sample. Most researchers hope to extend their results beyond just the participants in their research. Therefore, it is important that the people or objects in a study be representative of the larger group for which conclusions are to be drawn. We call those who are actually studied a **sample** and the larger group from which they were chosen a **population.** (In Chapter 4 we will learn some ways to select a proper sample.) For comparing pulse rates, it may be convenient to use the members of your class. But this sample would not be valid if there were something about your class that would relate pulse rates and sex, such as if the entire men's track team happened to be in the class. It would also be unacceptable if you wanted to extend your results to an age group much different from the distribution of ages in your class. Often researchers are constrained to using such "convenience" samples, and we will discuss the implications of this later in the book.

2. Get a large enough sample. Even experienced researchers often fail to recognize the importance of this concept. In Part 4 of this book, you will learn how to detect the problem of a sample that is too small; you will also learn that such a sample can sometimes lead to erroneous conclusions. In comparing pulse rates, collecting one pulse rate from each sex obviously does not tell us much. Is two enough? Four? One hundred? The answer to that question depends on how much *variability* there is among pulse rates. If all men had pulse rates of 65 and all women had pulse rates of 75, it wouldn't take long before you recognized a difference. However, if men's pulse rates ranged from 50 to 80 and women's pulse rates ranged from 52 to 82, it would take many more measurements to convince you of a difference. The question of how large is "large enough" is closely tied to how diverse the measurements are likely to be within each group. The more diverse, or variable, the individuals within each group, the larger the sample needs to be to detect a real difference between the groups.

3. Decide whether the study should be an observational study or an experiment. For comparing pulse rates, it would be sufficient to measure or "observe" both the pulse rate and the sex of the people in our sample. When we merely observe things about our sample, we are conducting an **observational study.** However, if we were interested in whether frequent use of aspirin would help prevent heart attacks (which has been suggested as a likely possibility), it would not be sufficient to simply observe whether people frequently took aspirin and then whether they had a heart attack. It could be that people who were more concerned with their health were both more likely to take aspirin and less likely to have a heart attack, or vice versa.

To be able to make a causal connection, we would have to conduct an **experiment** in which we *randomly* assigned people to one of two groups. **Random assignments** are made by doing something akin to flipping a coin to determine the group membership for each person. In one group, people would be given aspirin and, in the other, they would be given a dummy pill that looked like aspirin. So as not to influence people with our expectations, we would not tell people which one they were taking until the experiment was concluded. In Case Study 1.2, we briefly examine such an experiment; in Chapter 5 we discuss these ideas in much more detail.

CASE STUDY 1.2   Does Aspirin Prevent Heart Attacks?

In 1988, the Steering Committee of the Physicians' Health Study Research Group released the results of a 5-year experiment conducted using 22,071 male physicians between the ages of 40 and 84. The physicians had been randomly assigned to two groups. One group took an ordinary aspirin tablet every other day, whereas the other group took a "placebo," a pill designed to look just like an aspirin but with no active ingredients. Neither group knew whether they were taking the active ingredient.

The results, shown in Table 1.1, support the conclusion that taking aspirin does indeed help reduce the risk of having a heart attack. The rate of heart attacks in the group taking aspirin was only 55% of the rate of heart attacks in the placebo group, or just slightly more than half as big. Because the men were randomly assigned to the two conditions, other factors, such as amount of exercise, should have been similar for both groups. The only substantial difference in the two groups should have been whether they took the aspirin or the placebo. Therefore, we can conclude that taking aspirin caused the lower rate of heart attacks for that group.

Notice that, because the participants were all male physicians, these conclusions may not apply to the general population of men. They may not apply to women at all because no women were included in the study. More recent evidence has provided even more support for this effect, however, something we will examine in more detail in an example in Chapter 25. ■

TABLE 1.1	The Effect of Aspirin on Heart Attacks		
Condition	Heart Attack	No Heart Attack	Attacks per 1000
Aspirin	104	10,933	9.42
Placebo	189	10,842	17.13

1.3   DON'T BE DECEIVED BY IMPROPER USE OF STATISTICS

Let's look at some examples representative of the kinds of abuses of statistics you may see in the media. In the first example, the simple principles we have been discussing were violated; in the second example, the statistics have been taken out of their proper context; and in the third and fourth examples, you will see how to stop short of making too strong a conclusion on the basis of an observational study.

EXAMPLE 1

In 1986, a business-oriented magazine published in Washington, D.C., conducted a survey that concluded that Chrysler president Lee Iacocca would beat Vice-President George Bush in a Republican primary by a margin of 54% to 47% [sic]. Further reading revealed that the poll was based on questionnaires mailed to 2000 of the magazine's readers, surely a biased sample of American voters. To make matters worse, the results were compiled from only the first 200 respondents. It should not surprise you to learn that those who feel strongly about an issue, especially those who would like to see a change, are most likely to respond to a survey received in the mail. Therefore, the "sample" was not at all representative of the "population" of all people likely to vote in a Republican primary election. (In the next election year, 1988, George Bush not only won the Republican primary, but went on to win the presidential election with 54% of the popular vote.) ■

EXAMPLE 2

When a federal air report ranked the state of New Jersey as 22nd in the nation in its release of toxic chemicals, the New Jersey Department of Environmental Protection happily took credit (Wang, 1993, p. 170). The statistic was based on a reliable source, a study by the U.S. Environmental Protection Agency. However, the ranking had been made based on total pounds released, which was 38.6 million for New Jersey. When this total was turned into pounds per square mile in the state, New Jersey, it became apparent, was one of the worst—fourth on the list. Because New Jersey is one of the smallest states by area, the figures were quite misleading until adjusted for size. ■

FIGURE 1.1

*Don't Make Causal
Connections from
Observational Studies*

Source: Davis (CA)
Enterprise, 11 Feb. 1994.

STUDY: SMOKING MAY LOWER KIDS' IQS

ROCHESTER, N.Y. (AP)—Secondhand smoke has little impact on the intelligence scores of young children, researchers found.

But women who light up while pregnant could be dooming their babies to lower IQs, according to a study released Thursday.

Children ages 3 and 4 whose mothers smoked 10 or more cigarettes a day during pregnancy scored about 9 points lower on the intelligence tests than the offspring of nonsmokers, researchers at Cornell University and the University of Rochester reported in this month's *Pediatrics* journal.

That gap narrowed to 4 points against children of nonsmokers when a wide range of interrelated factors were controlled. The study took into account secondhand smoke as well as diet, education, age, drug use, parents' IQ, quality of parental care and duration of breast feeding.

"It is comparable to the effects that moderate levels of lead exposure have on children's IQ scores," said Charles Henderson, senior research associate at Cornell's College of Human Ecology in Ithaca.

EXAMPLE 3

Read the article in Figure 1.1, and then read the headline again. Notice that the headline stops short of making a causal connection between smoking during pregnancy and lower IQs in children. Reading the article, you can see that the results are based on an observational study and not an experiment—with good reason: It would clearly be unethical to randomly assign pregnant women to either smoke or not. With studies like this, the best that can be done is to try to measure and statistically adjust for other factors that might be related to both smoking behavior and children's IQ scores. Notice that when the researchers did so, the gap in IQ between the children of smokers and nonsmokers narrowed from 9 points down to 4 points. There may be even more factors that the researchers did not measure that would account for the remaining 4-point difference. Unfortunately, with an observational study, we simply cannot make causal conclusions. We will explore this particular example in more detail in Chapter 6. ■

EXAMPLE 4

An article headlined "New study confirms too much pot impairs brain" read as follows:

*More evidence that chronic marijuana smoking impairs mental ability:
Researchers at the University of Iowa College of Medicine say a test shows
those who smoke seven or more marijuana joints per week had lower math,*

verbal and memory scores than non-marijuana users. Scores were particularly reduced when marijuana users held a joint's smoke in their lungs for longer periods.— San Francisco Examiner, *13 March 1993, p. D-1.*

This research was clearly based on an observational study because people cannot be randomly assigned to either smoke marijuana or not. The headline is misleading because it implies that there is a causal connection between smoking marijuana and brain functioning. All we can conclude from an observational study is that there is a relationship. It could be the case that people who choose to smoke marijuana are those who would score lower on the tests anyway. ∎

CASE STUDY 1.3   ## A Mistaken Accusation of Cheating

Klein (1992) described a situation in which two students were accused of cheating on a multiple-choice medical licensing exam. They had been observed whispering during one part of the 3-day exam and their answers to the questions they got wrong very often matched each other. The licensing board determined that the statistical evidence for cheating was overwhelming. They estimated that the odds of two people having answers as close as these two did were less than 1 in 10,000. Further, the students were husband and wife. Their tests were invalidated.

The case went to trial, and upon further investigation the couple was exonerated. They hired a statistician who was able to show that the agreement in their answers during the session in which they were whispering was no higher than it was in the other sessions. What happened? The board assumed students who picked the wrong answer were simply guessing among the other choices. This couple had grown up together and had been educated together in India. Answers that would have been correct for their culture and training were incorrect for the American culture (for example, whether a set of symptoms was more indicative of tuberculosis or a common cold). Their common mistakes often would have been the right answers for India. So, the licensing board erred in calculating the odds of getting such a close match by using the assumption that they were just guessing. And, according to Klein, "with regard to their whispering, it was very brief and had to do with the status of their sick child" (p. 26). ∎

1.4   SUMMARY AND CONCLUSIONS

In this chapter, we have just begun to examine both the advantages and the dangers of using statistical methods. We have seen that it is not enough to know the results of a study, survey, or experiment. We also need to know how those numbers were collected and who was asked. In the upcoming chapters, you will learn much more about how to collect and process this kind of information properly and how to detect problems in what others have done. You will learn that a relationship between two characteristics (such as smoking marijuana and lower grades) does not necessarily mean that one causes the other, and you will learn how to determine other plausible explanations. In short, you will become an educated consumer of statistical information.

EXERCISES

1. Explain why the relationship shown in Table 1.1, concerning the use of aspirin and heart attack rates, can be used as evidence that aspirin actually prevents heart attacks.

2. "People who often attend cultural activities, such as movies, sports events and concerts, are more likely than their less cultured cousins to survive the next eight to nine years, even when education and income are taken into account, according to a survey by the University of Umea in Sweden" (*American Health,* April 1997, p. 20).

   a. Can this claim be tested by conducting a randomized experiment? Explain.

   b. On the basis of the study that was conducted, can we conclude that attending cultural events causes people to be likely to live longer? Explain.

   c. The article continued "No one's sure how Mel Gibson and Mozart help health, but the activities may enhance immunity or coping skills." Comment on the validity of this statement.

   d. The article notes that education and income were taken into account. Give two other factors about the people surveyed that you think should also have been taken into account.

3. Explain why the number of people in a sample is an important factor to consider when designing a study.

4. Explain what problems arise in trying to make conclusions based on a survey mailed to the subscribers of a specialty magazine. Find or construct an example.

5. "If you have borderline high blood pressure, taking magnesium supplements may help, Japanese researchers report. Blood pressure fell significantly in subjects who got 400–500 milligrams of magnesium a day for four weeks, but not in those getting a placebo" (*USA Weekend,* 22–24 May 1998, p. 11).

   a. Do you think this was a randomized experiment or an observational study? Explain.

   b. Do you think the relationship found in this study is a causal one, in which taking magnesium actually causes blood pressure to be lowered? Explain.

6. Refer to Case Study 1.1. When Salk measured the results, he divided the babies into three groups based on whether they had low (2510 to 3000 g), medium (3010 to 3500 g), or high (3510 g and over) birthweights. He then compared the infants from the heartbeat and silent nurseries separately within each birthweight group. Why do you think he did that? (*Hint:* Remember that it would be easier to detect a difference in male and female pulse rates if all males measured 65 beats per minute and all females measured 75 than it would be if both groups were quite diverse.)

7. A psychology department is interested in comparing two methods for teaching introductory psychology. Four hundred students plan to enroll for the course at 10:00 A.M. and another 200 plan to enroll for the course at 4:00 P.M. The registrar will allow the department to assign students to multiple sections at each time slot, if they so desire. Design a study to compare the two teaching methods. For example, would it be a good idea to use one method on all of the 10:00 sections and the other method on all of the 4:00 sections? Explain your reasoning.

8. Suppose you have a choice of two grocery stores in your neighborhood. Because you hate waiting, you want to choose the one for which there is generally a shorter wait in the checkout line. How would you gather information to determine which one is faster? Would it be sufficient to visit each store once and time how long you had to wait in line? Explain.

9. Suppose researchers want to know whether smoking cigars increases the risk of esophageal cancer.

   a. Could they conduct a randomized experiment to test this? Explain.

   b. If they conducted an observational study and found that cigar smokers had a higher rate of esophageal cancer than those who did not smoke cigars, could they conclude that smoking cigars increases the risk of esophageal cancer? Explain why or why not.

10. Universities are sometimes ranked for prestige according to the amount of research funding their faculty members are able to obtain from outside sources. Explain why it would not be fair to simply use total dollar amounts for each university, and describe what should be used instead.

11. Refer to Case Study 1.3, in which two students were accused of cheating because the licensing board determined the odds of such similar answers were less than 1 in 10,000. Further investigation revealed that over 20% of all pairs of students had matches giving low odds like these (Klein, 1992, p. 26). Clearly, something was wrong with the method used by the board. Read the case study and explain what erroneous assumption they made in their determination of the odds. (*Hint:* Use your own experience with answering multiple-choice questions.)

12. Suppose the officials in the city or town where you live would like to ask questions of a "representative sample" of the adult population. Explain some of the characteristics this sample should have. For example, would it be sufficient to include only homeowners?

13. Suppose you have 20 tomato plants and want to know if fertilizing them will help them produce more fruit. You randomly assign ten of them to receive fertilizer and the remaining ten to receive none. You otherwise treat the plants in an identical manner.

   a. Explain whether this would be an observational study or an experiment.

   b. If the fertilized plants produce 30% more fruit than the unfertilized plants, can you conclude that the fertilizer caused the plants to produce more? Explain.

14. Give an example of a decision in your own life, such as which route to take to school, for which you think statistics would be useful in making the decision. Explain how you could collect and process information to help make the decision.

15. National polls are often conducted by asking the opinions of a few thousand adults nationwide and using them to infer the opinions of all adults in the nation. Explain who is in the sample and who is in the population for such polls.

16. Sometimes television news programs ask viewers to call and register their opinions about an issue. One number is to be called for a "yes" opinion and another number for a "no" vote. Do you think viewers who call are a representative sample of all viewers? Explain.

17. Suppose a study first asked people whether they meditate regularly and then measured their blood pressures. The idea would be to see if those who meditate have lower blood pressure than those who do not do so.

    a. Explain whether this would be an observational study or an experiment.

    b. If it were found that meditators had lower than average blood pressures, can we conclude that meditation *causes* lower blood pressure? Explain.

18. Suppose a researcher would like to determine whether one grade of gasoline produces better gas mileage than another grade. Twenty cars are randomly divided into two groups, with ten cars receiving one grade and ten receiving the other. After many trips, average mileage is computed for each car.

    a. Would it be easier to detect a difference in gas mileage for the two grades if the 20 cars were all the same size, or would it be easier if they covered a wide range of sizes and weights? Explain.

    b. What would be one disadvantage to using cars that were all the same size?

19. Suppose the administration at your school wants to know how students feel about a policy banning smoking on campus. Because they can't ask all students, they must rely on a sample.

    a. Give an example of a sample they could choose that would *not* be representative of all students.

    b. Explain how you think they could get a representative sample.

20. A newspaper headline read "Study finds walking a key to good health: Six brisk outings a month cut death risk." Comment on what type of study you think was done and whether this is a good headline.

MINI-PROJECTS

1. Design and carry out a study to test the proposition that men have lower resting pulse rates than women.

2. Find a newspaper article that discusses a recent study. Identify it as either an observational study or an experiment. Comment on how well the simple concepts discussed in this chapter have been applied in the study. Comment on whether the news article, including the headline, accurately reports the conclusions that can legitimately be made from the study. Finally, discuss whether any information is missing from the news article that would have helped you answer the previous questions.

REFERENCES

Klein, Stephen P. (1992). Statistical evidence of cheating on multiple-choice tests. *Chance* 5, no. 3–4, pp. 23–27.

Salk, Lee. (May 1973). The role of the heartbeat in the relations between mother and infant. *Scientific American,* pp. 26–29.

Steering Committee. Physicians' Health Study Research Group. (28 January 1988). Preliminary report: Findings from the aspirin component of the ongoing Physicians' Health Study. *New England Journal of Medicine* 318, no. 4, pp. 262–264.

Wang, Chamont. (1993). *Sense and nonsense of statistical inference.* New York: Marcel Dekker.

CHAPTER TWO

Reading the News

THOUGHT QUESTIONS

1. Advice columnist Ann Landers often asks readers to write and tell her their feelings about certain topics, such as whether they think engineers make good husbands. Do you think the responses she gets are representative of public opinion? Explain why or why not.

2. Taste tests of new products are often done by having people taste both the new product and an old familiar standard. Do you think the results would be biased if the person handing the products to the respondents knew which was which? Explain why or why not.

3. Nicotine patches are patches attached to the arm of someone who is trying to quit smoking that dispense nicotine into the blood. Suppose you read about a study showing that nicotine patches were twice as effective in getting people to quit smoking as "control" patches (made to look like the real thing). Further, suppose you are a smoker trying to quit. What questions would you want answered about the study before you decided whether to try the patches yourself?

4. For a door-to-door survey on opinions about various political issues, do you think it matters who conducts the interviews? Give an example of how it might make a difference.

2.1   THE EDUCATED CONSUMER OF DATA

Pick up any newspaper or newsmagazine and you are almost certain to find a story containing conclusions based on data. Should you believe what you read? Not always. It depends on how the data were collected, measured, and summarized. In this chapter, we discuss seven critical components of statistical studies. We examine the kinds of questions you should ask before you believe what you read. We go into further detail about these issues in subsequent chapters. The goal in this chapter is to give you an overview of how to be a more educated consumer of the data you encounter in your everyday life.

What Are Data?

In statistical parlance, **data** is a plural word referring to a collection of numbers or other pieces of information to which meaning has been attached. For example, the numbers 1, 3, and 10 are not necessarily data, but they become so when we are told that these were the weight gains in grams of three of the infants in Salk's heartbeat study, discussed in Chapter 1. In Case Study 1.2, the data consisted of two pieces of information measured on each participant: (1) whether they took aspirin or a placebo, and (2) whether they had a heart attack.

Don't Always Believe What You Read

When you read the results of a study in the newspaper, you are rarely presented with the actual data. Someone has usually summarized the information for you, and he or she has probably already drawn conclusions and presented them to you. Don't always believe them. The meaning we can attach to data, and to the resulting conclusions, depends on how well the information was acquired and summarized.

In the remaining chapters of Part 1, we look at proper ways to obtain data. In Part 2, we turn our attention to how it should be summarized. In Part 4, we learn the power as well as the limitations of using the data collected from a sample to make conclusions about the larger population. In the rest of this chapter, we address seven features of statistical studies that you should think about when you read a news article. You will begin to be able to think critically and make your own conclusions about what you read.

2.2   HOW TO BE A STATISTICS SLEUTH: SEVEN CRITICAL COMPONENTS

Reading and interpreting the results of surveys or experiments is not much different from reading and interpreting the results of other events of interest, such as sports competitions or criminal investigations. If you are a sports fan, then you know what information should be included in reports of competitions, and you know when crucial information is missing. If you have ever been involved in an event that was later reported in the newspaper, you know that missing information can lead readers to erroneous conclusions.

In this section, you are going to learn what information should be included in news reports of statistical studies. Unfortunately, crucial information is often missing. With some practice, you can learn to figure out what's missing, as well as how to interpret what's reported. You will no longer be at the mercy of someone else's conclusions. You will be able to determine them yourself.

To provide structure to our examination of news reports, let's list Seven Critical Components that determine the soundness of statistical studies. A good news report should provide you with information about all of the components that are relevant to that study.

Component 1: The *source* of the research and of the *funding*.

Component 2: The *researchers* who had *contact* with the participants.

Component 3: The *individuals* or objects studied and how they were *selected*.

Component 4: The exact nature of the *measurements* made or *questions* asked.

Component 5: The *setting* in which the measurements were taken.

Component 6: The *extraneous differences* between groups being compared.

Component 7: The *magnitude* of any claimed effects or differences.

Before delving into some examples, let's examine each component more closely. You will find that most of the problems with studies are easy to identify. Listing these components simply provides a framework for using your common sense.

Component 1: The *source* of the research and of the *funding* Studies are conducted for three major reasons. First, governments and private companies need to have data in order to make wise policy decisions. Information such as unemployment rates and consumer spending patterns are measured for this reason. Second, researchers at universities and other institutes are paid to ask and answer interesting questions about the world around us. The curious questioning and experimentation of such researchers have resulted in many social, medical, and scientific advances. Much of this research is funded by government agencies, such as the National Institutes of Health. Third, companies want to convince consumers that their programs and products work better than the competition, or special-interest groups want to prove that their point of view is held by the majority.

Unfortunately, it is not always easy to discover who funded research. Many university researchers are now funded by private companies. In her book *Tainted Truth* (1994), Cynthia Crossen warns us:

> *Private companies, meanwhile, have found it both cheaper and more prestigious to retain academic, government, or commercial researchers than to set up in-house operations that some might suspect of fraud. Corporations, litigants,*

political candidates, trade associations, lobbyists, special interest groups—all can buy research to use as they like. (p. 19)

If you discover that a study was funded by an organization that would be likely to have a strong preference for a particular outcome, it is especially important to be sure that correct scientific procedures were followed. In other words, be sure the remaining components have sound explanations.

Component 2: The *researchers* who had *contact* with the participants It is important to know who actually had contact with the participants and what message those people conveyed. Participants often give answers or behave in ways to comply with the desires of the researchers. Consider, for example, a study done at a shopping mall to compare a new brand of a certain product to an old familiar brand. Shoppers are asked to taste each brand and state their preference. It is crucial that both the person presenting the two brands and the respondents be kept entirely blind as to which is which until after the preferences have been selected. Any clues might bias the respondent to choose the old familiar brand. Or, if the interviewer is clearly eager to have them choose one brand over the other, the respondents will most likely oblige in order to please. As another example, if you discovered that a study on the prevalence of illegal drug use was conducted by sending uniformed police officers door to door, you would probably not have much faith in the results. We will discuss other ways in which researchers influence participants in Chapters 4 and 5.

Component 3: The *individuals* or objects studied and how they were *selected* It is important to know to whom the results can be extended. For example, until recently, many medical studies included men only, so the results were of little value to women. Many studies rely on volunteers recruited through the newspaper, who are usually paid a small amount for their participation. People who would respond to such recruitment efforts may differ in relevant ways from those who would not. Studies relying on voluntary responses are likely to be biased because only those who feel strongly about the issues are likely to respond.

Component 4: The exact nature of the *measurements* made or *questions* asked As you will see in Chapter 3, precisely defining and measuring most of the things researchers study isn't easy. For example, if you wanted to measure whether people "eat breakfast," how would you do so? What if they just have juice? What if they work until midmorning and then eat a meal that satisfies them until dinner? You need to understand exactly what the various definitions mean when you read about someone else's measurements.

In polls and surveys, the "measurements" are usually answers to specific questions. Both the wording and the ordering of the questions can influence answers. For example, a question about "street people" would probably elicit different responses than a question about "families who have no home." Ideally, you should be given the exact wording that was used in a survey or poll,

Component 5: The *setting* in which the measurements were taken The setting in which measurements were taken includes factors such as when and where they were taken and whether respondents were contacted by phone, mail, or in person. A study can be easily biased by timing. For example, opinions on whether criminals should be locked away for life may change drastically following a highly publicized murder or kidnapping case. If a study is conducted by telephone and calls are made only in the evening, certain groups of people would be excluded, such as those who work the evening shift or who routinely eat dinner in restaurants.

Where the measurements were taken can also influence the results. Questions about sensitive topics, such as sexual behavior or income, might be more readily answered over the phone, where respondents feel more anonymous. Sometimes research is done in a laboratory or university office, and the results may not readily extend to a natural setting. For example, studies of communication between two people are sometimes done by asking them to conduct a conversation in a university office with a tape recorder present. Such conditions almost certainly produce more limited conversation than would occur in a more natural setting.

Component 6: The *extraneous differences* between groups being compared Often, when groups exhibit a difference in some feature of interest, researchers try to attribute the difference to membership in the group. For example, if the group of people who smoke marijuana has lower test scores than the group of people who doesn't, researchers may conclude that the lower test scores are due to smoking marijuana. Often, however, other disparities in the groups can explain the observed difference just as well. For example, people who smoke marijuana may simply be the type of people who are less motivated to study and thus would score lower on tests whether they smoked or not. Reports of research should include an explanation of any such extraneous differences that might account for the results. We will explore the issue of extraneous factors, and how to control for them, in much more detail in Chapter 5.

Component 7: The *magnitude* of any claimed effects or differences Media reports about statistical studies often fail to tell you how large the observed effects were. Without that knowledge, it is hard for you to assess whether you think the results are of any practical importance. For example, if, based on Case Study 1.2, you were told simply that taking aspirin every other day reduced the risk of heart attacks, you would not be able to determine whether it would be worthwhile to take aspirin. You should instead be told that for the men in the study, the rate was reduced from about 17 heart attacks per 1000 participants without aspirin to about 9.4 heart attacks per 1000 with aspirin. Often news reports simply report that a treatment had an effect or that a difference was observed. We will investigate this issue in great detail in Part 4 of this book.

2.3   FOUR HYPOTHETICAL EXAMPLES OF BAD REPORTS

Throughout this book, you will see numerous examples of real studies and news reports. So that you can get some practice finding problems without having to read

HYPOTHETICAL
NEWS ARTICLE 1

STUDY SHOWS PSYCHOLOGY MAJORS ARE SMARTER THAN CHEMISTRY MAJORS

A fourth-year psychology student, for her senior thesis, conducted a study to see if students in her major were smarter than those majoring in chemistry. She handed out questionnaires in five advanced psychology classes and five advanced chemistry labs. She asked the students who were in class to record their grade-point averages (GPAs) and their majors. Using the data only from those who were actually majors in these fields in each set of classes, she found that the psychology majors had an average GPA of 3.05, whereas the chemistry majors had an average GPA of only 2.91. The study was conducted last Wednesday, the day before students went home to enjoy Thanksgiving dinner.

unnecessarily long news articles, let's examine some hypothetical reports. These are admittedly more problematic than many real reports because they serve to illustrate several difficulties at once.

Read each article and see if your common sense gives you some reasons why the headline is misleading. Then proceed to read the commentary about the Seven Critical Components.

Hypothetical News Article 1: "Study Shows Psychology Majors Are Smarter Than Chemistry Majors"

Component 1: The *source* of the research and of the *funding* The study was a senior thesis project conducted by a psychology major. Presumably, it was cheap to run and was paid for by the student. One could argue that she would have a reason to want the results to come out as they did, although with a properly conducted study, the motives of the experimenter should be minimized. As we shall see, there were additional problems with this study.

Component 2: The *researchers* who had *contact* with the participants Presumably, only the student conducting the study had contact with the respondents. Crucial missing information is whether she told them the purpose of the study. Even if she did not tell them, many of the psychology majors may have known her and known what she was doing. Any clues as to desired outcomes on the part of experimenters can bias the results.

Component 3: The *individuals* or objects studied and how they were *selected* The individuals selected are the crux of the problem here. The measurements were taken on advanced psychology and chemistry students, which would have been fine if they had been sampled correctly. However, only those who were in the psychology classes or in the chemistry labs that day were actually measured. Less conscientious students are more likely to leave early before a holiday, but a missed class is probably easier to make up than a missed lab. Therefore, perhaps a larger proportion of the students with low grade-point averages were absent from the psychology classes than from the chemistry labs. Due to the missing students, the investigator's results would overestimate the average GPA for psychology students more so than for chemistry students.

Component 4: The exact nature of the *measurements* made or *questions* asked Students were asked to give a "self-report" of their grade-point averages. A more accurate method would have been to obtain this information from the registrar at the university. Students may not know their exact grade-point average. Also, one group may be more likely to know the exact value than the other. For example, if many of the chemistry majors were planning to apply to medical school in the near future, they may be only too aware of their grades. Further, the headline implies that GPA is a measure of intelligence. Finally, the research assumes that GPA is a standard measure. Perhaps grading is more competitive in the chemistry department.

Component 5: The *setting* in which the measurements were taken Notice that the article specifies that the measurements were taken on the day before a major holiday. Unless the university consisted mainly of commuters, many students may have left early for the holiday, further aggravating the problem that the students with lower grades were more likely to be missing from the psychology classes than from the chemistry labs. Further, because students turned in their questionnaires anonymously, there was presumably no accountability for incorrect answers.

Component 6: The *extraneous differences* between groups being compared It is difficult to know what differences might exist without knowing more about the particular university. For example, because psychology is such a popular major, at some universities students are required to have a certain GPA before they are admitted to the major. A university with a separate premedical major might have the best of the science students enrolled in that major instead of chemistry. Those kinds of extraneous factors would be relevant to interpreting the results of the study.

Component 7: The *magnitude* of any claimed effects or differences The news report does present this information. Additional useful information would be to know how many students were included in each of the averages given, what percentage of all students in each major were represented in the sample, and how much variation there was among GPAs within each of the two groups.

Hypothetical News Article 2: "Per Capita Income of U.S. Shrinks Relative to Other Countries"

Component 1: The *source* of the research and of the *funding* We are told nothing except the name of the group that conducted the study, which should be fair warning.

PER CAPITA INCOME OF U.S. SHRINKS RELATIVE TO OTHER COUNTRIES

An independent research group, the Institute for Foreign Investment, has noted that the per capita income of Americans has been shrinking relative to some other countries. Using per capita income figures from the *World Almanac* and exchange rates from last Friday's financial pages, the organization warned that per capita income for the United States has risen only 10% during the past 5 years, whereas per capita income for certain other countries has risen 50%. The researchers concluded that more foreign investment should be allowed in the United States to bolster the sagging economy.

Being called "an independent research group" in the story does not mean that it is an unbiased research group. In fact, the last line of the story illustrates the probable motive for their research.

Component 2: The *researchers* who had *contact* with the participants This component is not relevant because there were no participants in the study.

Component 3: The *individuals* or objects studied and how they were *selected*
The objects in this study were the countries used for comparison with the United States. We should have been told which countries were used, and why.

Component 4: The exact nature of the *measurements* made or *questions* asked
This is the major problem with this study. First, as mentioned, we are not even told which countries were used for comparison. Second, current exchange rates but older per capita income figures were used. If the rate of inflation in a country had recently been very high, so that a large rise in per capita income did not reflect a concomitant rise in spending power, then we should not be surprised to see a large increase in per capita income in terms of actual dollars. In order to make a valid comparison, all figures would have to be adjusted to comparable measures of spending power, taking inflation into account. We will learn how to do that in Chapter 13.

**Components 5, 6, and 7: The *setting* in which the measurements were taken.
The *extraneous differences* between groups being compared. The *magnitude* of
any claimed effects or differences** These issues are not relevant here, except as they have already been discussed. For example, although the magnitude of the difference between the United States and the other countries is reported, it is meaningless without an inflation adjustment.

RESEARCHERS FIND DRUG TO CURE EXCESSIVE BARKING IN DOGS

Barking dogs can be a real problem, as anyone who has been kept awake at night by the barking of a neighbor's canine companion will know. Researchers at a local university have tested a new drug that they hope will put all concerned to rest. Twenty dog owners responded to a newspaper article asking for volunteers with problem barking dogs to participate in a study. The dogs were randomly assigned to two groups. One group of dogs was given the drug, administered as a shot, and the other dogs were not. Both groups were kept overnight at the research facility and frequency of barking was observed. The researchers deliberately tried to provoke the dogs into barking by doing things like ringing the doorbell of the facility and having a mail carrier walk up to the door. The two groups were treated on separate weekends because the facility was only large enough to hold ten dogs. The researchers left a tape recorder running and measured the amount of time during which any barking was heard. The dogs who had been given the drug spent only half as much time barking as did the dogs in the control group.

Hypothetical News Article 3: "Researchers Find Drug to Cure Excessive Barking in Dogs"

Component 1: The *source* of the research and of the *funding* We are not told why this study was conducted. Presumably it was because the researchers were interested in helping to solve a societal problem, but perhaps not. It is not uncommon for drug companies to fund research to test a new product or a new use for a current product. If that were the case, the researchers would have added incentive for the results to come out favorable to the drug. If everything were done correctly, such an incentive wouldn't be a major factor; however, when research is funded by a private source, that information should be announced when the results are announced.

Component 2: The *researchers* who had *contact* with the participants We are not given any information about who actually had contact with the dogs. One important question is whether the same handlers were used with both groups of dogs. If not, the difference in handlers could explain the results. Further, we are not told whether the dogs were primarily left alone or were attended most of the time. If researchers were present most of the time, their behavior toward the dogs could have had a major impact on the amount of barking.

Component 3: The *individuals* or objects studied and how they were *selected* We are told that the study used dogs whose owners volunteered them as problem dogs for the study. Although the report does not mention payment, it is quite common for volunteers to receive monetary compensation for their participation. The volunteers presumably lived in the area of the university. The dog owners had to be willing to be separated from their pets for the weekend. These and other factors mean that the owners and dogs who participated may differ from the general population. Further, the initial reasons for the problem behavior may vary from one participant to the next, yet the dogs were measured together. Therefore, there is no way to ascertain if, for example, dogs who bark only because they are lonely would be helped. In any case, we cannot extend the results of this study to conclude that the drug would work similarly on all dogs or even on all problem dogs. Because the dogs were randomly assigned to the two groups—and if there were no other problems—we would be able to extend the results to all dogs similar to those who participated.

Component 4: The exact nature of the *measurements* made or *questions* asked The researchers measured each group of dogs as a group, by listening to a tape and recording the amount of time during which there was any barking. Because dogs are quite responsive to group behavior, one barking dog could set the whole group barking for a long time. Therefore, just one particularly obnoxious dog in the control group alone could explain the results. It would have been better to separate the dogs and measure each one individually.

Component 5: The *setting* in which the measurements were taken The groups were measured on separate weekends. This creates another problem. First, the researchers knew which group was which, and may have unconsciously provoked the control group slightly more than the group receiving the drug. Further, conditions differed over the 2 weeks. Perhaps it was sunny one weekend and raining the next, or there were other subtle differences, such as more traffic one weekend than the next, small planes overhead, and so on. All of these could change the behavior of the dogs but might go unnoticed or unreported by the experimenters.

The measurements were also taken outside of the dogs' natural environments. The dogs in the experimental group in particular would have reason to be upset because they were first given a shot and then put together with nine other dogs in the research facility. It would have been better to put them back into their natural environment because that's where the problem barking was known to occur.

Component 6: The *extraneous differences* between groups being compared The dogs were randomly assigned to the two groups, which should have minimized overall differences in size, temperament, and so on. However, differences were induced between the two groups by the way the experiment was conducted. Recall that the two groups were measured on different weekends—this could have created the difference in behavior. Also, the treated dogs were given a shot to administer the drug, whereas the control group was given no shot. It could be that the very act of getting a shot made the drug group lethargic. A better design would have been to administer a placebo shot—that is, a shot with an inert substance—to the control group.

SURVEY FINDS MOST WOMEN UNHAPPY IN THEIR CHOICE OF HUSBANDS

A popular women's magazine, in a survey of its subscribers, found that over 90% of them are unhappy in their choice of whom they married. Copies of the survey were mailed to the magazine's 100,000 subscribers. Surveys were returned by 5000 readers. Of those responding, 4520, or slightly over 90%, answered no to the question: "If you had it to do over again, would you marry the same man?" To keep the survey simple so that people would return it, only two other questions were asked. The second question was, "Do you think being married is better than being single?" Despite their unhappiness with their choice of spouse, 70% answered yes to this. The final question, "Do you think you will outlive your husband?" received a yes answer from 80% of the respondents. Because women generally live longer than men, and tend to marry men somewhat older than themselves, this response was not surprising. The magazine editors were at a loss to explain the huge proportion of women who would choose differently. The editor could only speculate: "I guess finding Mr. Right is much harder than anyone realized."

Component 7: The *magnitude* of any claimed effects or differences We are told only that the treated group barked half as much as the control group. We are not told how much time either group spent barking. If one group barked 8 hours a day but the other group only 4 hours a day, that would not be a satisfactory solution to the problem of barking dogs.

Hypothetical News Article 4: "Survey Finds Most Women Unhappy in Their Choice of Husbands"

Components 1 through 7 We don't even need to consider the details of this study because it contains a fatal flaw from the outset. The survey is an example of what is called "volunteer response." Of the 100,000 who received the survey, only 5% responded. The people who are most likely to respond to such a survey are those who have a strong emotional response to the question. In this case, it would be women who are unhappy with their current situation who would probably respond. Notice that the other two questions are more general and thus not likely to arouse much emotion either way. Thus, it is the strong reaction to the first question that would drive people to respond. The results would certainly not be representative of "most women" or even of most subscribers to the magazine.

2.4   PLANNING YOUR OWN STUDY: DEFINING THE COMPONENTS IN ADVANCE

Although you may never have to design your own survey or experiment, it will help you understand how difficult it can be if we illustrate the Seven Critical Components for a very simple hypothetical study you might want to conduct. Suppose you are interested in determining which of three local supermarkets has the best prices so you can decide where to shop. Because you obviously can't record and summarize the prices for all available items, you would have to use some sort of sample.

To obtain meaningful data, you would need to make many decisions. Some of the Components need to be reworded because they are being answered in advance of the study, and obviously not all of the Components are relevant for this simple example. However, by going through them for such a simple case, you can see how many ambiguities and decisions can arise when designing a study.

Component 1: The *source* of the research and of the *funding* Presumably you would be funding the study yourself, but before you start you need to decide why you are doing the study. Are you only interested in items you routinely buy, or are you interested in comparing the stores on the multitude of possible items?

Component 2: The *researchers* who had *contact* with the participants In this example, the question would be who is going to visit the stores and record the prices. Will you personally visit each store and record the prices? Will you send friends to two of the stores and visit the third yourself? If you use other people, you would need to train them so there would be no ambiguities.

Component 3: The *individuals* or objects studied and how they were *selected* In this case, the "objects studied" are items in the grocery store. The correct question is, "On what items should prices be recorded?" Do you want to use exactly the same items at all stores? What if one store offers its own brand but another only offers name brands? Do you want to choose a representative sampling of items you are likely to buy or choose from all possible items? Do you want to include nonfood items? How many items should you include? How should you choose which ones to select? If you are simply trying to minimize your own shopping bill, it is probably best to list the 20 or 30 items you buy most often. However, if you are interested in sharing your results with others, you might prefer to choose a representative sample of items from a long list of possibilities.

Component 4: The exact nature of the *measurements* made or *questions* asked You may think that the cost of an item in a supermarket is a well-defined measurement. But if a store is having a sale on a particular item on your list, should you use the sale price or the regular price? Should you use the price of the smallest possible size of the product? The largest? What if a store always has a sale on one brand or another of something, such as laundry soap, and you don't really care which brand you buy? Should you then record the price of the brand on sale that week? Should you record the prices listed on the shelves or actually purchase the items and see if the prices listed were accurate?

Component 5: The *setting* in which the measurements were taken When will you conduct the study? Supermarkets in university towns may offer sale prices on items typically bought by students at certain times of the year—for example, just after students have returned from vacation. Many stores also offer sale items related to certain holidays, such as ham or turkey just before Christmas or eggs just before Easter. Should you take that kind of timing into account?

Component 6: The *extraneous differences* between groups being compared There should be no extraneous differences related to the direct costs of the items. However, if you were conducting the study in order to minimize your shopping costs, you might ask if there are hidden costs for shopping at one store versus another. For example, do you always have to wait in line at one store and not at another, and should you therefore put a value on your time? Does one store make mistakes at the cash register more often than another? Does one store charge a higher fee to use your cash card for payment? Does it cost more to drive to one store than another?

Component 7: The *magnitude* of any claimed effects or differences This component should enter into your decision about where to shop after you have finished the study. Even if you find that one store costs less than another, the magnitude of the difference may not convince you to shop there. You would probably want to figure out approximately how much shopping in a particular store would save you over the course of a year. You can see why knowing the amount of a difference found in a study is an important component for using that study to make future decisions.

CASE STUDY 2.1   # Brooks Shoes Brings Flawed Study to Court

SOURCE: Gastwirth, 1988, pp. 517–520.

In 1981, Brooks Shoe Manufacturing Company sued Suave Shoe Corporation for manufacturing shoes incorporating a "V" design used in Brooks's athletic shoes. Brooks claimed that the design was an unregistered trademark that people used to identify Brooks shoes. According to Gastwirth (1988, p. 517), it was the role of the court to determine "the distinctiveness or strength of the mark as well as its possible secondary meaning (similarity of product or mark might confuse prospective purchasers of the source of the item)."

To show that the design had "secondary meaning" to buyers, Brooks conducted a survey of 121 spectators and participants at three track meets. Interviewers approached people and asked them a series of questions that included showing them a Brooks shoe with the name masked and asking them to identify it. Of those surveyed, 71% were able to identify it as a Brooks shoe, and 33% of those people said it was because they recognized the "V." When shown a Suave shoe, 39% of them thought it was a Brooks shoe, with 48% of those people saying it was because of the "V" design on the Suave shoe. Brooks Company argued that this was sufficient evidence that people might be confused and think Suave shoes were manufactured by Brooks.

Suave had a statistician as an expert witness, who pointed out a number of flaws in the Brooks survey. Let's examine those using the Seven Critical Components as a guide. First, the survey was funded and conducted by Brooks, and the company's lawyer was instrumental in designing it. Second, the court determined that the interviewers who had contact with the respondents were inadequately trained in how to conduct an unbiased survey. Third, the individuals asked were not selected to be representative of the general public in the area (Baltimore/Washington, D.C.). For example, 78% had some college education, compared with 18.4% in Baltimore and 37.7% in Washington, D.C. Further, the settings for the interviews were track meets, where people were likely to be more familiar with athletic shoes. The questions asked were biased. For example, the exact wording used when a person was handed the shoes was: "I am going to hand you a shoe. Please tell me what brand you think it is." The way the question is framed would presumably lead respondents to think the shoe has a well-known brand name. Later in the questioning, respondents were asked, "How long have you known about Brooks Running Shoes?" Because of the setting, respondents could have informed others at the track meet that Brooks was probably conducting the survey, and those informed could have subsequently been interviewed.

Suave introduced its own survey conducted on 404 respondents properly sampled from the population of all people who had purchased any type of athletic shoe during the previous year. Of those, only 2.7% recognized a Brooks shoe on the basis of the "V" design. The combination of the poor survey methods by Brooks and the proper survey by Suave convinced the court that the public did not make enough of an association between Brooks and the "V" design to allow Brooks to claim legal rights to the design. ■

EXERCISES

1. Suppose that a television network wants to know how daytime television viewers feel about a new soap opera the network is broadcasting. A staff member suggests that just after the show ends they give two phone numbers, one for viewers to call if they like the show and the other to call if they don't. Give two reasons why this method would not produce the desired information.

2. The April 24, 1997, issue of "UCDavis Lifestyle Newstips" reported that a professor of veterinary medicine was conducting a study to see if a drug called clomipramine, an anti-anxiety medication used for humans, could reduce "canine aggression toward family members." The newsletter said, "Dogs demonstrating this type of aggression are needed to participate in the study. . . .

Half of the participating dogs will receive clomipramine, while the others will be given a placebo." A phone number was given for dog owners to call to volunteer their dogs for the study. To what group could the results of this study be applied? Explain.

3. A prison administration wants to know whether the prisoners think the guards treat them fairly. Explain how each of the following components could be used to produce biased results, versus how each could be used to produce unbiased results:

  a. Component 2: The *researchers* who had *contact* with the participants.

  b. Component 4: The exact nature of the *measurements* made or *questions* asked.

4. According to Cynthia Crossen (1994, p. 106): "It is a poller's business to press for an opinion whether people have one or not. 'Don't knows' are worthless to pollers, whose product is opinion, not ignorance. That's why so many polls do not even offer a 'don't know' alternative." Explain how this problem might lead to bias in a survey and how the problem would be uncovered by answering the Seven Critical Components.

5. The student who conducted the study in "Hypothetical News Article 1" in this chapter collected two pieces of data from each participant. What were the two pieces of data?

6. Many research organizations give their interviewers an exact script to follow when conducting interviews to measure opinions on controversial issues. Why do you think they do so?

7. Is it necessary that "data" consist of numbers? Explain.

8. Refer to Case Study 1.1, "Heart or Hypothalamus?" Discuss each of the following Components, including whether you think the way it was handled would detract from Salk's conclusion:

  a. Component 3

  b. Component 4

  c. Component 5

  d. Component 6

9. Suppose a tobacco company is planning to fund a telephone survey of attitudes about banning smoking in restaurants. In each of the following phases of the survey, should the company disclose who is funding the study? Explain your answer in each case.

  a. When respondents answer the phone, before they are interviewed.

  b. When the survey results are reported in the news.

  c. When the interviewers are trained and told how to conduct the interviews.

10. Suppose a study were to find that twice as many users of nicotine patches quit smoking than nonusers. Suppose you are a smoker trying to quit. Which version of an answer to each of the following Components would be more compelling evidence for you to try the nicotine patches? Explain.

    a. Component 3. Version 1 is that the nicotine patch users were lung cancer patients, whereas the nonusers were healthy. Version 2 is that participants were randomly assigned to use the patch or not after answering an advertisement in the newspaper asking for volunteers who wanted to quit smoking.

    b. Component 7. Version 1 is that 25% of nonusers quit, whereas 50% of users quit. Version 2 is that 1% of nonusers quit, whereas 2% of users quit.

11. In most studies involving human participants, researchers are required to fully disclose the purpose of the study to the participants. Do you think people should always be informed about the purpose *before* they participate? Explain.

12. Explain why news reports should give the magnitude of the claimed effects or differences from a study instead of just reporting that an effect or difference was found.

13. Suppose a study were to find that drinking coffee raised cholesterol levels. Further, suppose you drink two cups of coffee a day and have a family history of heart problems related to high cholesterol. Which three of the Seven Critical Components would interest you most, in terms of deciding whether to change your coffee-drinking habits? Explain.

14. Holden (1991, p. 934) discusses the methods used to rank high school math performance among various countries. She notes that: "According to the International Association for the Evaluation of Educational Achievement, Hungary ranks near the top in 8th-grade math achievement. But by the 12th grade, the country falls to the bottom of the list because it enrolls more students than any other country—50%—in advanced math. Hong Kong, in contrast, comes in first, but only 3% of its 12th graders take math."

    Explain which of the Seven Critical Components should be considered when interpreting the results of rankings of high school math performance in various countries, and describe how your interpretation of the results would be affected.

15. Moore (1991, p. 19) reports the following contradictory evidence: "The advice columnist Ann Landers once asked her readers, 'If you had it to do over again, would you have children? She received nearly 10,000 responses, almost 70% saying 'No!'. . . A professional nationwide random sample commissioned by *Newsday* polled 1373 parents and found that 91% would have children again." Using the Seven Critical Components, explain the contradiction in the two sources of answers.

16. An advertisement for a cross-country ski machine, NordicTrack, claimed, "In just 12 weeks, research shows that *people who used a NordicTrack lost an average of 18 pounds.*" Explain how each of the following Components should have been addressed if the research results are fair and unbiased.

    a. Component 3: The *individuals* or objects studied and how they were *selected.*

    b. Component 4: The exact nature of the *measurements* made or *questions* asked.

    c. Component 5: The *setting* in which the measurements were taken.

    d. Component 6: The *extraneous differences* between groups being compared.

MINI-PROJECTS

1. Scientists publish their findings in technical magazines called journals. Most university libraries have hundreds of journals available for browsing. Find out where the medical journals are located. Browse the shelves until you find an article with a study that sounds interesting to you. (The *New England Journal of Medicine* and the *Journal of the American Medical Association* often have articles of broad interest, but there are also numerous specialized journals on pediatrics, cancer, AIDS, and so on.) Read the article and write a report that discusses each of the Seven Critical Components for that particular study. Argue for or against the believability of the results on the basis of your discussion. Be sure you find an article discussing a single study and not a collection or "meta-analysis" of numerous studies.

2. Explain how you would design and carry out a study to find out how students at your school feel about an issue of interest to you. Be explicit enough that someone would actually be able to follow your instructions and implement the study. Be sure to consider each of the Seven Critical Components when you design and explain how to do the study.

3. Find an example of a statistical study reported in the news for which information about one of the Seven Critical Components is missing. Write two hypothetical reports addressing the missing component that would lead you to two different conclusions about the applicability of the results of the study.

REFERENCES

Crossen, Cynthia. (1994). *Tainted truth: The manipulation of fact in America.* New York: Simon & Schuster.

Gastwirth, Joseph L. (1988). *Statistical reasoning in law and public policy.* Vol. 2. *Tort law, evidence and health.* Boston: Academic Press.

Holden, Constance. (1991). Questions raised on math rankings. *Science 254,* p. 934.

Moore, David S. (1991). *Statistics: Concepts and controversies.* 3d ed. New York: W.H. Freeman.

CHAPTER THREE

Measurements, Mistakes, and Misunderstandings

THOUGHT QUESTIONS

1. Suppose you were interested in finding out what people felt to be the most important problem facing society today. Do you think it would be better to give them a fixed set of choices from which they must choose or an open-ended question that allowed them to specify whatever they wished? What would be the advantages and disadvantages of each approach?

2. You and a friend are each doing a survey to see if there is a relationship between height and happiness. You both attempt to measure the height and happiness of the same 100 people. Are you more likely to agree on your measurement of height or on your measurement of happiness? Explain, discussing how you would measure each characteristic.

3. A newsletter distributed by a politician to his constituents gave the results of a "nationwide survey on Americans' attitudes about a variety of educational issues." One of the questions asked was, "Should your legislature adopt a policy to assist children in failing schools to opt out of that school and attend an alternative school—public, private, or parochial—of the parents' choosing?" From the wording of this question, can you speculate on what answer was desired? Explain.

4. You are at a swimming pool with a friend and become curious about the width of the pool. Your friend has a 12-inch ruler, with which he sets about measuring the width. He reports that the width is 15.771 feet. Do you believe the pool is exactly that width? What is the problem?

5. If you were to have your intelligence, or IQ, measured twice using a standard IQ test, do you think it would be exactly the same both times? What factors might account for any changes?

3.1   SIMPLE MEASURES DON'T EXIST

In the last chapter, we listed seven critical components that need to be considered when someone conducts a study. You saw that many decisions need to be made and many potential problems can arise when you try to use data to answer a question. One of the hardest decisions is contained in Component 4—that is, in deciding exactly what to measure or what questions to ask. In this chapter, we focus on problems with defining measurements and on the subsequent misunderstandings and mistakes that can result. When you read the results of a study, it is important that you understand exactly how the information was collected and what was measured or asked. Consider something as apparently simple as trying to measure your own height. Try it a few times and see if you get the measurement to within a quarter of an inch from one time to the next. Now imagine trying to measure something much more complex, such as the amount of fat in someone's diet or the degree of happiness in someone's life. Researchers routinely attempt to measure these kinds of factors.

3.2   IT'S ALL IN THE WORDING

You may be surprised at how much answers to questions can change based on simple changes in wording. Here is one example. Loftus and Palmer (1974; quoted in Plous, 1993, p. 32) showed college students films of an automobile accident, after which they asked them a series of questions. One group was asked the question: "About how fast were the cars going when they contacted each other?" The average response was 31.8 miles per hour. Another group was asked: "About how fast were the cars going when they collided with each other?" In that group, the average response was 40.8 miles per hour. Simply changing from the word *contacted* to the word *collided* increased the estimates of speed by 9 miles per hour, or 28%, even though the respondents had witnessed the same film.

Many pitfalls can be encountered when asking questions in a survey or experiment. Here are some of them; each will be discussed in turn:

1. Deliberate bias
2. Unintentional bias
3. Desire to please
4. Asking the uninformed
5. Unnecessary complexity
6. Ordering of questions
7. Confidentiality and anonymity

Deliberate Bias

Sometimes, if a survey is being conducted to support a certain cause, questions are deliberately worded in a biased manner. Be careful about survey questions that begin with phrases like "Do you agree that. . . ." Most people want to be agreeable and will be inclined to answer yes unless they have strong feelings the other way. For example, suppose an anti-abortion group and a pro-choice group each wanted to conduct a survey in which they would find the best possible agreement with their position. Here are two questions that would each produce an estimate of the proportion of people who think abortion should be completely illegal. Each question is almost certain to produce a different estimate:

1. Do you agree that abortion, the murder of innocent beings, should be outlawed?
2. Do you agree that there are circumstances under which abortion should be legal, to protect the rights of the mother?

Appropriate wording should not indicate a desired answer. For instance, a Gallup Poll conducted in June 1998 contained the question "Do you think it was a good thing or a bad thing that the atomic bomb was developed?" Notice that the question does not indicate which answer is preferable. In fact 61% of the respondents said "bad," whereas 36% said "good" and 3% were undecided.

Unintentional Bias

Sometimes questions are worded in such a way that the meaning is misinterpreted by a large percentage of the respondents. For example, if you were to ask people whether they use drugs, you would need to specify if you mean prescription drugs, illegal drugs, over-the-counter drugs, or common substances such as caffeine. If you were to ask people to recall the most important date in their life, you would need to clarify if you meant the most important calendar date or the most important social engagement with a potential partner. (It is unlikely that anyone would mistake the question as being about the shriveled fruit, but you can see that the same word can have multiple meanings.)

Desire to Please

Most survey respondents have a desire to please the person who is asking the question. They tend to understate their responses about undesirable social habits and opinions, and vice versa. For example, in recent years estimates of the prevalence of cigarette smoking based on surveys do not match those based on cigarette sales. Either people are not being completely truthful or lots of cigarettes are ending up in the garbage.

Asking the Uninformed

People do not like to admit that they don't know what you are talking about when you ask them a question. Crossen (1994, p. 24) gives an example: "When the American Jewish Committee studied Americans' attitudes toward various ethnic groups, almost 30% of the respondents had an opinion about the fictional Wisians, rating them in social standing above a half-dozen other real groups, including Mexicans, Vietnamese, and African blacks." Political pollsters, who are interested in surveying only those who will actually vote, learned long ago that it is useless to simply ask people if they plan to vote. Most of them will say yes. Instead, they ask questions to establish a history of voting, such as "Where did you go to vote in the last election?"

Unnecessary Complexity

If questions are to be understood, they must be kept simple. A question such as "Shouldn't former drug dealers not be allowed to work in hospitals after they are released from prison?" is sure to lead to confusion. Does a yes answer mean they should or they shouldn't be allowed to work in hospitals? It would take a few readings to figure that out.

Another way in which a question can be unnecessarily complex is to actually ask more than one question at once. An example would be a question such as, "Do you support the president's health care plan because it would ensure that all Americans receive health coverage?" If you agree with the idea that all Americans should receive health coverage, but disagree with the remainder of the plan, do you answer yes or no? Or what if you support the President's plan, but not for that reason?

Ordering of Questions

If one question requires respondents to think about something that they may not have otherwise considered, then the order in which questions are presented can change the results. For example, suppose a survey were to ask, "To what extent do you think teenagers today worry about peer pressure related to drinking alcohol?" and then ask, "Name the top five pressures you think face teenagers today." It is quite likely that respondents would use the idea they had just been given and name peer pressure related to drinking alcohol as one of the five choices.

Confidentiality and Anonymity

People sometimes answer questions differently based on the degree to which they believe they are anonymous. Because researchers often need to perform follow-up surveys, it is easier to try to ensure confidentiality than true anonymity. In ensuring confidentiality, the researcher promises not to release identifying information about respondents. In a truly anonymous survey, the researcher does not know the identity of the respondents.

Questions on issues such as sexual behavior and income are particularly difficult because people consider those to be private matters. A variety of techniques have been developed to help ensure confidentiality, but surveys on such issues are hard to conduct accurately.

CASE STUDY 3.1   # No Opinion of Your Own? Let Politics Decide

SOURCE: Morin, 10–16 April 1995, p. 36.

This is an excellent example of how people will respond to survey questions, even when they do not know about the issues, and how the wording of questions can influence responses. In 1995, the *Washington Post* decided to expand on a 1978 poll taken in Cincinnati, Ohio, in which people were asked whether they "favored or opposed repealing the 1975 Public Affairs Act." There was no such act, but about one-third of the respondents expressed an opinion about it.

In February 1995, the *Washington Post* added this fictitious question to its weekly poll of 1000 randomly selected respondents: "Some people say the 1975 Public Affairs Act should be repealed. Do you agree or disagree that it should be repealed?" Almost half (43%) of the sample expressed an opinion, with 24% agreeing that it should be repealed and 19% disagreeing. The *Post* then tried another trick that produced even more disturbing results. This time, they polled two separate groups of 500 randomly selected adults. The first group was asked: "President Clinton [a Democrat] said that the 1975 Public Affairs Act should be repealed. Do you agree or disagree?" The second group was asked: "The Republicans in Congress said that the 1975 Public Affairs Act should be repealed. Do you agree or disagree?" Respondents were also asked about their party affiliation. Overall, 53% of the respondents expressed an opinion about repealing this fictional act. The results by party affiliation were striking: For the "Clinton" version, 36% of the Democrats but only 16% of the Republicans agreed that the Act should be repealed. For the "Republicans in Congress" version, 36% of the Republicans but only 19% of the Democrats agreed that the act should be repealed. ∎

3.3   OPEN OR CLOSED QUESTIONS: SHOULD CHOICES BE GIVEN?

An **open question** is one in which respondents are allowed to answer in their own words, whereas a **closed question** is one in which they are given a list of alternatives from which to choose their answer. Usually the latter form offers a choice of "other," in which the respondent is allowed to fill in the blank.

Problems with Closed Questions

To show the limitation of closed questions, Schuman and Scott (22 May 1987) asked about "the most important problem facing this country today." Half of the sample, 171 people, were given this as an open question. The most common responses were

Unemployment (17%)

General economic problems (17%)

Threat of nuclear war (12%)

Foreign affairs (10%)

In other words, one of these four choices was volunteered by over half of the respondents.

The other half of the sample was given this as a closed question. Following is the list of choices and the percentage of respondents who chose them:

The energy shortage (5.6%)

The quality of public schools (32.0%)

Legalized abortion (8.4%)

Pollution (14.0%)

These four choices combined were mentioned by only 2.4% of respondents in the open-question survey; yet they were selected by 60% when they were the only specific choices given. Further, respondents in this closed-question survey were given an open choice. In addition to the list of four, they were told: "If you prefer, you may name a different problem as most important." On the basis of the closed-form questionnaire, policymakers would have been seriously misled about what is important to the public.

It is possible to avoid this kind of astounding discrepancy. If closed questions are preferred, they first should be presented as open questions to a test sample before the real survey is conducted. Then the most common responses should be included in the list of choices for the closed question. This kind of exercise is usually done as part of what's called a "pilot survey," in which various aspects of a study design can be tried before it's too late to change them.

Problems with Open Questions

The biggest problem with open questions is that the results can be difficult to summarize. If a survey includes thousands of respondents, it can be a major chore to categorize their responses.

Another problem, found by Schuman and Scott (22 May 1987), is that the wording of the question might unintentionally exclude answers that would have been appealing had they been included in a list of choices (such as in a closed question). To test this, they asked 347 people to "name one or two of the most important national or world event(s) or change(s) during the past 50 years." The most common choices and the percentage who mentioned them were

World War II (14.1%)

Assassination of John F. Kennedy (4.6%)

The Vietnam War (10.1%)

Don't know (10.6%)

All other responses (53.7%)

The same question was then repeated in closed form to a new group of 354 people. Five choices were given: the first four choices in the preceding list plus "invention of the computer." Of the 354 respondents, the percentage of those who selected each choice was:

> World War II (22.9%)
>
> Exploration of space (15.8%)
>
> Assassination of John F. Kennedy (11.6%)
>
> The Vietnam War (14.1%)
>
> Invention of the computer (29.9%)
>
> Don't know (0.3%)
>
> All other responses (5.4%)

The most frequent response was "invention of the computer," which had been mentioned by only 1.4% of respondents in the open question. Clearly the wording of the question led respondents to focus on "events" rather than "changes," and the invention of the computer did not readily come to mind. When it was presented as an option, however, people realized that it was indeed one of the most important events or changes during the past 50 years. In summary, there are advantages and disadvantages to both approaches. One compromise is to ask a small test sample to list the first several answers that come to mind, and then use the most common of those in a closed-question survey. These choices could be supplemented with additional answers such as "invention of the computer," which may not readily come to mind.

Remember that, as the reader, you have an important role in interpreting the results. You should always be informed as to whether questions were asked in open or closed form, and if the latter, you should be told what the choices were. You should also be told whether "don't know" or "no opinion" was offered as a choice in either case.

3.4 DEFINING WHAT IS BEING MEASURED

EXAMPLE 1 **TEENAGE SEX**

To understand the results of a survey or an experiment, we need to know exactly what was measured. Consider this example. A few years ago, a letter to advice columnist Ann Landers stated: "According to a report from the University of California at San Francisco . . . sexual activity among adolescents is on the rise. There is no indication that this trend is slowing down or reversing itself." The letter went on to explain that these results were based on a national survey (*Davis* (CA) *Enterprise*, 19 February 1990, p. B-4). On the same day, in the same newspaper, an article entitled "Survey: Americans conservative with sex" reported that "teenage boys are not living up to their reputations. [A study by the Urban Institute in Washington] found that adolescents seem to be having sex less often, with fewer girls and at a later age than teenagers did a decade ago" (p. A-9).

Here we have two apparently conflicting reports on adolescent sexuality, both reported on the same day in the same newspaper. One indicated that teenage sex was on the rise; the other indicated that it was on the decline. Although neither report specified exactly what was measured, the letter to Ann Landers proceeded to note that "national statistics show the average age of first intercourse is 17.2 for females and 16.5 for males." The article stating that adolescent sex was on the decline measured it in terms of *frequency*. The result was based on interviews with 1880 boys between the ages of 15 and 19, in which "the boys said they had had six sex partners, compared with seven a decade earlier. They reported having had sex an average of three times during the previous month, compared with almost five times in the earlier survey."

Thus, it is not enough to note that both surveys were measuring adolescent or teenage sexual behavior. In one case, the author was, at least partially, discussing the *age* of first intercourse, whereas in the other case the author was discussing the *frequency*. ∎

EXAMPLE 2   **THE UNEMPLOYED**

Ask people whether they know anyone who is unemployed; they will invariably say yes. But most people don't realize that in order to be officially unemployed, and included in the unemployment statistics given by the U.S. government, you must meet very stringent criteria. These criteria exclude "discouraged workers" who are "identified as persons who, though not working or seeking work, have a current desire for a job, have looked for one within the past year [but not within the past four weeks], and are currently available for work" (U.S. Dept. of Labor, September 1992, p. 10). If you know someone who fits that definition, you would undoubtedly think of that person as unemployed. However, he or she would not be included in the official statistics. You can see that the true number of people who are not working is higher than government statistics would lead you to believe. ∎

These two examples illustrate that when you read about measurements taken by someone else, you should not automatically assume you are speaking a common language. A precise definition of what is meant by "adolescent sexuality" or "unemployment" should be provided.

Some Concepts Are Hard to Define Precisely

Sometimes it is not the language but the concept itself that is ill-defined. For example, there is still not universal agreement on what should be measured with intelligence, or IQ, tests. The tests were originated at the beginning of the 20th century in order to determine the mental level of school children. The intelligence quotient (IQ) of a child was found by dividing the child's "mental level" by his or her chronological age. The "mental level" was determined by comparing the child's performance on the test with that of a large group of "normal" children, to find the age group the individual's performance matched. Thus, if an 8-year-old child performed as well on the test as a "normal" group of 10-year-old children, he or she would have an IQ of $100 \times (10/8) = 125$.

IQ tests have been expanded and refined since the early days, but they continue to be surrounded by controversy. One reason is that it is very difficult to define what is meant by intelligence. It is difficult to measure something if you can't even agree on what it is you are trying to measure. If you are interested in knowing more about these tests and the surrounding controversies, you can find numerous books on the subject. Anastasi and Urbina (1997) provide a detailed discussion of a large variety of psychological tests, including IQ tests.

Measuring Attitudes and Emotions

Similar problems exist with trying to measure attitudes and emotions such as self-esteem and happiness. The most common method for trying to measure such things is to have respondents read statements and determine the extent to which they agree with the statement. For example, a test for measuring happiness might ask respondents to indicate their level of agreement, from "strongly disagree" to "strongly agree," with statements such as "I generally feel optimistic when I get up in the morning." To produce agreement on what is meant by characteristics such as "introversion," psychologists have developed standardized tests that claim to measure those attributes.

CASE STUDY 3.2 ## Questions in Advertising

Advertisements commonly present results without telling the listener or reader what choices were given to the respondents of a survey. Here are two examples:

EXAMPLE 3 Levi Strauss released a marketing package presented as "Levi's 501 Report, a fall fashion survey conducted annually on 100 U.S. campuses." As part of the report, it was noted that 90% of college students chose Levi's 501 jeans as being "in" on campus. What the resulting advertising failed to reveal was the list of choices, which noticeably omits blue jeans except for Levi's 501 jeans:

Levi's 501 jeans	T-shirts with graphics
1960s-inspired clothing	Lycra/spandex clothing
Overalls	Patriotic-themed clothing
Decorated denim	Printed, pull-on beach pants
Long-sleeved, hooded T-shirts	Neon-colored clothing

EXAMPLE 4 An advertisement for Triumph cigarettes boasted: "TRIUMPH BEATS MERIT—an amazing 60% said Triumph tastes as good or better than Merit." In truth, three choices were offered to respondents, including "no preference." The results were: 36% preferred Triumph, 40% preferred Merit, and *24% said the brands were equal.* So, although the wording of the advertisement is not false, it is also true that 64% said Merit tastes as good as or better than Triumph. Which brand do you think wins?

SOURCE: Crossen, 1994, pp. 74–75. ■

3.5   DEFINING A COMMON LANGUAGE

So that we're all speaking a common language for the rest of this book, we need to define some terms. We can perform different manipulations on different types of data, so we need a common understanding of what those types are. Other terms defined in this section are those that are well-known in everyday usage but that have a slightly different technical meaning.

Categorical versus Measurement Variables

Thus far in this book, we have seen examples of measuring opinions (such as what you think is the most important problem facing society), numerical information (such as weight gain in infants), and attributes that can be transformed into numerical information (such as IQ). To understand what we can do with these measurements, we need definitions to distinguish numerical measures from qualitative ones. Although statisticians make numerous fine distinctions among types of measurements, for our purposes it will be sufficient to distinguish between just two main types: categorical variables and measurement variables.

Categorical Variables

Categorical variables are those we can place into a category but that may not have any logical ordering. For example, you could be categorized as male or female. You could also be categorized based on what you name as the most important problem facing society. Notice that we are limited in how we can manipulate this kind of information numerically. For example, we cannot talk about the average problem facing society in the same way as we can talk about the average weight gain of infants during the first few days of life.

Measurement Variables

Measurement variables are those for which we can record a numerical value and then order respondents according to those values. For example, IQ is a measurement variable because it can be expressed as a single number. An IQ of 130 is higher than an IQ of 100. Age, height, and number of cigarettes smoked per day are other examples of measurement variables. Notice that these can be worked with numerically. Of course, not all numerical summaries will make sense even with measurement variables. For example, if one person in your family smokes 20 cigarettes a day and the remaining three members smoke none, it is accurate but misleading to say that the average number of cigarettes smoked by your family per day is 5 per person. We will learn about reasonable numerical summaries in Chapter 7.

Continuous versus Discrete Measurement Variables

Even when we can measure something with a number, we may need to distinguish further whether it can fall on a continuum. A **discrete variable** is one for which you could actually count the possible responses. For example, if we measure the number of automobile accidents on a certain stretch of highway, the answer could be zero, one, two, three, and so on. It could not be 2 1/2 or 3.8. Conversely, a **continuous variable** can be anything within a given interval. Age, for example, falls on a continuum.

Something of a gray area exists between these definitions. For example, if we measure age to the nearest year, it may seem as though it should be called a discrete variable. But the real difference is conceptual. With a discrete variable you can count the possible responses without having to round off. With a continuous variable you can't. In case you are confused by this, note that long ago you probably figured out the difference between the phrase "the number of" and "the amount of." You wouldn't say, "the amount of cigarettes smoked," nor would you say, "the number of water consumed." Discrete variables are analogous to numbers of things, and continuous variables are analogous to amounts. You still need to be careful about wording, however, because we have a tendency to express continuous variables in discrete units. Although you wouldn't say, "the number of water consumed," you might say, "the number of glasses of water consumed." That's why it's the *concept* of number versus amount that you need to think about.

Validity, Reliability, Bias, and Variability

The words we define in this section are commonly used in the English language, but they also have specific definitions when applied to measurements. Although these definitions are close to the general usage of the words, to avoid confusion we will spell them out.

Validity

When you talk about something being *valid*, you generally mean that it makes sense to you; it is sound and defensible. The same can be said for a measurement. A **valid measurement** is one that actually measures what it claims to measure. Thus, if you tried to measure happiness with an IQ test, you would not get a valid measure of happiness.

A more realistic example would be trying to determine the selling price of a home. Getting a valid measurement of the actual sales price of a home is tricky because the purchase often involves bargaining on what items are to be left behind by the old owners, what repairs will be made before the house is sold, and so on.

These items can change the recorded sales price by thousands of dollars. If we were to define the "selling price" as the price recorded in public records, it may not actually reflect the price the buyer and seller had agreed was the true worth of the home. To determine whether a measurement is valid, you need to know exactly what was measured. For example, many readers, once they are informed of the definition, do not think the unemployment figures provided by the U.S. government are a valid measure of unemployment, as the term is generally understood. Remember that the figures do not include "discouraged workers." However, the government statistics are a valid measure of the percentage of the "civilian labor force" that is currently "unemployed," according to the precise definitions supplied by the Bureau of Labor Statistics. The problem is that most people do not understand exactly what the government has measured.

Reliability

When we say something or someone is *reliable,* we mean that that thing or person can be depended upon time after time. A reliable car is one that will start every time and get us where we are going without worry. A reliable friend is one who is always there for us, not one who is sometimes too busy to bother with us. Similarly, a **reliable measurement** is one that will give you or anyone else approximately the same result time after time, when taken on the same object or individual. For example, a reliable way to define the selling price of a home would be the officially recorded amount. This may not be valid, but it would give us a consistent figure without any ambiguity.

Reliability is a useful concept in psychological and aptitude testing. An IQ test is obviously not much use if it measures the same person's IQ to be 80 one time and 130 the next. Whether we agree that the test is measuring what we really mean by "intelligence" (that is, whether it is really valid), it should at least be reliable enough to give us approximately the same number each time. Commonly used IQ tests are fairly reliable: About two-thirds of the time, taking the test a second time gives a reading within 2 or 3 points of the first test, and, most of the time, it gives a reading within about 5 points.

The most reliable measurements are physical ones taken with a precise measuring instrument. For example, it is much easier to get a reliable measurement of height than of happiness, assuming you have an accurate tape measure.

However, you should be cautious of measurements given with greater precision than you think the measuring tool would be capable of providing. The degree of precision probably exceeds the reliability of the measurement. For example, if your friend measures the width of a swimming pool with a ruler and reports that it is 15.771 feet wide, which is 15' 9 1/4", you should be suspicious. It would be very difficult to measure a distance that large reliably with a 12-inch ruler. A second measuring attempt would undoubtedly give a different number.

Bias

A systematic prejudice in one direction is called a *bias.* Similarly, a measurement that is systematically off the mark in the same direction is called a **biased measurement.** If you were trying to weigh yourself with a scale that was not satisfactorily adjusted at the factory, and was always a few pounds under, you would get a biased view of your own weight. When we used the term earlier in discussing the wording of questions, we noted that either intentional or unintentional bias could enter into the responses of a poorly worded survey question. Notice that a biased measurement differs from an unreliable measurement because it is consistently off the mark in the same direction.

Variability

If someone has *variable* moods, we mean that that person has unpredictable swings in mood. When we say the weather is quite variable, we mean it changes without any consistent pattern. Most measurements are prone to some degree of **variability.** By that, we mean that they include unpredictable errors or discrepancies that are not readily explained. If you tried to measure your height as instructed at the beginning of this chapter, you probably found some unexplainable variability from one time to the next. If you tried to measure the length of a table by laying a ruler end to end, you would undoubtedly get a slightly different answer each time.

Unlike the other terms we have defined, which are used to characterize a single measurement, variability is a concept used when we talk about two or more measurements in relation to each other. Sometimes two measurements vary because the measuring device produces unreliable results—for example, when we try to measure a large distance with a small ruler. Other times variability results from changes in the system being measured. For example, even with a very precise measuring device your recorded blood pressure will differ from one moment to the next.

Natural Variability

A related concept, that of **natural variability,** is crucial to understanding modern statistical methods. When we measure the same quantity across several individuals, such as the weight gain of newborn babies, we are bound to get some variability. Although some of this may be due to our measuring instrument, most of it is simply due to the fact that everyone is different. Variability is simply inherent in nature. Babies all gain weight at their own pace. If we want to compare the weight gain of a group of babies who have consistently listened to a heartbeat to the weight gain of a group of babies who have not, we first need to know how much variability to expect due to natural causes.

We encountered the idea of natural variability when we discussed comparing resting pulse rates of men and women in Chapter 1. If there were no variability within each sex, it would be easy to detect a difference between males and females. The more variability there is within each group, the more difficult it is to detect a difference between groups.

In Part 4, we will learn how to sort out differences due to natural variability from differences due to features we can define, measure, and possibly manipulate, such as amount of salt consumed or time spent exercising. In this way, we can study the effects of diet or medical treatments on disease, of advertising campaigns on consumer choices, of exercise on weight loss, and so on.

This one basic idea, comparing natural variability to the variability induced by different behaviors, interventions, or group memberships, forms the heart of modern statistics. It has allowed Salk to conclude that heartbeats are soothing to infants and the medical community to conclude that aspirin helps prevent heart attacks. We will see numerous other conclusions based on this idea throughout this book.

EXERCISES

1. Give an example of a measure that is:
   a. Valid and categorical
   b. Reliable but biased
   c. Unbiased but not reliable

2. Give an example of a survey question that is:
   a. Deliberately biased
   b. Unintentionally biased
   c. Unnecessarily complex
   d. Likely to cause respondents to lie

3. Give an example of a survey question that is:
   a. Most appropriately asked as an open question
   b. Most appropriately asked as a closed question

4. Explain which (one or more) of the seven pitfalls listed in Section 3.2 applies to each of the following potential survey questions:
   a. Do you support banning prayers in schools so that teachers have more time to spend teaching?
   b. Do you agree that marijuana should be legal?
   c. Studies have shown that consuming one alcoholic drink daily helps reduce heart disease. How many alcoholic drinks do you consume daily?

5. Refer to Question 4. Reword each question so it avoids the seven pitfalls.

6. Specify whether each of the following is a categorical or measurement variable. If you think the variable is ambiguous, discuss why.
   a. Years of formal education
   b. Highest level of education completed (grade school, high school, college, higher than college)
   c. Brand of car owned
   d. Price paid for the last car purchased
   e. Type of car owned (subcompact, compact, mid-size, full-size, sports, pickup)

7. Specify whether each of the following measurements is discrete or continuous. If you think the measurement is ambiguous, discuss why.
   a. The number of floors in a building
   b. The height of a building measured as precisely as possible

    c. The number of words in this book

    d. The weight of this book

8. If we were interested in knowing whether the average price of homes in a certain county had gone up or down this year in comparison with last year, would we be more interested in having a valid measure or a reliable measure of sales price? Explain.

9. In Chapter 1, we discussed Lee Salk's experiment in which he exposed one group of infants to the sound of a heartbeat and compared their weight gain to that of a group not exposed. Do you think it would be easier to discover a difference in weight gain between the group exposed to the heartbeat and the "control group" if there were a lot of natural variability among babies, or if there were only a little? Explain.

10. Do you think the crime statistics reported by the police are a valid measure of the amount of crime in a given city? Are they a reliable measure? Discuss.

11. Refer to Case Study 2.1, "Brooks Shoes Brings Flawed Study to Court." Discuss the study conducted by Brooks Shoe Manufacturing Company in the context of the seven pitfalls that can be encountered when asking questions in a survey, listed in Section 3.2.

12. An advertiser of a certain brand of aspirin (let's call it Brand B) claims that it is the preferred painkiller for headaches, based on the results of a survey of headache sufferers. The choices given to respondents were: Tylenol, Extra-Strength Tylenol, Brand B aspirin, Advil.

    a. Is this an open- or closed-form question? Explain.

    b. Comment on the variety of choices given to respondents.

    c. Comment on the advertiser's claim.

13. Schuman and Presser (1981, p. 277) report a study in which one set of respondents was asked question A, and the other set was asked question B:

    A. Do you think the United States should forbid public speeches against democracy?

    B. Do you think the United States should allow public speeches against democracy?

    For one version of the question, only about one-fifth of the respondents were against such freedom of speech, whereas for the other version almost half were against such freedom of speech. Which question do you think elicited which response? Explain.

14. Give an example of two questions in which the order in which they are presented would determine whether the responses were likely to be biased.

15. In February 1998, U.S. president Bill Clinton was under investigation for allegedly having had an extramarital affair. A Gallup Poll asked the following two questions: "Do you think most presidents have or have not had extramarital affairs while they were president?" and then "Would you describe Bill Clinton's faults as worse than most other presidents, or as no worse than most other presidents?" For the first question, 59% said "have had," 33% said "have not," and the remaining 8% had no opinion. For the second question, 24% said "worse," 75% said "no worse," and only 1% had no opinion. Do you think the order of these two questions influenced the results? Explain.

16. Sometimes medical tests, such as those for detecting HIV, are so sensitive that people do not want to give their names when they take the test. Instead, they are given a number or code, which they use to obtain their results later. Is this procedure anonymous testing or is it confidential testing? Explain.

17. Give three versions of a question to determine whether people think smoking should be banned on all airline flights. Word the question to be as follows:

   a. As unbiased as possible

   b. Likely to get people to respond that smoking should be forbidden

   c. Likely to get people to respond that smoking should not be forbidden

18. Explain the difference between a discrete variable and a categorical variable. Give an example of each type.

19. Suppose you were to compare two routes to school or work by timing yourself on each route for five days. Suppose the times on one route were (in minutes) 10, 12, 13, 15, 20, and on the other route they were 10, 15, 16, 18, 21.

   a. The average times for the two routes are 14 minutes and 16 minutes. Would you be willing to conclude that the first route is faster, on average, based on these sample measurements?

   b. Give an example of two sets of times, where the first has an average of 14 minutes and the second an average of 16 minutes, for which you *would* be willing to conclude that the first route is faster.

   c. Explain how the concept of natural variability entered into your conclusions in parts a and b.

20. Give an example of a characteristic that could be measured as either a discrete or a continuous variable, depending on the types of units used.

21. Airlines compute the percentage of flights that are on time to be the percentage that arrive no later than 15 minutes after their scheduled arrival time. Is this a valid measure of on-time performance? Is it a reliable measure? Explain.

....

MINI-PROJECTS

1. Measure the heights of five males and five females. Draw a line to scale, starting at the lowest height in your group and ending at the highest height, and mark each male with an M and each female with an F. It should look something like this:

F		F	M		FMF		F	MM	M
5'									6'2"

   Explain exactly how you measured the heights, and then answer each of the following:

   a. Are your measures valid?

    b. Are your measures reliable?

    c. How does the variability in the measurements within each group compare to the difference between the two groups? For example, are all of your men taller than all of your women? Are they completely intermixed?

    d. Do you think your measurements would convince an alien being that men are taller, on average, than women? Explain. Use your answer to part c as part of your explanation.

2. Design a survey with three questions to measure attitudes toward something of interest to you. Now design a new version by changing just a few words in each question to make it deliberately biased. Choose 20 people to whom you will administer the survey. Put their names in a hat (or a box or a bag) and draw out ten names. Administer the first (unbiased) version of the survey to them and the second (biased) version to the remaining ten people. Compare the responses and discuss what happened.

REFERENCES

Anastasi, Anne, and Susana Urbina. (1997). *Psychological testing.* 7th ed. New York: Macmillan.

Crossen, Cynthia (1994). *Tainted truth.* New York: Simon and Schuster.

Loftus, E. F., and J. C. Palmer. (1974). Reconstruction of automobile destruction: An example of the interaction between language and memory. *Journal of Verbal Learning and Verbal Behavior* 13, pp. 585–589.

Morin, Richard. (10–16 April 1995). What informed public opinion? *Washington Post,* National Weekly Edition.

Plous, Scott. (1993). *The psychology of judgment and decision making.* New York: McGraw-Hill.

Schuman, H., and S. Presser. (1981). *Questions and answers in attitude surveys.* New York: Academic Press.

Schuman, H., and J. Scott. (22 May 1987). Problems in the use of survey questions to measure public opinion. *Science* 236, pp. 957–959.

U.S. Department of Labor. Bureau of Labor Statistics. (September 1992). *BLS handbook of methods.* Bulletin 2414.

CHAPTER FOUR

How to Get a Good Sample

THOUGHT QUESTIONS

1. What do you think is the major difference between a *survey* (such as a public opinion poll) and an *experiment* (such as the heartbeat experiment in Case Study 1.1)?

2. Suppose a properly chosen sample of 1600 people across the United States was asked if they regularly watch a certain television program, and 24% said yes. How close do you think that is to the percentage of the entire country who watch the show? Within 30%? 10%? 5%? 1%? Exactly the same?

3. Many television stations conduct polls by asking viewers to call one phone number if they feel one way about an issue and a different phone number if they feel the opposite. Do you think the results of such a poll represent the feelings of the community? Do you think they represent the feelings of all those watching the TV station at the time or the feelings of some other group? Explain.

4. Suppose you had a telephone directory listing all the businesses in a city, alphabetized by type of business. If you wanted to phone 100 of them to get a representative sampling of opinion on some issue, how would you select which 100 to phone? Why would it not be a good idea to simply use the first 100 businesses listed?

5. There are many professional polling organizations, such as Gallup and Roper. They often report on surveys they have done, announcing that they have sampled 1243 adults, or some such number. How do you think they select the people to include in their samples?

4.1

COMMON RESEARCH STRATEGIES

In Chapters 1 to 3, we discussed scientific studies in general, without differentiating them by type. In this chapter and the next, we are going to look at proper ways to conduct specific types of studies. When you read the results of a scientific study, the first thing you need to do is determine which research strategy was used. You can then see whether or not the study used the proper methods for that strategy. In this chapter and the next, you will learn about potential difficulties and outright disasters that can befall each type of study, as well as some principles for executing them correctly. First, let's examine the common types of research strategies.

Sample Surveys

You are probably quite familiar with sample surveys, at least in the form of political and opinion polls. In a **sample survey,** a subgroup of a large population is questioned on a set of topics. The results from the subgroup are used as if they were representative of the larger population, which they will be if the sample was chosen correctly. There is no intervention or manipulation of the respondents in this type of research, they are simply asked to answer some questions. We examine sample surveys in more depth later in this chapter.

Experiments

An **experiment** measures the effect of manipulating the environment in some way. For example, the manipulation may include receiving a drug or medical treatment, going through a training program, agreeing to a special diet, and so on. Most experiments on humans use volunteers because you can't force someone to accept a manipulation. You then measure the result of the feature being manipulated, called the **explanatory variable,** on an outcome, called the **outcome variable.** Examples of outcome variables are cholesterol level (after taking a new drug), amount learned (after a new training program), or weight loss (after a special diet).

As an example, recall Case Study 1.2, an experiment that investigated the relationship between aspirin and heart attacks. The explanatory variable, manipulated by the researchers, was whether a participant took aspirin or a placebo. The variable was then used to help explain the outcome variable, which was whether a participant had a heart attack or not. Notice that the explanatory and outcome variables are both categorical in this case, with two categories each (aspirin/placebo and heart attack/no heart attack).

Experiments are important because, unlike most other studies, they often allow us to determine cause and effect. The participants in an experiment are usually randomly assigned to either receive the manipulation or take part in a control group. The purpose of the random assignment is to make the two groups approximately equal in all respects except for the explanatory variable, which is purposely manipulated. Differences in the outcome variable between the groups, if large enough to rule out natural chance variability, can then be attributed to the manipulation of the explanatory variable.

For example, suppose we flip a coin to assign each of a number of new babies into one of two groups. Without any intervention, we should expect both groups to gain about the same amount of weight, on average. If we then expose one group to the sound of a heartbeat and that group gains significantly more weight than the other group, we can be reasonably certain that the weight gain was due to the sound of the heartbeat.

Observational Studies

As we noted in Chapter 1, an **observational study** resembles an experiment except that the manipulation occurs naturally rather than being imposed by the experimenter. For example, we can observe what happens to people's weight when they quit smoking, but we can't experimentally manipulate them to quit smoking. We must rely on naturally occurring events.

This reliance on naturally occurring events leads to problems with establishing a causal connection because we can't arrange to have a similar control group. For instance, people who quit smoking may do so because they are on a "health kick" that also includes better eating habits, a change in coffee consumption, and so on. In this case, if we were to observe a weight loss (or gain) after cessation of smoking, we would not know if it were caused by the changes in diet or the lack of cigarettes. In an observational study, you cannot assume that the explanatory variable of interest to the researchers is the only one that may be responsible for any observed differences in the outcome variable.

A special type of observational study is frequently used in medical research. Called a **case-control study,** it is an attempt to include an appropriate control group. In Chapter 5, we will explore the details of how these and other observational studies are conducted, and in Chapter 6, we will cover some examples in depth.

Observational studies do have one advantage over experiments. Researchers are not required to induce artificial behavior. Participants are simply observed doing what they would do naturally; therefore, the results can be more readily extended to the real world.

Meta-Analyses

A **meta-analysis** is a quantitative review of a collection of studies all done on a similar topic. Combining information from various researchers may result in the emergence of patterns or effects that weren't conclusively available from the individual studies.

It is becoming quite common for the results of meta-analyses to appear in newspapers and magazines. For example, the top headline in the November 24, 1993, *San Jose Mercury News* was, "Why mammogram advice keeps changing: S.F. study contradicts cancer society's finding." The article explained that, in addition to the results of new research in San Francisco, "A recent analysis of eight international studies did not find any clear benefit to women getting routine mammograms while in their 40s."

When you see wording indicating that many studies were analyzed together, the report is undoubtedly referring to a meta-analysis. In this case, the eight studies in question had been conducted over a 30-year period with 500,000 women. Unfortunately, that information was missing from the newspaper article. Missing information is one of the problems with trying to evaluate a news article based on meta-analysis. In Chapter 24, we will examine meta-analyses and the role they play in science.

Case Studies

A **case study** is an in-depth examination of one or a small number of individuals. The researcher observes and interviews that individual and others who know about the topic of interest. For example, to study a purported psychic healer, a researcher might observe her at work, interview her about techniques, and interview clients who had been treated by the healer. We do not cover case studies in this book because they are descriptive and do not require statistical methods. We will issue one warning, though. Be careful not to assume you can extend the findings of a case study to any person or situation other than the one studied. In fact, case studies may be used to investigate situations precisely because they are rare and unrepresentative.

4.2   DEFINING A COMMON LANGUAGE

In the remainder of this chapter, we explore the methods used in sample surveys. To make our discussion of sampling methods clear, let's establish a common language. As we have seen before, statisticians borrow words from everyday language and attach specialized meaning to them.

The first thing you need to know is that researchers sometimes speak synonymously of the individuals being measured and the measurements themselves. You can usually figure this out from the context. The relevant definitions cover both meanings.

- A **unit** is a single individual or object to be measured.
- The **population** (or **universe**) is the entire collection of units about which we would like information or the entire collection of measurements we would have if we could measure the whole population.
- The **sample** is the collection of units we actually measure or the collection of measurements we actually obtain.
- The **sampling frame** is a list of units from which the sample is chosen. Ideally, it includes the whole population.
- In a **sample survey**, measurements are taken on a subset, or sample, of units from the population.
- A **census** is a survey in which the entire population is measured.

EXAMPLE 1

DETERMINING MONTHLY UNEMPLOYMENT IN THE UNITED STATES

In the United States, the Bureau of Labor Statistics (BLS) is responsible for determining monthly unemployment rates. To do this, the BLS does not collect information on all adults; that is, it does not take a census. Instead, employees visit approximately 60,000 households, chosen from a list of all known households in the country, and obtain information on the approximately 116,000 adults living in them. They classify each person as employed, unemployed, or "not in the labor force." The last category includes the "discouraged workers" discussed in Chapter 3. The unemployment rate is the number of unemployed persons divided by the sum of the employed and unemployed. Those "not in the labor force" are not included at all. (See the U.S. Department of Labor's *BLS Handbook of Methods*, referenced at the end of Chapter 3, for further details.)

Before reading any further, try to apply the definitions you have just learned to the way the BLS calculates unemployment. In other words, specify the units, the population, the sampling frame, and the sample. Be sure to include both forms of each definition when appropriate.

The *units* of interest to the BLS are adults in the labor force, meaning adults who meet their definitions of employed and unemployed. Those who are "not in the labor force" are not relevant units. The *population of units* consists of all adults who are in the labor force. The *population of measurements,* if we could obtain it, would consist of the employment status (working or not working) of everyone in the labor force. The *sampling frame* is the list of all known households in the country. The people who actually get asked about their employment status by the BLS constitute the *units in the sample,* and their actual employment statuses constitute the *measurements in the sample.* ■

4.3   THE BEAUTY OF SAMPLING

Here is some information that may astound you. If you use commonly accepted methods to sample 1500 adults from an entire population of millions of adults, you can almost certainly gauge, to within 3%, the percentage of the entire population who have a certain trait or opinion. (There is nothing magical about 1500 and 3%, as you will soon see.) Even more amazing is the fact that this result doesn't depend on how big the population is; it depends only on how many are in the sample. Our sample of 1500 would do equally well at estimating, to within 3%, the percentage of a population of 10 billion. Of course, you have to use a proper sampling method—but we address that later.

You can see why researchers are content to rely on public opinion polls rather than trying to ask everyone for their opinion. It is much cheaper to ask 1500 people than several million, especially when you can get an answer that is almost as accurate. It also takes less time to conduct a sample survey than a census, and because fewer interviewers are needed, there is better quality control.

Accuracy of a Sample Survey: Margin of Error

Most sample surveys are used to estimate the proportion or percentage of people who have a certain trait or opinion. For example, the Nielsen ratings, used to determine the percentage of American television sets tuned to a particular show, are based on a sample of a few thousand households. Newspapers and magazines routinely conduct surveys of a few thousand people to determine public opinion on current topics of interest. As we have said, these surveys, if properly conducted, are amazingly accurate. The measure of accuracy is a number called the **margin of error.** The sample proportion differs from the population proportion by more than the margin of error less than 5% of the time, or in fewer than 1 in 20 surveys. To express results in terms of percentages instead of proportions, simply multiply everything by 100.

> *As a general rule, the amount by which the proportion obtained from the sample will differ from the true population proportion rarely exceeds 1 divided by the square root of the number in the sample.* This is expressed by the simple formula $1/\sqrt{n}$, where the letter n represents the number of people in the sample.

For example, with a sample of 1600 people, we usually get an estimate that is accurate to within $1/40 = 0.025 = 2.5\%$ of the truth because the square root of 1600 is 40. You might see results such as, "Fifty-five percent of respondents support the president's economic plan. The margin of error for this survey is plus or minus 2.5 percentage points." This means that it is almost certain that between 52.5% and 57.5% of the entire population support the plan. In other words, add and subtract the margin of error to the sample value, and the resulting interval almost surely covers the true population value. If you were to follow this method every time you read the results of a properly conducted survey, the interval would only miss covering the truth about 1 in 20 times.

Other Advantages of Sample Surveys

When a Census Isn't Possible

Suppose you needed a laboratory test to see if your blood had too high a concentration of a certain substance. Would you prefer that the lab measure the entire population of your blood, or would you prefer to give a sample? Similarly, suppose a manufacturer of firecrackers wanted to know what percentage of their products were duds. They would not make much of a profit if they tested them all, but they could get a reasonable estimate of the desired percentage by testing a properly selected sample. As these examples illustrate, there are situations where measurements destroy the units being tested and thus a census is not feasible.

Speed

Another advantage of a sample survey over a census is the amount of time it takes to conduct a sample survey. For example, it takes several years to successfully plan and execute a census of the entire population of the United States. Getting monthly unemployment rates would be impossible with a census; the results would be quite out of date by the time they were released. It is much faster to collect a sample than a census if the population is large.

Accuracy

A final advantage of a sample survey is that you can devote your resources to getting the most accurate information possible from the sample you have selected. It is easier to train a small group of interviewers than a large one, and it is easier to track down a small group of nonrespondents than the larger one that would inevitably result from trying to conduct a census.

4.4   SIMPLE RANDOM SAMPLING

The ability of a relatively small sample to accurately reflect the opinions of a huge population does not happen haphazardly. It works only if proper sampling methods are used. Everyone in the population must have a specified chance of making it into the sample. Methods with this characteristic are called **probability sampling plans.**

The simplest way of accomplishing this goal is to use a **simple random sample.** With a simple random sample, every conceivable group of people of the required size has the same chance of being the selected sample.

To actually produce a simple random sample, you need only two things. First, you need a list of the units in the population. Second, you need a source of **random numbers.** Random numbers can be found in tables designed for that purpose, called "tables of random digits," or they can be generated by computers and calculators. If the population isn't too large, physical methods can be used, as illustrated in the next hypothetical example.

EXAMPLE 2   **HOW TO SAMPLE FROM YOUR CLASS**

Suppose you are taking a class with 200 students and are unhappy with the teaching method. To substantiate that a problem exists so that you can complain to higher powers, you decide to collect a simple random sample of 25 students and ask them for their opinions.

Notice that a sample of this size would have a margin of error of about 20% because $1/\sqrt{25} = 1/5 = 0.20$. Thus, the percentage of those 25 people who were dissatisfied would almost surely be within 20% of the percentage of the

entire class who were dissatisfied. If 60% of the sample said they were dissatisfied, you could tell the higher powers that somewhere between 40% and 80% of
the entire class was probably dissatisfied. Although that's not a very precise statement, it is certainly enough to show major dissatisfaction.

To collect your sample, you would proceed as follows:

STEP 1: Obtain a list of the students in the class, numbered from 1 to 200.

STEP 2: Obtain 25 random numbers between 1 and 200. One simple way to
do this would be to write each of the numbers from 1 to 200 on equally sized
slips of paper, put them in a bag, mix them very well, and draw out 25. However, we will instead use a computer program called *Minitab* to select the 25
numbers. Here is what the program and the results look like:

```
MTB > set c1
DATA > 1:200
DATA > end
MTB > sample 25 c1 c2
MTB > print c2
C2
31 141 35 69 100 182 61 116 191 161 129 120 150 15 84
   194 135 101 44 163 152 39 99 110 36
```

STEP 3: Locate and interview the people on your list whose numbers were
selected.

Notice that it is important to try to locate the actual 25 people resulting from this
process. If you tried to phone someone only once and gave up when you could
not reach that person, you would bias your results toward people who were home
more often. If you collected your sample correctly, as described, you would have
legitimate data to present to the higher powers. ■

4.5   OTHER SAMPLING METHODS

By now you may be asking yourself how polling organizations could possibly get a
numbered list of all voters or of all adults in the country. In truth, they don't. Instead,
they rely on more complicated sampling methods. Here we describe a few other
sampling methods, all of which are good substitutes for simple random sampling in
most situations. In fact, they often have advantages over simple random sampling.

Stratified Random Sampling

Sometimes the population of units falls into natural groups, called **strata.** For example, public opinion pollsters often take separate samples from each region of the
country so they can spot regional differences as well as measure national trends.

Political pollsters may sample separately from each political party to compare opinions by party.

A **stratified random sample** is collected by first dividing the population of units into groups (strata) and then taking a simple random sample from each. For example, the strata might be regions of the country or political parties. You can often recognize this type of sampling when you read the results of a survey because the results will be listed separately for each of the strata.

Stratified sampling has other advantages besides the fact that results are available separately by strata. One is that different interviewers may work best with different people. For example, people from separate regions of the country (South, Northeast, and so on) may feel more comfortable with interviewers from the same region. It may also be more convenient to stratify before sampling. If we were interested in opinions of college students across the country, it would probably be easier to train interviewers at each college rather than to send the same interviewer to all campuses.

So far we have been focusing on the collection of categorical variables, such as opinions or traits people might have. Surveys are also used to collect measurement variables, such as age at first intercourse or number of cigarettes smoked per day. We are often interested in the population average for such measurements. The accuracy with which we can estimate the average depends on the natural variability among the measurements. The less variable they are, the more precisely we can assess the population average on the basis of the sample values. For instance, if everyone in a relatively large sample reports that his or her age at first intercourse was between 16 years 3 months and 16 years 4 months, then we can be relatively sure that the average age in the population is close to that. However, if reported ages range from 13 years to 25 years, then we cannot pinpoint the average age for the population nearly as accurately.

Stratified sampling can help to solve the problem of great natural variability. Suppose we could figure out how to stratify in a way that allowed little natural variability in the answers within each strata. We could then get an accurate estimate for each stratum and combine estimates to get a much more precise answer for the group than if we measured everyone together. For example, if we wanted to estimate the average weight gain of newborn babies during the first four days of life, we could do so more accurately by dividing the babies into groups based on their initial birth weight. Very heavy newborns actually tend to lose weight during the first few days, whereas very light ones tend to gain more weight.

Cluster Sampling

Cluster sampling is often confused with stratified sampling, but it is actually a radically different concept and can be much easier to accomplish. The population units are again divided into groups, called **clusters,** but rather than sampling within each group, we select a random sample of clusters and measure only those clusters. One obvious advantage of cluster sampling is that you need only a list of clusters, instead of a list of all individual units.

For example, suppose we wanted to sample students living in the dormitories at a college. If the college had 30 dorms and each dorm had 6 floors, we could consider the 180 floors to be 180 clusters of units. We could then randomly select the desired number of floors and measure everyone on those floors. Doing so would probably be much cheaper and more convenient than obtaining a simple random sample of all dormitory residents.

If cluster sampling is used, the analysis must proceed differently because similarities may exist among the members of the clusters, and these must be taken into account. Numerous books are available that describe proper analysis methods based on which sampling plan was employed. [See, for example, B. Williams, *A sampler on sampling* (New York: Wiley, 1978).]

Stratified sampling is sometimes used instead of simple random sampling for the following reasons:

1. We can find individual estimates for each stratum.

2. If the variable measured gives more consistent values within each of the strata than within the whole population, we can get more accurate estimates of the population values.

3. If strata are geographically separated, it may be cheaper to sample them separately.

4. We may want to use different interviewers within each of the strata.

Systematic Sampling

Suppose you had a list of 5000 names and telephone numbers from which you wanted to select a sample of 100. That means you would want to select 1 of every 50 people on the list. The first idea that might occur to you is to simply choose every 50th name on the list. If you did so, you would be using a **systematic sampling plan.** With this plan, you divide the list into as many consecutive segments as you need, randomly choose a starting point in the first segment, then sample at that same point in each segment. In our example, you would randomly choose a starting point in the first 50 names, then sample every 50th name after that. When you were finished, you would have selected one person from each of 100 segments, equally spaced throughout the list.

Systematic sampling is often a good alternative to random sampling. In a few instances, however, it can lead to a biased sample, and common sense must be used to avoid those. As an example, suppose you were doing a survey of potential noise problems in a high-rise college dormitory. Further, suppose a list of residents was provided, arranged by room number, with 20 rooms per floor and two people per room. If you were to take a systematic sample of, say, every 40th person on the list, you would get people who lived in the same location on every floor—and thus a biased sampling of opinions about noise problems.

Random Digit Dialing

Most of the national polling organizations in the United States now use a method of sampling called **random digit dialing.** This method results in a sample that approximates a simple random sample of all households in the United States that have telephones. The method proceeds as follows. First, they make a list of all possible telephone *exchanges,* where the exchange consists of the area code and the next three digits. Using numbers listed in the white pages, they can approximate the proportion of all households in the country that have each exchange. They then use a computer to generate a sample that has approximately those same proportions. Next, they use the same method to randomly sample *banks* within each exchange, where a bank consists of the next two numbers. Phone companies assign numbers using banks, so that certain banks are mainly assigned to businesses, certain ones are held for future neighborhoods, and so on. Finally, to complete the number, the computer randomly generates two digits from 00 to 99.

Once a phone number has been determined, a well-conducted poll will make multiple attempts to reach someone at that household. Sometimes they will ask to speak to a male because females are more likely to answer the phone and would thus be overrepresented.

Multistage Sampling

Many large surveys, especially those that are conducted in person rather than over the telephone, use a combination of the methods we have discussed. They might stratify by region of the country; then stratify by urban, suburban, and rural; and then choose a random sample of communities within those strata. They would then divide those communities into city blocks or fixed areas, as clusters, and sample some of those. Everyone on the block or within the fixed area may then be sampled. This is called a **multistage sampling plan.**

4.6   DIFFICULTIES AND DISASTERS IN SAMPLING

Difficulties
1. Using the wrong sampling frame
2. Not reaching the individuals selected
3. Getting no response or getting a volunteer response

Disasters
4. Getting a volunteer sample
5. Using a convenience or haphazard sample

In theory, designing a good sampling plan is easy and straightforward. However, the real world rarely cooperates with well-designed plans, and trying to collect a proper sample is no exception. Difficulties that can occur in practice need to be considered when you evaluate a study. If a proper sampling plan is never implemented, the conclusions can be misleading and inaccurate.

Difficulties in Sampling

Following are some problems that can occur even when a sampling plan has been well designed.

Using the Wrong Sampling Frame

Remember that the sampling frame is the list of the population of units from which the sample is drawn. Sometimes a sampling frame will either include unwanted units or exclude desired units. For example, using a list of registered voters to predict election outcomes includes those who are not likely to vote as well as those who are likely to do so. Using a telephone directory to survey the general population excludes those who move often, those with unlisted home numbers (such as many physicians and teachers), and those who cannot afford a telephone.

Common sense can often lead to a solution for this problem. In the example of registered voters, interviewers may try to first ascertain the voting history of the person contacted by asking where he or she votes and then continuing the interview only if the person knows the answer. Instead of using a telephone directory, surveys use random digit dialing. This solution still excludes those without phones but not those who didn't happen to be in the last printed directory.

Not Reaching the Individuals Selected

Even if a proper sample of units is selected, the units may not be reached. For example, *Consumer Reports* magazine mails a lengthy survey to its subscribers to obtain information on the reliability of various products. If you were to receive such a survey, and you had a close friend who had been having trouble with a highly rated automobile, you may very well decide to pass the questionnaire on to your friend to answer. That way, he would get to register his complaints about the car, but *Consumer Reports* would not have reached the intended recipient.

Telephone surveys tend to reach a disproportionate number of women because they are more likely to answer the phone. To try to counter that problem, researchers sometimes ask to speak to the oldest adult male at home. Surveys are also likely to have trouble contacting people who work long hours and are rarely home or those who tend to travel extensively.

In recent years, news organizations have been pressured to produce surveys of public opinion quickly. When a controversial story breaks, people want to know how others feel about it. This pressure results in what *Wall Street Journal* reporter Cynthia Crossen calls "quickie polls." As she notes, these are "most likely to be wrong because questions are hastily drawn and poorly pretested, and it is almost impossible to get a random sample in one night" (Crossen, 1994, p. 102). Even with the computer randomly generating phone numbers for the sample, many people are not likely to be home that night—and they may have different opinions from those who are likely to be home. Most responsible reports about polls include information about the dates during which they were conducted. If a poll was done in one night, beware!

It is important that once a sample has been selected, those individuals are the ones who are actually measured. It is better to put resources into getting a smaller sample than to get one that has been biased because the survey takers moved on to the next person on the list when a selected individual was initially unavailable.

Getting No Response or a Volunteer Response

Even the best surveys are not able to contact everyone on their list, and not everyone contacted will respond. The General Social Survey (GSS), run by the prestigious National Opinion Research Center (NORC) at the University of Chicago, noted in its September 1993 *GSS News:*

> *In 1993 the GSS achieved its highest response rate ever, 82.4%. This is five percentage points higher than our average over the last four years. Given the long length of the GSS (90 minutes), the high response rates of the GSS are testimony to the extraordinary skill and dedication of the NORC field staff.*

Beyond having a dedicated staff, not much can be done about getting everyone in the sample to respond. Response rates should simply be reported in research summaries. As a reader, remember that the lower the response rate, the less the results can be generalized to the population as a whole. Responding to a survey (or not) is voluntary, and those who respond are likely to have stronger opinions than those who do not.

With mail surveys, it may be possible to compare those who respond immediately with those who need a second prodding, and in telephone surveys you could compare those who are home on the first try with those who require numerous callbacks. If those groups differ on the measurement of interest, then those who were never reached are probably different as well.

In a mail survey, it is best not to rely solely on "volunteer response." In other words, don't just accept that those who did not respond the first time can't be cajoled into it. Often, sending a reminder with a brightly colored stamp or following up with a personal phone call will produce the desired effect. Surveys that simply use those who respond voluntarily (such as the one conducted by *Consumer Reports*) are sure to be biased in favor of those with strong opinions or with time on their hands.

EXAMPLE 3   **WHICH SCIENTISTS TRASHED THE PUBLIC?**

According to a poll taken among scientists and reported in the prestigious journal *Science* (Mervis, 1998), scientists don't have much faith in either the public or the media. The article reported that, based on the results of a "recent survey of 1400 professionals" in science and in journalism, 82% of scientists "strongly or somewhat agree" with the statement "The U.S. public is gullible and believes in miracle cures or easy solutions," and 80% agreed that "the public doesn't understand the importance of federal funding for research." About the same percentage (82%) also trashed the media, agreeing with the statement "The media do not understand statistics well enough to explain new findings." It isn't until the end of the article that we learn who responded: "The study reported a 34% response rate among scientists, and the typical respondent was a white, male physical scientist over the age of 50 doing basic research." Remember that those who feel strongly about the issues in a survey are the most likely to respond. With only about a third of those contacted responding, it is inappropriate to generalize these findings and conclude that most scientists have so little faith in the public and the media. This is especially true because we were told that the respondents represented only a narrow subset of scientists. ■

Disasters in Sampling

A few sampling methods are so bad that they don't even warrant a further look at the study or its results.

Getting a Volunteer Sample

Although relying on volunteer *responses* presents somewhat of a difficulty in determining the extent to which surveys can be generalized, relying on a volunteer *sample* is a complete waste of time. If a magazine or television station runs a survey and asks any readers or viewers who are interested to respond, the results reflect only the opinions of those who decide to volunteer. As noted earlier, those who have a strong opinion about the question are more likely to respond than those who do not. Thus, the responding group is simply not representative of any larger group. Most media outlets now acknowledge that such polls are "unscientific" when they report the results, but most readers are not likely to understand how misleading the results can be. The next example illustrates the contradiction that can result between a scientific poll and one relying solely on a volunteer sample.

EXAMPLE 4　　**A MEANINGLESS POLL**

On February 18, 1993, shortly after Bill Clinton became president of the United States, a television station in Sacramento, California, asked viewers to respond to the question: "Do you support the president's economic plan?" The next day, the results of a properly conducted study asking the same question were published in the newspaper. Here are the results:

	TELEVISION POLL	SURVEY
Yes (support plan)	42%	75%
No (don't support plan)	58%	18%
Not sure	0%	7%

As you can see, those who were dissatisfied with the president's plan were much more likely to respond to the television poll than those who supported it, and no one who was "Not sure" called the television station because they were not invited to do so. Trying to extend those results to the general population is misleading. It is irresponsible to publicize such studies, especially without a warning that they result from an unscientific survey and are not representative of general public opinion. You should never interpret such polls as anything other than a count of who bothered to go to the telephone and call. ■

Using a Convenience or Haphazard Sample

Another worthless sampling technique for surveys is to use the most convenient group available or to decide on the spot who to sample. Again, the group does not represent any larger population.

EXAMPLE 5   **HAPHAZARD SAMPLING**

A few years ago, the student newspaper at a California university announced as a front page headline: "Students ignorant, survey says." The article explained that a "random survey" indicated that American students were less aware of current events than international students were. However, the article quoted the undergraduate researchers, who were international students themselves, as saying that "the students were randomly sampled on the quad." The quad is an open-air, grassy area where students relax, eat lunch, and so on. There is simply no proper way to collect a random sample of students by selecting them in an area like that. In such situations, the researchers are likely to approach people whom they think will support the results they intended for their survey. Or, they are likely to approach friendly looking people who look as though they will easily cooperate. This is called a haphazard sample, and it cannot be expected to be representative at all. ■

You have seen the proper way to collect a sample and have been warned about the many difficulties and dangers inherent in the process. We finish the chapter with a famous example that helped researchers learn some of these pitfalls.

CASE STUDY 4.1 | # The Infamous <u>Literary Digest</u> Poll of 1936

Before the election of 1936, a contest between Democratic incumbent Franklin Delano Roosevelt and Republican Alf Landon, the magazine *Literary Digest* had been extremely successful in predicting the results in U.S. presidential elections. But 1936 turned out to be the year of their downfall, when they predicted a 3-to-2 victory for Landon. To add insult to injury, young pollster George Gallup, who had just founded the American Institute of Public Opinion in 1935, not only correctly predicted Roosevelt as the winner of the election, he also predicted that the *Literary Digest* would get it wrong. He did this before they even conducted their poll. And Gallup surveyed only 50,000 people, whereas the *Literary Digest* sent questionnaires to 10 million people (Freedman, Pisani, Purves, and Adhikari, 1991, p. 307).

The *Literary Digest* made two classic mistakes. First, the lists of people to whom they mailed the 10 million questionnaires were taken from magazine subscribers, car owners, telephone directories, and, in just a few cases, lists of registered voters. In 1936, those who owned telephones or cars, or subscribed to magazines, were more likely to be wealthy individuals who were not happy with the Democratic incumbent.

Despite what many accounts of this famous story conclude, the bias produced by the more affluent list was not likely to have been as severe as the second problem (Bryson, 1976). *The main problem was volunteer response.* The magazine received 2.3 million responses, a response rate of only 23%. Those who felt strongly about the outcome of the election were most likely to respond. And that included a majority of those who wanted a change, the Landon supporters. Those who were happy with the incumbent were less likely to bother to respond.

Gallup, however, knew the value of random sampling. He was able not only to predict the election, but to predict the results of the *Literary Digest* poll within 1%. How did he do this? According to Freedman and colleagues (1991, p. 308), "he just chose 3000 people at random from the same lists the *Digest* was going to use, and mailed them all a postcard asking them how they planned to vote."

This example illustrates the beauty of random sampling and the idiocy of trying to base conclusions on nonrandom and biased samples. The *Literary Digest* went bankrupt the following year, and so never had a chance to revise its methods. The organization founded by George Gallup has flourished, although not without making a few sampling blunders of its own (see, for example, Exercise 11 on page 65). ■

....

EXERCISES

1. For each of the following situations, state which type of sampling plan was used. Explain whether you think the sampling plan would result in a biased sample.

   a. To survey the opinions of its customers, an airline company made a list of all its flights and randomly selected 25 flights. All of the passengers on those flights were asked to fill out a survey.

b. A pollster interested in opinions on gun control divided a city into city blocks, then surveyed the third house to the west of the southeast corner of each block. If the house was divided into apartments, the westernmost ground floor apartment was selected. The pollster conducted the survey during the day, but left a notice for those who were not at home to phone her so she could interview them.

c. To learn how its employees felt about higher student fees imposed by the legislature, a university divided employees into three categories: staff, faculty, and student employees. A random sample was selected from each group and they were telephoned and asked for their opinions.

d. A large variety store wanted to know if consumers would be willing to pay slightly higher prices to have computers available throughout the store to help them locate items. The store posted an interviewer at the door and told her to collect a sample of 100 opinions by asking the next person who came in the door each time she had finished an interview.

2. Explain the difference between a proportion and a percentage as used to present the results of a sample survey.

3. Construct an example in which a systematic sampling plan would result in a biased sample.

4. In the March 8, 1994, edition of the *Scotsman,* a newspaper published in Edinburgh, Scotland, a headline read, "Reform study finds fear over schools." The article described a survey of 200 parents who had been asked about proposed education reforms and indicated that most parents felt uninformed and thought the reforms would be costly and unnecessary. The report did not clarify whether a random sample was chosen, but make that assumption in answering the following questions.

a. What is the margin of error for this survey?

b. It was reported that "about 80 percent added that they were satisfied with the current education set-up in Scotland." What is the range of values that almost certainly covers the percentage of the population of parents who were satisfied?

c. The article quoted Lord James Douglas-Hamilton, the Scottish education minister, as saying, "If you took a similar poll in two years' time, you would have a different result." Comment on this statement.

5. An article in the *Sacramento Bee* (12 January 1998, p. A4) was titled "College freshmen show conservative side" and reported the results of a fall 1997 survey "based on responses from a representative sample of 252,082 full-time freshmen at 464 two- and four-year colleges and universities nationwide." The article did not explain how the schools or students were selected.

a. For this survey, explain what a unit is, what the population is, and what the sample is.

b. Assuming a random sample of students was selected at each of the 464 schools, what type of sample was used in this survey? Explain.

    c. Now assume that the 464 schools were randomly selected from all eligible colleges and universities and that all first-year students at those schools were surveyed. Explain what type of sample was used in the survey.

    d. Why would one of the two sampling methods described in parts b and c have been simpler to implement than a simple random sample of all first-year college students in the United States?

6. The survey in Exercise 5 has been conducted annually by the Higher Education Research Institute at UCLA since 1966. One of the results reported was that "Students' disengagement from politics continues. The percentage of freshmen believing that 'keeping up to date with political affairs' is important fell to 26.7 percent, down from 29.4 percent a year ago [in 1996] and a high of 57.8 percent in 1966." In 1966, college students were in the midst of protesting the Vietnam War, and in 1996 there was a presidential election. Do you think the results of this survey indicate that first-year college students have become more apathetic in general? Explain.

7. Specify the population and the sample, being sure to include both units and measurements, for the situation described in:

    a. Exercise 1a

    b. Exercise 1b

    c. Exercise 1c

    d. Exercise 1d

8. Give an example in which:

    a. A sample would be preferable to a census

    b. A cluster sample would be the easiest method to use

    c. A systematic sample would be the easiest to use and would not be biased

9. Explain whether a survey or an experiment would be most appropriate to find out about each of the following:

    a. Who is likely to win the next presidential election

    b. Whether the use of nicotine gum reduces cigarette smoking

    c. Whether there is a relationship between height and happiness

    d. Whether a public service advertising campaign has been effective in promoting the use of condoms

10. Find a newspaper or magazine article describing a survey that is obviously biased. Explain why you think it is biased.

11. Despite his success in 1936, George Gallup failed miserably in trying to predict the winner of the 1948 U.S. presidential election. His organization, as well as two others, predicted that Thomas Dewey would beat incumbent Harry Truman. All three used what is called "quota sampling." The interviewers were told to find a certain number, or quota, of each of several types of people. For example, they might have been told to interview six women under age 40, one of whom was black and the other five of whom were white. Imagine that you are one of their interviewers trying to follow these instructions. Who would you ask? Now explain why you think these polls failed to predict the true winner and why quota sampling is not a good method.

12. Explain the difference between a volunteer *response* and a volunteer *sample*. Explain which is worse, and why.

13. Explain why the main problem with the *Literary Digest* poll is described as "volunteer response" and not "volunteer sample."

14. Gastwirth (1988, p. 507) describes a court case in which Bristol-Myers was ordered by the Federal Trade Commission to stop advertising that "twice as many dentists use Ipana as any other dentifrice" and that more dentists recommended it than any other dentifrice. Bristol-Myers had based its claim on a survey of 10,000 randomly selected dentists from a list of 66,000 subscribers to two dental magazines. They received 1983 responses, with 621 saying they used Ipana and only 258 reporting that they used the second most popular brand. As for the recommendations, 461 respondents recommended Ipana, compared with 195 for the second most popular choice.

   a. Specify the sampling frame for this survey, and explain whether you think "using the wrong sampling frame" was a difficulty here, based on what Bristol-Myers was trying to conclude.

   b. Of the remaining four "difficulties and disasters in sampling" listed in Section 4.6 (other than "using the wrong sampling frame"), which do you think was the most serious in this case? Explain.

   c. What could Bristol-Myers have done to improve the validity of the results after it had mailed the 10,000 surveys and received 1983 back? Assume the company kept track of who had responded and who had not.

15. A survey in *Newsweek* (14 November 1994, p. 54) asked: "Does the Senate generally pay too much attention to personal lives of people nominated to high office, or not enough?" Fifty-six percent of the respondents said "too much attention." It was also reported that "for this *Newsweek* poll, Princeton Survey Research Associates telephoned 756 adults Nov. 3–4. The margin of error is ±4 percentage points."

   a. Verify that the margin of error reported by *Newsweek* is consistent with the rule given in this chapter for finding the approximate margin of error.

   b. Based on these sample results, are you convinced that a majority of the population (that is, over 50%) think that the Senate pays too much attention? Explain.

16. The student newspaper at a university in California reported a debate between two student council members, revolving around a survey of students *(California Aggie,* 8 November 1994, p. 3). The newspaper reported that "according to an AS [Associated Students] Survey Unit poll, 52 percent of the students surveyed said they opposed a diversity requirement." The report said that one council member "claimed that the roughly 500 people polled were not enough to guarantee a statistically sound cross section of the student population." Another council member countered by saying that "three percent is an excellent random sampling, so there's no reason to question accuracy." (Note that the 3% figure is based on the fact that there were about 17,000 undergraduate students currently enrolled.)

a. Comment on the remark attributed to the first council member, that the sample size is not large enough to "guarantee a statistically sound cross section of the population." Is the size of the sample the relevant issue to address his concern?

b. Comment on the remark by the second council member that "three percent is an excellent random sampling, so there's no reason to question accuracy." Is she correct in her use of terminology and in her conclusion?

c. Produce an interval that almost certainly covers the true percentage of the population of students who oppose the diversity requirement. Use your result to comment on the debate. In particular, do these results allow a conclusion as to whether the majority of students on campus oppose the requirement?

17. Identify each of the following studies as a survey, an experiment, an observational study, or a case study. Explain your reasoning.

a. A doctor claims to be able to cure migraine headaches. A researcher administers a questionnaire to each of the patients the doctor claims to have cured.

b. Patients who visit a clinic to help them stop smoking are given a choice of two treatments: undergoing hypnosis or applying nicotine patches. The percentages who quit are compared for the two methods.

c. A large company wants to compare two incentive plans for increasing sales. The company randomly assigns a number of its sales staff to receive each kind of incentive and compares the average change in sales of the employees under the two plans.

18. Is using a convenience sample an example of a probability sampling plan? Explain why or why not.

19. What role does natural variability play when trying to determine the population average of a measurement variable from a sample?

20. Suppose that a gourmet food magazine wants to know how its readers feel about serving beer with various types of food. The magazine sends surveys to 1000 randomly selected readers. Explain which one of the "difficulties and disasters" in sampling the magazine is most likely to face.

21. Suppose you had a student telephone directory for your local college and wanted to sample 100 students. Explain how you would obtain each of the following:

a. A simple random sample

b. A systematic sample

22. Suppose you have a telephone directory for your local college from which you randomly select 100 names. To find out how students feel about a new pub on campus, you call the 100 numbers and interview the person who answers the phone. Explain which one of the "difficulties and disasters" in sampling you are most likely to encounter and how it could bias your results.

23. The U.S. government uses a multitude of surveys to measure opinions, behaviors, and so on. Yet, every 10 years it takes a census. What can the government learn from a census that it could not learn from a sample survey?

MINI-PROJECTS

1. For this project, you will use the telephone directory for your community to estimate the percentage of households that list their phone number but not their address. Use two different sampling methods, chosen from simple random sampling, stratified sampling, cluster sampling, or systematic sampling. In each case, sample about 100 households and figure out the proportion who do not list an address.

   a. Explain exactly how you chose your samples.

   b. Explain which of your two methods was easier to use.

   c. Do you think either of your methods produced biased results? Explain.

   d. Report your results, including a margin of error. Are your estimates from the two methods in agreement with each other?

2. Go to a large parking lot or a large area where bicycles are parked. Choose a color or a manufacturer. Design a sampling scheme you can use to estimate the percentage of cars or bicycles of that color or model. In choosing the number to sample, consider the margin of error that will accompany your sample result. Now go through the entire area, *actually taking a census,* and compute the population percentage of cars or bicycles of that type.

   a. Explain your sampling method and discuss any problems or biases you encountered in using it.

   b. Construct an interval from your sample that almost surely covers the true population percentage with that characteristic. Does your interval cover the true population percentage you found when you took the census?

   c. Use your experience with taking the census to give one practical difficulty with taking a census. (*Hint:* Did all the cars or bicycles stay put while you counted?)

REFERENCES

Bryson, M. C. (1976). The *Literary Digest* poll: Making of a statistical myth. *American Statistician* 30, pp. 184–185.

Crossen, Cynthia. (1994). *Tainted truth.* New York: Simon and Schuster.

Freedman, D., R. Pisani, R. Purves, and A. Adhikari. (1991). *Statistics.* 2d ed. New York: W. W. Norton.

Gastwirth, Joseph L. (1988). *Statistical reasoning in law and public policy.* Vol. 2. *Tort law, evidence and health.* Boston: Academic Press.

Mervis, Jeffrey (1998). Report deplores science-media gap. *Science* 279, p. 2036.

CHAPTER FIVE

Experiments and Observational Studies

THOUGHT QUESTIONS

1. In conducting a study to relate two conditions (activities, traits, and so on), researchers often define one of them as the *explanatory variable* and the other as the outcome or *response variable*. In a study to determine whether surgery or chemotherapy results in higher survival rates for a certain type of cancer, whether the patient survived is one variable, and whether he or she received surgery or chemotherapy is the other. Which is the explanatory variable and which is the response variable?

2. In an experiment, researchers assign "treatments" to participants, whereas in an observational study, they simply observe what the participants do naturally. Give an example of a situation where an experiment would not be feasible for ethical reasons.

3. Suppose you are interested in determining whether a daily dose of vitamin C helps prevent colds. You recruit 20 volunteers to participate in an experiment. You want half of them to take vitamin C and the other half to agree not to take it. You ask them each which they would prefer, and ten say they would like to take the vitamin and the other ten say they would not. You ask each of them to record how many colds he or she gets during the next ten weeks. At the end of that time, you compare the results reported from the two groups. Give three reasons why this is not a good experiment.

4. When experimenters want to compare two treatments, such as an old and a new drug, they use *randomization* to assign the participants to the two conditions. If you had 50 people participate in such a study, how would you go about randomizing them? Why do you think randomization is necessary? Why shouldn't the experimenter decide which people should get which treatment?

5. "Graduating is good for your health," according to a headline in the *Boston Globe* (3 April 1998, p. A25). The article noted, "According to the Center for Disease Control, college graduates feel better emotionally and physically than do high school dropouts." Do you think the headline is justified based on this statement? Explain why or why not.

5.1   DEFINING A COMMON LANGUAGE

In this chapter, we focus on studies that attempt to detect relationships between variables. Connections others have found, which we examine in this chapter and the next, include a relationship between baldness and heart attacks in men, between smoking during pregnancy and subsequent lower IQ in the child, between listening to Mozart and scoring higher on an IQ test, and between handedness and age at death. We will see that some of these connections are supported by properly conducted studies, whereas other connections are not as solid.

Explanatory Variables, Response Variables, and Treatments

In most studies, we imagine that if there is a causal relationship, it occurs in a particular direction. For example, if we found that left-handed people die at a younger age than right-handed people, we could envision reasons why their handedness might be responsible for the earlier death, such as accidents resulting from living in a right-handed world. It would be more difficult to argue that they were left-handed because they were going to die at an earlier age.

Explanatory Variables versus Response Variables

We define an **explanatory variable** to be one that attempts to explain or is purported to cause (at least partially) differences in a **response variable** (sometimes called an **outcome variable**). In the previous example, handedness would be the explanatory variable and age at death the response variable. In the Salk experiment described in Chapter 1, whether the baby listened to a heartbeat was the explanatory variable and weight gain was the response variable. In a study comparing chemotherapy to surgery for cancer, the medical treatment is the explanatory variable and surviving (usually measured as surviving for five years) or not surviving is the response variable. Many studies have more than one explanatory variable, but usually there is only one response variable.

We can usually distinguish which is which, but occasionally we simply examine relationships in which there is no conceivable causal connection. An example is the apparent relationship between baldness and heart attacks. Because the level of baldness was measured at the time of the heart attack, the heart attack could not have caused the baldness. It would be farfetched to assume that baldness results in such stress that men are led to have heart attacks. Instead, a third variable may be causing both the baldness and the heart attack. In such cases, we simply refer to the variables generically and do not assign one to be the explanatory variable and one to be the response variable.

Treatments

Sometimes the explanatory variable takes the form of a manipulation applied by the experimenter, such as when Salk played the sound of a heartbeat for some of the babies. A **treatment** is one or a combination of explanatory variables assigned by the experimenter. Notice that the term incorporates a collection of conditions. In Salk's experiment there were two conditions: Some babies received the heartbeat treatment and others received the silent treatment.

Experiment versus Observational Study

Ideally, if we were trying to ascertain the connection between the explanatory and response variables, we would keep everything constant except the explanatory variable. We would then manipulate the explanatory variable and notice what happened to the response variable as a consequence. We rarely reach this ideal, but we can come closer with an experiment than with an observational study.

In an **experiment,** we create differences in the explanatory variable and then examine the results. In an **observational study** we observe differences in the explanatory variable and then notice whether these are related to differences in the response variable.

For example, suppose we wanted to detect the effects of the explanatory variable "smoking during pregnancy" on the response variable "child's IQ at 4 years of age." In an experiment, we would randomly assign half of the mothers to smoke during pregnancy and the other half to not smoke. In an observational study, we would merely record smoking behavior. This example demonstrates why we can't always perform an experiment.

> *Two reasons why we must sometimes use an observational study instead of an experiment:*
> 1. It is unethical or impossible to assign people to receive a specific treatment.
> 2. Certain explanatory variables, such as handedness, are inherent traits and cannot be randomly assigned.

Confounders and Interactions

Confounding Variables

A **confounding variable** is one whose effect on the response variable cannot be separated from the effect of the explanatory variable. For instance, if we notice that women who smoke during pregnancy have children with lower IQs than the children of women who don't smoke, it could be because women who smoke also have poor nutrition. In that case, nutrition would be confounded with smoking in terms of its influence on subsequent IQ of the child.

Confounding variables are a bigger problem in observational studies than in experiments. In fact, one of the major advantages of an experiment over an observational study is that, in an experiment, the researcher attempts to control for confounding variables.

Interactions Between Variables

When you read the results of a study, you should also be aware that there may be **interactions** between explanatory variables. An interaction occurs when the effect of one explanatory variable on the response variable depends on what's happening with another explanatory variable. For example, if smoking during pregnancy reduces IQ when the mother does not exercise, but raises or does not influence IQ when the mother does exercise, then we would say that smoking interacts with exercise to produce an effect on IQ. Notice that if two variables do interact, it is important that the results be given separately for each combination. To simply say that smoking lowers IQ when, in fact, it only did so for those who didn't exercise would be a misleading conclusion.

Experimental Units, Subjects, and Volunteers

In addition to humans, it is common for studies to be performed on plants, animals, machine parts, and so on. To have a generic term for this conglomeration of possibilities, we define **experimental units** to be the smallest basic objects to which we can assign different treatments in an experiment and **observational units** to be the objects or people measured in any study.

The terms **participants** or **subjects** are commonly used when the observational units are people. In most cases, the participants in studies are **volunteers.** Sometimes they are passive volunteers, such as when all patients treated at a particular medical facility are asked to sign a consent form agreeing to participate in a study.

Often researchers recruit volunteers through the newspaper. For example, a weekly newspaper in a small town in California recently ran an article with the head line, "Volunteers sought for silicone study" (*Winters* (CA) *Express,* 16 December 1993, p. 8). The article explained that researchers at a local medical school were "seeking 100 women with silicone breast implants and 100 without who are willing to fast overnight and then give a blood sample." The article also explained who was doing the research and its purpose.

Notice that by recruiting volunteers for studies, the results cannot necessarily be extended to the larger population. For example, if volunteers are enticed to participate by receiving a small payment or free medical care, as is often the case, those who respond are more likely to be from lower socioeconomic backgrounds. Common sense should enable you to figure out if this is likely to be a problem, but researchers should always report the source of their participants so you can judge this for yourself.

5.2   DESIGNING A GOOD EXPERIMENT

Designing a flawless experiment is extremely difficult, and carrying one out is probably impossible. Nonetheless, there are ideals to strive for, and in this section, we investigate those first. We then explore some of the pitfalls that are still quite prevalent in research today.

Randomization: The Crucial Element

Experiments are supposed to reduce the effects of confounding variables and other sources of bias that are necessarily present in observational studies. They do so by using a simple principle called **randomization.**

We discussed the idea of randomization in Chapter 4, when we described how to choose a sample for a survey. There, we were concerned that everyone in the population had a specified probability of making it into the sample. In experimentation, we are concerned that each of the experimental units (people, animals, and so on) has a specified probability of receiving any of the potential treatments. For example, Salk should have ensured that each group of babies available for study had an equal chance of being assigned to hear the heartbeat or to go into the silent nursery. Otherwise, he could have chosen the babies who looked healthier to begin with to receive the heartbeat treatment.

Randomizing the Order of Treatments

In some experiments, all treatments are applied to each unit. In that case, randomization should be used to determine the *order* in which they are applied. For example, suppose an experiment is conducted to determine the extent to which drinking alcohol or smoking marijuana impairs driving ability. Because drivers are all so different, it makes sense to test the same drivers under all three conditions (alcohol, marijuana, and sober) rather than using different drivers for each condition. But if everyone were tested under alcohol, then marijuana, then sober, by the time they were traveling the course for the second and third times, their performance would improve just from having learned something about the course.

A better method would be to randomly assign some drivers to each of the possible orderings so the learning effect would average out over the three treatments. Notice that it is important that the assignments be made *randomly.* If we let the experimenter decide which drivers to assign to which ordering, assignments could be made to give an unfair advantage to one of the treatments.

Randomizing the Type of Treatments

Randomly assigning the treatments to the experimental units also helps protect against hidden or unknown biases. For example, suppose that in the experiment in Case Study 1.2, approximately the first 11,000 physicians who enrolled were given aspirin and the remainder were given placebos. It could be that the healthier, more energetic physicians enrolled first, thus giving aspirin an unfair advantage.

In statistics, "random" is not synonymous with "haphazard"—despite what your thesaurus might say. Random assignments may not be possible or ethical under some circumstances. But in situations where randomization is feasible, it is usually not difficult to accomplish. It can easily be done with a table of random digits, a computer, or even—if done carefully—by physical means such as flipping a coin or drawing numbers from a hat. The important feature, ensured by proper randomization, is that the chances of being assigned to each condition are the same for each participant.

Control Groups, Placebos, and Blinding

Control Groups

To determine whether a drug, heartbeat sound, meditation technique, and so on, has an effect, we need to know what would have happened to the response variable if the treatment had not been applied. To find that out, experimenters create **control groups,** which are treated identically in all respects, except that they don't receive the active treatment.

Placebos

A special kind of control group is usually used in studies of the effectiveness of drugs. A substantial body of research shows that people respond not only to active drugs but also to **placebos.** A placebo looks like the real drug but has no active ingredients. Placebos can be amazingly effective; studies have shown that they can help up to 62% of headache sufferers, 58% of those suffering from seasickness, and 39% of those with postoperative wound pain.

Because the **placebo effect** is so strong, drug research is generally conducted by randomly assigning half of the volunteers to receive the drug and the other half to receive a placebo, without telling them which they are receiving. The placebo looks just like the real thing, so the participants will not be able to distinguish between it and the actual drug and thus will not be influenced by belief biases.

Blinding

The patient isn't the only one who can be affected by knowing whether he or she has received an active drug. If the researcher who is measuring the reaction of the patients were to know which group was which, the researcher might take the measurements in a biased fashion.

To avoid these biases, good experiments use **double-blind** procedures. A double-blind experiment is one in which neither the participant nor the researcher taking the measurements knows who had which treatment. A **single-blind** experiment is one in which only the participants do not know which treatment they have been assigned. An experiment would also be called single-blind if the participants knew the treatments but the researcher were kept blind.

Although double-blind experiments are preferable, they are not always possible. For example, in testing the effect of meditation on blood pressure, the subjects would obviously know if they were in the meditation group or the control group. In this case, the only possible experiment would be a single-blind study in which the person taking the blood pressure measurement would not know who was in which group.

Pairing and Blocking

It is sometimes easier and more efficient to have each person in a study serve as his or her own control. That way, differences that inherently exist among individuals don't obscure the treatment effects. We encountered this idea when we discussed how to compare driving ability when under the influence of alcohol and marijuana and when sober.

Sometimes, instead of using the same individual for the treatments, researchers will match people on traits that are likely to be related to the outcome, such as age, IQ, or weight. They then randomly assign each of the treatments to one member of each matched pair or grouping. For example, in a study comparing chemotherapy to surgery to treat cancer, patients might be matched by sex, age, and level of severity of the illness. One from each pair would then be randomly chosen to receive the chemotherapy and the other to receive surgery. (Of course, such a study would only be ethically feasible if there were no prior knowledge that one treatment was superior to the other. Patients in such cases are always required to sign an informed consent.)

Matched-Pair Designs

Experimental designs that use either two matched individuals or the same individual to receive each of two treatments are called **matched-pair designs.** The important feature of these designs is that randomization is used to assign the order of the two treatments. Of course, it is still important to try to conduct the experiment in a double-blind fashion so that neither the participant nor the researcher knows which order was used.

Block Designs

An extension of the matched-pair design to three or more treatments is called a **block design.** The method discussed for comparing drivers under three conditions was a block design. Each driver is called a **block.** This somewhat peculiar terminology results from the fact that these ideas were first used in agricultural experiments, where the experimental units were plots of land that had been subdivided into "blocks." In the social sciences, designs such as these, in which the same participants are measured repeatedly, are referred to as **repeated-measures designs.**

CASE STUDY 5.1 ## Quitting Smoking with Nicotine Patches

There is no longer any doubt that smoking cigarettes is hazardous to your health and to those around you. Yet, for someone addicted to smoking, quitting is no simple matter. One promising technique for helping people to quit smoking is to apply a patch to the skin that dispenses nicotine into the blood. These "nicotine patches" have become one of the most frequently prescribed medications in the United States. To test the effectiveness of these patches on the cessation of smoking, Dr. Richard Hurt and his colleagues (Hurt et al., 23 February 1994) recruited 240 smokers at Mayo Clinics in Rochester, Minnesota; Jacksonville, Florida; and Scottsdale, Arizona. Volunteers were required to be between the ages of 20 and 65, have an expired carbon monoxide level of 10 ppm or greater (showing that they were indeed smokers), be in good health, have a history of smoking at least 20 cigarettes per day for the past year, and be motivated to quit.

Volunteers were randomly assigned to receive either 22-mg nicotine patches or placebo patches for 8 weeks. They were also provided with an intervention program recommended by the National Cancer Institute, in which they received counseling before, during, and for many months after the 8-week period of wearing the patches.

After the 8-week period of patch use, almost half (46%) of the nicotine group had quit smoking, whereas only one-fifth (20%) of the placebo group had. Having quit was defined as "self-reported abstinence (not even a puff) since the last visit and an expired air carbon monoxide level of 8 ppm or less" (p. 596).

After a year, rates in both groups had declined, but the group that had received the nicotine patch still had a higher percentage who had successfully quit than did the placebo group: 27.5% versus 14.2%.

The study was double-blind, so neither the participants nor the nurses taking the measurements knew who had received the active nicotine patches. The study was funded by a grant from Lederle Laboratories and was published in the *Journal of the American Medical Association.* ∎

5.3   DIFFICULTIES AND DISASTERS IN EXPERIMENTS

We have already introduced some of the problems that can be encountered with experiments, such as biases introduced by lack of randomization. However, many of the complications that result from poorly conducted experiments can be negated with proper planning and execution.

Here are some potential complications:

1. Confounding variables
2. Interacting variables
3. Placebo, Hawthorne, and experimenter effects
4. Ecological validity and generalizability

Confounding Variables

The Problem

Variables that are connected with the explanatory variable can distort the results of an experiment because they—and not the explanatory variable—may be the agent actually causing a change in the response variable.

The Solution

Randomization is the solution. If experimental units are randomly assigned to treatments, then the effects of the confounding variables should apply equally to each treatment. Thus, observed differences between treatments should not be attributable to the confounding variables.

EXAMPLE 1   **NICOTINE PATCH THERAPY**

The nicotine patch therapy in Case Study 5.1 was more effective when there were no other smokers in the participant's home. Suppose the researchers had assigned the first 120 volunteers to the placebo group and the last 120 to the nicotine group. Further, suppose that those with no other smokers at home were more eager to volunteer. Then the treatment would have been confounded with whether there were other smokers at home. The observed results showing that the active patches were more effective than the placebo patches could have merely represented a difference between those with other smokers at home and those without. By using randomization, approximately equal numbers in each group should have come from homes with other smokers. Thus, any impact of that variable would be spread equally across the two groups. ∎

Interacting Variables

The Problem

Sometimes a second variable interacts with the explanatory variable, but the results are reported without taking that interaction into account. The reader is then misled into thinking the treatment works equally well, no matter what the condition is for the second variable.

The Solution

Researchers should measure and report variables that may interact with the main explanatory variable(s).

| EXAMPLE 2 | **OTHER SMOKERS AT HOME** |

In the study we discussed in Case Study 5.1, there was an interaction between the treatment and whether there were other smokers at home. The researchers measured and reported this interaction. After the 8-week patch therapy, the proportion of the nicotine group who had quit smoking was only 31% if there were other smokers at home, whereas it was 58% if there were not. In the placebo group, the proportions who had quit were the same whether there were other smokers at home or not. Therefore, it would be misleading to merely report that 46% of the nicotine recipients had quit, without also providing the information about the interaction. ■

Placebo, Hawthorne, and Experimenter Effects

The Problem

We have already discussed the strong effect that a placebo can have on experimental outcomes because the power of suggestion is somehow able to affect the result. A related idea is that participants in an experiment respond differently than they otherwise would, just because they *are* in the experiment. This is called the "Hawthorne effect" because it was first detected in 1924 during a study of factory workers at the Hawthorne, Illinois, plant of the Western Electric Company. (The phrase was not actually coined until much later; see French, 1953.)

Related to these effects are numerous ways in which the experimenter can bias the results. These include recording the data erroneously to match the desired outcome, treating subjects differently based on which condition they are receiving, or subtly letting the subjects know the desired outcome.

The Solution

As we have seen, most of these problems can be overcome by using double-blind designs and by including a placebo group or a control group that receives identical handling except for the active part of the treatment. Other problems, such as incorrect data recording, should be addressed by allowing data to be automatically entered into a computer as it is collected, if possible. Depending on the experiment, there may still be subtle ways in which experimenter effects can sneak into the results. You should be aware of these possibilities when you read the results of a study.

| EXAMPLE 3 | **DULL RATS** |

In an experiment designed to test whether the expectations of the experimenter could really influence the results, Rosenthal and Fode (1963) deliberately conned

12 experimenters. They gave each one five rats that had been taught to run a maze. They told six of the experimenters that the rats had been bred to do well (that is, that they were "maze bright") and told the other six that their rats were "maze dull" and should not be expected to do well. Sure enough, the experimenters who had been told they had bright rats found learning rates far superior to those found by the experimenters who had been told they had dull rats. Hundreds of other studies have since confirmed the "experimenter effect." ■

Ecological Validity and Generalizability

The Problem

Suppose you wanted to compare three assertiveness training methods to see which was more effective in teaching people how to say no to unwanted requests on their time. Would it be realistic to give them the training, then measure the results by asking them to role-play in situations in which they would have to say no? Probably not because everyone involved would know it was only role-playing. The usual social pressures to say yes would not be as striking. This is an example of an experiment with little **ecological validity.** In other words, the variables have been removed from their natural setting and are measured in the laboratory or in some other artificial setting. Thus, the results do not accurately reflect the impact of the variables in the real world or in everyday life. A related problem is one we have already mentioned—namely, if volunteers are used for a study, can the results be generalized to any larger group?

The Solution

There are no ideal solutions to these problems, other than trying to design experiments that can be performed in a natural setting with a random sample from the population of interest. In most experimental work, these idealistic solutions are impossible. A partial solution is to measure variables for which the volunteers might differ from the general population, such as income, age, or health, and then try to determine the extent to which those variables would make the results less general than desired. In any case, when you read the results of a study, you should question its ecological validity and its generalizability.

EXAMPLE 4   **REAL SMOKERS WITH A DESIRE TO QUIT**

The researchers in Case Study 5.1 did many things to help ensure ecological validity and generalizability. First, they used a standard intervention program available from and recommended by the National Cancer Institute instead of inventing their own so that other physicians could follow the same program. Next, they used participants at three different locations around the country, rather than in one community only, and they involved a wide range of ages (20 to 65). They included individuals who lived in households with other smokers as well as those who did not. Finally, they recorded numerous other variables (sex, race, education, marital status, psychological health, and so on) and checked to make sure these were not related to the response variable or the patch assignment. ■

Exercise Yourself to Sleep

SOURCE: King et al., 1 January 1997, pp. 32–37.

According to the *UC Davis Health Journal* (November–December 1997, p. 8), older adults constitute only 12% of the population but receive almost 40% of the sedatives prescribed. The purpose of this randomized experiment was to see if regular exercise could help reduce sleep difficulties in older adults. The 43 subjects were sedentary volunteers between the ages of 50 and 76, with moderate sleep problems but no heart disease. They were randomly assigned to either participate in a moderate community-based exercise program four times a week for 16 weeks, or continue to be sedentary. For ethical reasons, the control group was admitted to the program when the experiment was complete. The results were striking. The exercise group fell asleep an average of 11 minutes faster and slept an average of 42 minutes longer than the control group. Note that this could not be a double-blind experiment because subjects obviously knew whether they were exercising. Because sleep patterns were self-reported, there could have been a tendency to err in reporting, in the direction desired by the experimenters. However, this experiment is an example of a well-designed experiment, given the practical constraints, and, as the authors conclude, it does allow the finding that "older adults with moderate sleep complaints can improve self-rated sleep quality by initiating a regular moderate intensity exercise program" (p. 32). ■

5.4 DESIGNING A GOOD OBSERVATIONAL STUDY

In trying to establish causal links, observational studies start with a distinct disadvantage compared to experiments: The researchers observe, but cannot control, the explanatory variables. However, these researchers do have the advantage that they are more likely to measure participants in their natural setting. Before looking at some complications that can arise, let's look at an example of a well-designed observational study.

Baldness and Heart Attacks

On March 8, 1993, *Newsweek* announced, "A really bad hair day: Researchers link baldness and heart attacks" (p. 62). The article reported that "men with typical male pattern baldness . . . are anywhere from 30 to 300 percent more likely to suffer a heart attack than men with little or no hair loss at all." Pattern baldness is the type affecting the crown or vertex and is not the same as a receding hairline; it affects approximately one-third of middle-aged men.

The report was based on an observational study conducted by researchers at Boston University School of Medicine, in which they compared 665 men who had been admitted to the hospital with their first heart attack to 772 men in the same age group (21- to 54-years-old) who had been admitted to the same hospitals for other reasons. Thirty-five hospitals were involved, all in eastern Massachusetts and Rhode Island. The full results were reported by Lesko, Rosenberg, and Shapiro (23 February 1993).

The study found that the percentage of men who showed some degree of pattern baldness was substantially higher for those who had had a heart attack (42%) than for those who had not (34%). Further, when they used sophisticated statistical tests to ask the question in the reverse direction, they found an increased risk of heart attack for men with any degree of pattern baldness. The analysis methods included adjustments for age and other heart attack risk factors. The increase in risk was more severe with increasing severity of baldness, after adjusting for age and other risk factors.

The authors of the study speculated that there may be a third variable, perhaps a male hormone, that both increases the risk of heart attacks and leads to a propensity for baldness. With an observational study such as this, scientists can establish a connection, and they can then look for causal mechanisms in future work. ■

Types of Observational Studies

Case-Control Studies

Some terms are used specifically for observational studies. Case Study 5.3 is an example of a **case-control study.** In such a study, "cases" who have a particular attribute or condition are compared with "controls" who do not. In this example, those who had been admitted to the hospital with a heart attack were the cases, and those who had been admitted for other reasons were the controls. The cases and controls are compared to see how they differ on the variable of interest, which in Case Study 5.3 was the degree of baldness.

Sometimes cases are matched with controls on an individual basis. This type of design is similar to a matched-pair experimental design. The analysis proceeds by first comparing the pair, then summarizing over all pairs. Unlike a matched-pair experiment, the researcher does not randomly assign treatments within pairs but is restricted to how they occur naturally. For example, to identify whether left-handed people die at a younger age, researchers might match each left-handed case with a right-handed sibling as a control, and compare their ages at death. Handedness could obviously not be randomly assigned to the two individuals, so confounding factors might be responsible for any observed differences.

Retrospective or Prospective Studies

Observational studies are also classified according to whether they are **retrospective,** in which participants are asked to recall past events, or **prospective,** in which participants are followed into the future, and events are recorded. The latter is a better procedure because people often do not remember past events accurately.

Advantages of Case-Control Studies

Case-control studies have become increasingly popular in medical research, and with good reason. Much more efficient than experiments, they do not suffer from the ethical considerations inherent in the random assignment of potentially harmful or beneficial treatments. The purpose of a case-control study is to find out whether one or more explanatory variables are related to a certain disease. For instance, in an example given later in this book, researchers were interested in whether owning a pet bird is related to incidence of lung cancer.

A case-control study begins with the identification of a suitable number of cases, or people who have just been diagnosed with the disease of interest. Researchers then identify a group of controls, who are as similar as possible to the cases, except that they don't have the disease. To achieve this similarity, researchers often use patients hospitalized for other causes as the controls.

Efficiency

The case-control design has some clear advantages over experiments as well as other observational studies. Case-control studies are very efficient in terms of time, money, and inclusion of enough people with the disease. Imagine trying to design an experiment to find out whether a relationship exists between owning a bird and getting lung cancer. You would randomly assign people to either own a bird or not and then wait to see how many in each group contracted lung cancer. The problem is that you would have to wait a long time, and even then, you would have very few cases of lung cancer in either group. In the end, you may not have enough cases for a valid comparison.

A case-control study, in contrast, would identify a large group of people who had just been diagnosed with lung cancer, and would then ask them whether they had owned a pet bird. A similar control group would be identified and asked the same question. A comparison would then be made between the proportion of cases (lung cancer patients) who had birds and the proportion of controls who had birds.

Reducing Potential Confounding Variables

Another advantage of case-control studies over other observational studies is that the controls are chosen to try to reduce potential confounding variables. For example, in Case Study 5.3, suppose it were the case that bald men were simply less healthy than other men and were therefore more likely to get sick in some way. An observational study that recorded only whether someone had baldness and whether they had had a heart attack would not be able to control for that fact. By using other hospitalized patients as the controls, the researchers were able to at least partially account for general health as a potential confounding factor.

You can see that careful thought is needed to choose controls that reduce potential confounding factors and do not introduce new ones. For example, suppose we wanted to know if heavy exercise induced heart attacks and, as cases, we used people as they were admitted to the hospital with a heart attack. We would certainly not want to use other newly admitted patients as controls. People who were sick enough to enter the hospital (for anything other than sudden emergencies) would probably not have been recently engaging in heavy exercise. When you read the results of a case-control study, you should pay attention to how the controls were selected.

5.5   DIFFICULTIES AND DISASTERS IN OBSERVATIONAL STUDIES

As with other types of research we have discussed, when you read the results of observational studies you need to watch for procedures that could negate the results of the study.

Here are some complications that can arise:
1. Confounding variables and the implications of causation
2. Extending the results inappropriately
3. Using the past as a source of data

Confounding Variables and the Implications of Causation

The Problem

Don't be fooled into thinking that a link between two variables established by an observational study implies that one causes the other. There is simply no way to separate out all potential confounding factors if randomization has not been used.

The Solution

A partial solution is achieved if researchers measure all the potential confounding variables they can imagine and include those in the analysis to see whether they are related to the response variable. Another partial solution can be achieved in case-control studies by choosing the controls to be as similar as possible to the cases. The other part of the solution is up to the reader: Don't be fooled into thinking a causal relationship exists.

EXAMPLE 5   **SMOKING DURING PREGNANCY**

In Chapter 1, we introduced a study showing that women who smoked during pregnancy had children whose IQs at age 4 were lower than those of similar women who had not smoked. The difference was as high as 9 points before accounting for confounding variables, such as diet and education, but was reduced to just over 4 points after accounting for those factors. However, other confounding variables could exist that are different for mothers who smoke and that were not measured and analyzed, such as amount of exercise the mother got during pregnancy. Therefore, we should not conclude that smoking during pregnancy necessarily caused the children to have lower IQs. ■

Extending the Results Inappropriately

The Problem

Many observational studies use convenience samples, which are generally not representative of any population. Results should be considered with that in mind. Case-control studies often use only hospitalized patients, for example.

The Solution

If possible, researchers should use an entire segment of the population rather than just a convenient sample. In studying the relationship between smoking during pregnancy and child's IQ, for example, the researchers included most of the women in a particular county in upstate New York who were pregnant during the right time period. Had they relied solely on volunteers recruited through the media, their results would not be as extendable.

EXAMPLE 6   **BALDNESS AND HEART ATTACKS**

The observational study relating baldness and heart attacks only used men who were hospitalized for some reason. Although that may make sense in terms of providing a more similar control group, you should consider whether the results should be extended to all men. ■

Using the Past as a Source of Data

The Problem

Retrospective observational studies can be particularly unreliable because they ask people to recall past behavior. Some medical studies, in which the response variable is whether someone has died, can be even worse because they rely on the memories of relatives and friends rather than on the actual participants. Retrospective studies also suffer from the fact that variables that confounded things in the past may no longer be similar to those that would currently be confounding variables, and researchers may not think to measure them.

The Solution

If at all possible, prospective studies should be used. That's not always possible. For example, researchers who first considered the potential causes of AIDS or Toxic Shock Syndrome had to start with those who were inflicted and try to find common factors from their pasts. If possible, retrospective studies should use authoritative sources such as medical records, rather than relying on memory.

EXAMPLE 7 — **DO LEFT-HANDERS DIE YOUNG?**

A few years ago, a highly publicized study pronounced that left-handed people did not live as long as right-handed people (Coren and Halpern, 1991). In one part of the study, the researchers had sent letters to next of kin for a random sample of recently deceased individuals, asking which hand the deceased had used for writing, drawing, and throwing a ball. They found that the average age of death for those who had been left-handed was 66, whereas for those who had been right-handed, it was 75. What they failed to take into account was that in the early part of the 20th century, many children were forced to write with their right hands, even if their natural inclination was to be left-handed. Therefore, people who died in their 70s and 80s during the time of this study were more likely to be right-handed than those who died in their 50s and 60s. The confounding factor of how long ago one learned to write was not taken into account. A better study would be a prospective one, following current left- and right-handers to see which group survived longer. ■

EXERCISES

1. Explain why it may be preferable to conduct an experiment rather than an observational study to determine the relationship between two variables. Support your argument with an example concerning something of interest to you.

2. In each of the following examples, explain whether the experiment was double-blind, single-blind, or neither, and explain whether it was a matched-pair or block design or neither.

   a. A utility company was interested in knowing if agricultural customers would use less electricity during peak hours if their rates were different during those hours. Customers were randomly assigned to continue to get standard rates or to receive the time-of-day rate structure. Special meters were attached that recorded usage during peak and off-peak hours; the technician who read the meter did not know what rate structure each customer had.

   b. To test the effects of drugs and alcohol on driving performance, 20 volunteers were each asked to take a driving test under three conditions: sober, after two drinks, and after smoking marijuana. The order under which they took these was randomized. An evaluator watched them drive on a test course and rated their accuracy on a scale from 1 to 10, without knowing which condition they were under each time.

   c. To compare four brands of tires, one of each brand was randomly assigned to the four locations on each of 50 cars. These tires were specially manufactured to remove any labels identifying the brand. After the tires had been on the cars for 30,000 miles, the researchers removed them and measured the remaining tread.

3. Designate the explanatory variable and the response variable for each of the three studies in Exercise 2.

4. Refer to Thought Question 5 at the beginning of this chapter. The headline was based on a study in which a representative sample of over 400,000 adults in the United States were asked a series of questions, including level of education and on how many of the past 30 days they felt physically and emotionally healthy.

   a. What were the intended explanatory variable and response variable for this study?

   b. Explain how each of the "difficulties and disasters in observational studies" (Section 5.5) applies to this study, if at all.

5. A study to see whether birds remember color was done by putting birdseed on a piece of red cloth and letting the birds eat the seed. Later, empty pieces of cloth of varying colors (red, purple, white, and blue) were displayed. The birds headed for the red cloth. The researcher concluded that the birds remembered the color.

   a. Using the terminology in this chapter, give an alternative explanation for the birds' behavior.

   b. Suppose 20 birds were available and they could each be tested separately. Suggest a better method for the study than the one used.

6. Suppose researchers were interested in determining the relationship, if any, between brain cancer and the use of cellular telephones. Would it be better to use an experiment or a case-control study? Explain.

7. Researchers have found that women who take oral contraceptives (birth control pills) are at higher risk of having a heart attack or stroke and that the risk is substantially higher if a woman smokes. In investigating the relationship between taking oral contraceptives (the explanatory variable) and having a heart attack or stroke (the response variable), would smoking be called a confounding variable or an interacting variable? Explain.

Each of the situations in Exercises 8 to 10 contains one of the complications listed as "difficulties and disasters" with designed experiments or observational studies. Explain the problem and suggest how it could have been either avoided or addressed. If you think more than one complication could have occurred, mention them all, but go into detail about only the most problematic.

8. To study the effectiveness of vitamin C in preventing colds, a researcher recruited 200 volunteers. She randomly assigned 100 of them to take vitamin C for 10 weeks and the remaining 100 to take nothing. The 200 participants recorded how many colds they had during the 10 weeks. The two groups were compared, and the researcher announced that taking vitamin C reduces the frequency of colds.

9. A researcher was interested in teaching couples to communicate more effectively. She had 20 volunteer couples, 10 of which were randomly assigned to receive the training program and 10 of which were not. After they had been trained (or not), she presented each of the 20 couples with a hypothetical problem situation and asked them to resolve it while she tape-recorded their conversation. She was blind as to which 10 couples had taken the training program until after she had analyzed the results.

10. Researchers ran an advertisement in a campus newspaper asking for sedentary volunteers who were willing to begin an exercise program. The volunteers were allowed to choose which of three programs they preferred: jogging, swimming or aerobic dance. After 5 weeks on the exercise programs, weight loss was measured. The joggers lost the most weight, and the researchers announced that jogging was better for losing weight than either swimming or aerobic dance.

11. Refer to Exercise 10. What are the explanatory and response variables?

12. Suppose you wanted to know if men or women students spend more money on clothes. You consider two different plans for carrying out an observational study:

Plan 1: Ask the participants how much they spent on clothes during the last 3 months; then compare the men and women.

Plan 2: Ask the participants to keep a diary in which they record their clothing expenditures for the next 3 months; then compare the men and women.

   a. Which of these plans is a retrospective study? What term is used for the other plan?

   b. Give one disadvantage of each plan.

13. Suppose an observational study finds that people who use public transportation to get to work have better knowledge of current affairs than those who drive to work, but that the relationship is weaker for well-educated people. What term from the chapter (for example, *response variable*) applies to each of the following variables?

   a. Method of getting to work

   b. Knowledge of current affairs

   c. Level of education

   d. Whether the participant reads a daily newspaper

14. A case-control study claimed to have found a relationship between drinking coffee and pancreatic cancer. The cases were people recently hospitalized with pancreatic cancer, and the controls were people hospitalized for other reasons. When asked about their coffee consumption for the past year, it was found that the cancer cases drank more coffee than the controls. Give a reasonable expla-

nation for this difference other than a relationship between coffee drinking and pancreatic cancer.

15. A headline in the *Sacramento Bee* (11 December 1997, p. A15) read, "Study: Daily drink cuts death," and the article began with the statement, "One drink a day can be good for health, scientists are reporting, confirming earlier research in a new study that is the largest to date of the effects of alcohol consumption in the United States." The article also noted that "most subjects were white, middle-class and married, and more likely than the rest of the U.S. population to be college-educated."

  a. Explain why this study could not have been a randomized experiment.

  b. Explain whether you think the headline is justified for this study.

  c. The study was based on recording drinking habits for the 490,000 participants in 1982, and then noting death rates for the next 9 years. Is this a retrospective or a prospective study?

  d. Comment on each of the "difficulties and disasters in observational studies" (Section 5.5) as applied to this study.

16. Is it possible to conduct an experiment to compare two conditions using volunteers recruited through the newspaper? If not, explain why not. If so, explain how it would be done and explain any "difficulties and disasters" that would be encountered.

17. Explain why a randomized experiment allows researchers to draw a causal conclusion whereas an observational study does not.

18. Refer to Case Study 5.2, "Exercise Yourself to Sleep."

  a. Discuss each of the "difficulties and disasters in experiments" (Section 5.3) as applied to this experiment.

  b. Explain whether the authors can conclude that exercise actually caused improvements in sleep.

19. Explain why each of the following is used in experiments:

  a. Placebo treatments

  b. Blinding

  c. Control groups

20. Is the "experimenter effect" most likely to be present in a double-blind experiment, a single-blind experiment, or an experiment with no blinding? Explain.

21. Give an example of an experiment that would have poor ecological validity.

22. Explain which of the "difficulties and disasters" is most likely to be a problem in each of the following experiments, and why:

  a. To see if eating just before going to bed causes nightmares, volunteers are recruited to spend the night in a sleep laboratory. They are randomly assigned to be given a meal before bed or not. Numbers of nightmares are recorded and compared for the two groups.

  b. A company wants to know if placing green plants in workers' offices will help reduce stress. Employees are randomly chosen to participate, and plants are delivered to their offices. One week later, all employees are given a stress questionnaire and those who received plants are compared with those who did not.

23. Explain which of the "difficulties and disasters" is most likely to be a problem in each of the following observational studies, and why:

    a. A study measured the number of writing courses taken by students and their subsequent scores on the quantitative part of the Graduate Record Exam. The students who had taken the largest number of writing courses scored lowest on the exam, so the researchers concluded that students who want to pursue graduate careers in quantitative areas should not take many writing courses.

    b. Successful female social workers and engineers were asked to recall whether they had any female professors in college who were particularly influential in their choice of career. More of the engineers than the social workers recalled a female professor who stood out in their minds.

MINI-PROJECTS

1. Design an experiment to test something of interest to you. Explain how your design addresses each of the four complications listed in Section 5.3, "Difficulties and Disasters in Experiments."

2. Design an observational study to test something of interest to you. Explain how your design addresses each of the three complications listed in Section 5.5, "Difficulties and Disasters in Observational Studies."

3. Go to the library or the Internet and locate a journal article that describes a randomized experiment. Explain what was done correctly and incorrectly in the experiment and whether you agree with the conclusions drawn by the authors.

4. Go to the library or the Internet and locate a journal article that describes an observational study. Explain how it was done using the terminology of this chapter and whether you agree with the conclusions drawn by the authors.

5. Design and carry out a single-blind study using ten participants. Your goal is to establish whether people write more legibly with their dominant hand. In other words, do right-handed people write more legibly with their right hand and vice versa for left-handed people? Explain exactly what you did, including how you managed to conduct a single-blind study. Mention things such as whether it was an experiment or an observational study and whether you used matched pairs or not.

REFERENCES

Coren, S., and D. Halpern. (1991). Left-handedness: A marker for decreased survival fitness. *Psychological Bulletin* 109, no. 1, pp. 90–106.

French, J. R. P (1953). Experiments in field settings. In L. Festinger and D. Katz (eds.), *Research method in the behavioral sciences* (pp. 98–135). New York: Holt.

Hurt, R., L. Dale, P. Fredrickson, C. Caldwell, G. Lee, K. Offord, G. Lauger, Z. Marŭisić, I. Neese, and T. Lundberg. (23 February 1994). Nicotine patch therapy for smoking cessation combined with physician advice and nurse follow-up. *Journal of the American Medical Association* 271, no. 8, pp. 595–600.

King, A. C., R. F. Oman, G. S. Brassington, D. L. Bliwise, and W. L. Haskell. (1 January 1997). Moderate-intensity exercise and self-rated quality of sleep in older adults. A randomized controlled trial. *Journal of the American Medical Association* 277, no. 1, pp. 32–37.

Lesko, S. M., L. Rosenberg, and S. Shapiro. (23 February 1993). A case-control study of baldness in relation to myocardial infarction in men. *Journal of the American Medical Association* 269, no. 8, pp. 998–1003.

Rosenthal, R., and K. L. Fode. (1963). The effect of experimenter bias on the performance of the albino rat. *Behavioral Science* 8, pp. 183–189.

CHAPTER SIX

Getting the Big Picture

6.1   FINAL QUESTIONS

By now, you should have a fairly clear picture of how data should be acquired in order to be useful. We have examined how to conduct a sample survey, an experiment and an observational study, and how to critically evaluate what others have done. In this chapter, we look at a few examples in depth and determine what conclusions can be drawn from them.

The final question you should ask when you read the results of research is whether you will make any changes in your lifestyle or beliefs as a result of the research. To reach that conclusion, you need to answer a series of questions—not all statistical—for yourself.

Here are some guidelines for how to evaluate a study:

STEP 1: Determine if the research was a sample survey, an experiment, an observational study, a combination, or based on anecdotes.

STEP 2: Consider the Seven Critical Components in Chapter 2 (p. 16) to familiarize yourself with the details of the research.

STEP 3: Based on the answer in step 1, review the "difficulties and disasters" inherent in that type of research and determine if any of them apply.

STEP 4: Determine if the information is complete. If necessary, see if you can find the original source of the report or contact the authors for missing information.

STEP 5: Ask if the results make sense in the larger scope of things. If they are counter to previously accepted knowledge, see if you can get a possible explanation from the authors.

STEP 6: Ask yourself if there is any alternative explanation for the results.

STEP 7: Determine if the results are meaningful enough to encourage you to change your lifestyle, attitudes, or beliefs on the basis of the research.

CASE STUDY 6.1

Mozart, Relaxation, and Performance on Spatial Tasks

Summary

The researchers performed a repeated-measures experiment on 36 college students. Each student participated in three listening conditions, each of which was followed by a set of abstract/visual reasoning tasks taken from the Stanford-Binet IQ test. The conditions each lasted for 10 minutes. They were

1. Listening to Mozart's *Sonata for Two Pianos in D Major*
2. Listening to a relaxation tape designed to lower blood pressure
3. Silence

The tasks were taken from the three abstract/visual reasoning parts of the Stanford-Binet IQ test that are suitable for adults: a pattern analysis test, a multiple-choice matrices test, and a multiple-choice paper folding and cutting test. The abstract/visual reasoning parts constitute one of four categories of the Stanford-Binet test; the others are verbal reasoning, quantitative reasoning, and short-term memory. None of those were tested in this experiment.

The scores on the abstract/visual reasoning tasks were translated into what the corresponding IQ score would have been for a full-fledged test. The results showed averages of 119, 111, and 110, respectively, for the three listening conditions. The results after listening to Mozart were significantly higher than those for the other two conditions—enough so that chance differences could be ruled out as an explanation.

The researchers tested some potential confounding factors and found that none of these could explain the results. First, they measured pulse rates before and after each listening session to be sure the results weren't simply due to arousal. They found no effects for pulse rate, nor any interactions between pulse rate and IQ test results. Next, they tested to see if the order of presentation or the use of different experimenters could have been confounded with the results, and again found no effect. They noted that the three different tests correlated strongly with each other, so they treated them as "equal measures of abstract reasoning ability."
SOURCE: Rauscher, Shaw, and Ky, 14 October 1993, p. 611.

Discussion

To evaluate the usefulness of this research, let's analyze it according to the seven steps listed at the beginning of this chapter.

STEP 1: Determine if the research was a sample survey, an experiment, an observational study, a combination, or based on anecdotes.

This is supposed to be an experiment although the authors do not provide information about how (or if) they randomly assigned the order of the three listening conditions. It qualifies as an experiment because they manipulated the environment of the participants. Notice that because the same people were tested under all three listening conditions, this was a *repeated-measures* experiment.

STEP 2: Consider the Seven Critical Components in Chapter 2 (p. 16) to familiarize yourself with the details of the research.

Based on the information given, you should be able to understand some, but not all, of the Seven Critical Components. The most important information missing relates to Component 2, the researchers who had contact with the participants. We were not told whether those who tested the participants knew the purpose of the experiment. Information about funding is missing as well (Component 1), but presumably the research was conducted at a university because the participants were college students. Finally, we were not told how the participants were selected (part of Component 3). The results might be interpreted differently if they were music majors, for example.

STEP 3: Based on the answer in step 1, review the "difficulties and disasters" inherent in that type of research and determine if any of them apply.

The four possible complications listed for an experiment include confounding variables; interacting variables; placebo, Hawthorne, and experimenter effects; and ecological validity and generalizability. In this experiment, all four could be problematic, but the most obvious is a possible experimenter effect. We were not told whether the subjects knew the intent of the experimenters, but even if they were not explicitly told, they may have figured out that they were expected to do better after listening to Mozart. This could create inflated scores after the Mozart condition or deflated scores after the other two. There is really no way to get around this because the subjects could not be blind to the listening condition and they were tested under all three conditions.

Another problem is generalizability. It is probably not true that results obtained after 10 minutes in a laboratory would extend directly to the real world.

There are also potential confounding variables. For example, we were not told whether the particular IQ task assigned after each listening condition was the same for each subject or was randomized among the three tasks. If it was the same, and if one task was easier for this particular group of volunteers, that condition would be confounded with the listening condition. We were also not told whether the experimenters had as much contact with the participants for the silent condition as for the two listening conditions. If they did not, then amount of contact could have interacted with the listening condition to produce the effect.

STEP 4: Determine if the information is complete. If necessary, see if you can find the original source of the report or contact the authors for missing information.

The summary at the beginning of this case study contains almost all of the information in the original report in *Nature,* which was probably shorter than the authors would have liked because it was contained in the "Scientific Correspondence" section of the magazine. Substantial information is missing, some of which would probably help determine whether the complications listed in step 3 were really problems.

STEP 5: Ask if the results make sense in the larger scope of things. If they are counter to previously accepted knowledge, see if you can get a possible explanation from the authors.

The authors gave references to "correlational, historical and anecdotal relationships between music cognition and other 'higher brain functions'" but did not otherwise attempt to justify how their results could be explained.

STEP 6: Ask yourself if there is an alternative explanation for the results.

As mentioned in step 3, the subjects were not blind to listening condition and could have performed better after the Mozart condition to satisfy the experimenters. Perhaps the particular IQ task assigned after the Mozart condition was easier for this group. Perhaps the experimenters interacted more (or less) with the participants under the listening conditions.

STEP 7: Determine if the results are meaningful enough to encourage you to change your lifestyle, attitudes, or beliefs on the basis of the research.

If these results are accurate, they indicate that listening to Mozart raises a certain type of IQ for at least a short period of time. That could be useful to you if you are about to take a test involving abstract or spatial reasoning. ■

Meditation and Aging

Summary

Meditation may have more to offer than a calm mind and lower blood pressure. Recent research reported in the *Journal of Behavioral Medicine* shows that a simple meditation practiced twice a day for a 20-minute period leads to marked changes in an age-associated enzyme, DHEA-S. Levels of DHEA-S in experienced meditators correspond to those expected of someone 5–10 years younger who does not meditate.

The enzyme is produced by the adrenal glands, and its level is closely correlated with age in humans. It has also been associated with measures of health and stress. Higher levels are specifically associated with lower incidences of heart disease and lower rates of mortality in general for males, and with less breast cancer and osteoporosis in women.

This study compared levels of DHEA-S in 270 men and 153 women who had practiced Transcendental Meditation (TM) or TM-Sidhi for a mean length of 10.3 years or 11.1 years, respectively. The meditation technique is a simple mental technique in which the meditator sits with eyes closed while remaining wakeful and alert, focusing without effort on a specific meaningless sound (mantram).

The results show that DHEA-S levels are higher in all age groupings and in both sexes among those who practice meditation than in nonmeditating controls, an effect independent of dietary habits and alcohol or drug consumption.

Are these salutary changes directly due to "sitting in meditation"? According to Jay Glaser, who headed the study, the effect may be because meditators learn to approach life with less physiological reaction to stress. Whatever the case, spending 20 minutes twice a day for a body that is measurably more youthful seems like a fair exchange.

ORIGINAL SOURCE: Glaser et al., 1992, pp. 327–341.

NEWS SOURCE: The effects of meditation on aging, *Noetic Sciences Review,* Summer 1993, p. 28.

Discussion

STEP 1: Determine if the research was a sample survey, an experiment, an observational study, a combination, or based on anecdotes.

This was an observational study because the researchers did not assign participants to either meditate or not; they simply measured meditators and nonmeditators.

STEP 2: Consider the Seven Critical Components in Chapter 2 (p. 16) to familiarize yourself with the details of the research.

Due to the necessary brevity of the news report, several pieces of information from the original report are missing. Therefore, based on the news report alone, you would not be able to consider all of the components. Following are some of the missing pieces, derived from the original report.

The first author on the study, J. L. Glaser, was a researcher at the Maharishi International University, widely known for teaching TM. No acknowledgments were given for funding, so presumably there were no outside funders. The control group consisted of 799 men and 453 women who "represented a healthy fraction of the patients of a large, well-known New York City practice specializing in cosmetic dermatology who visited the practice from 1980 to 1983 for cosmetic procedures such as hair transplants, dermabrasion, and removal of warts and moles" (Glaser et al., 1992, p. 329).

The TM group was recruited at the campus of the Maharishi International University (MIU) in Fairfield, Iowa. All of those under 45 years old and 28 of those over 45 were recruited from local faculty, staff, students, and meditating community members. The remainder of those over 45 were recruited "by public announcements during conferences for advanced TM practitioners held on campus from 1983 to 1987" (Ibid., p. 329). Ninety-two percent of the women and 93% of the men were practitioners of the more advanced TM-Sidhi program. More of the meditators than the controls were vegetarians, and the meditators were less likely to drink alcohol or smoke.

The measurements were made from blood samples drawn at office visits throughout the year for the control group, and at specified time periods for the TM group (men under 45 years of age: between 10:45 and 11:45 A.M. in September and October; women under 45 years of age: same hours in April and May; over 45 years of age: between 1:00 and 2:30 P.M., in December for men and in July for women). DHEA-S levels were measured using direct radioimmunoassay. The control and TM groups were assayed in different batches, but random samples from the control group were included in all batches to make sure there was no drift over time. Also, contrary to the news summary, the original report noted a difference in DHEA-S levels for all age groups of women, but DHEA-S levels for men varied only for those over 40.

STEP 3: Based on the answer in step 1, review the "difficulties and disasters" inherent in that type of research and determine if any of them apply.

The most obvious potential problem in this observational study (see p. 84) is complication 1, "confounding variables and implications of causation." Many differences between the TM and control groups could be confounding the results, and because there is no random assignment of treatments, a causal conclusion about the effects of meditation cannot be made. Also, the results cannot be extended to people other than those similar to the TM group measured. For example, there is no way to know if they would extend to practitioners of other relaxation or meditation techniques, or even to practitioners of TM who are not as heavily involved with it as those attending MIU. We consider other explanations in step 6.

STEP 4: Determine if the information is complete. If necessary, see if you can find the original source of the report or contact the authors for missing information.

Most of the necessary information was available, at least in the original report. One piece of missing information was whether those who drew and analyzed the blood knew the purpose of the study.

STEP 5: Ask if the results make sense in the larger scope of things. If they are counter to previously accepted knowledge, see if you can get a possible explanation from the authors.

In the news report, we are not given any information about prior medical knowledge of the effects of meditation. However, returning to the original report, we find that the authors do cite other evidence and give potential mechanistic explanations for why meditation may help reduce the level of the enzyme. They also cite a study showing that 2000 practitioners of TM who were enrolled in major medical plans made less use of the plans than nonmeditators in every category except obstetrics. They noted that the TM group had 55.5% fewer admissions for tumors and 87% fewer admissions for heart disease than the comparison group. Of course, these results do not imply a causal relationship either because they are based on an observational study. They simply add support to the idea that meditators are healthier than nonmeditators.

STEP 6: Ask yourself if there is an alternative explanation for the results.

The obvious explanation is that those who choose to practice TM, especially to the degree that they would be visiting or attending MIU, are somehow different, and probably healthier than those who do not. There is no way to test this assumption. Remember that more of the meditators were vegetarians and they were less likely to drink alcohol or smoke. The authors discuss these points, however, and cite evidence that shows that these factors would not influence the levels of this particular enzyme in the observed direction. Another potential confounding variable is the location of the test. The control group was in New York City and the TM group was in Iowa. Also, the control group consisted of people visiting plastic surgeons, who may be more likely to be those who show early signs of aging. Perhaps you can think of other potential explanations.

STEP 7: Determine if the results are meaningful enough to encourage you to change your lifestyle, attitudes, or beliefs on the basis of the research.

Because there is no way to establish a causal connection, these results must be taken only as support of a difference between the two groups in the study. Nonetheless, we cannot rule out the idea that meditation may be the cause of the observed differences between the TM and control groups and, if slowing down the aging process were crucial to you, these results might encourage you to learn to meditate. ∎

CASE STUDY 6.3   Drinking, Driving, and the Supreme Court

Summary

This case study doesn't require you to make a personal decision about the results. Rather, it involves a decision that was made by the Supreme Court based on statistical evidence and illustrates how laws can be affected by studies and statistics.

In the early 1970s, a young man between the ages of 18 and 20 challenged an Oklahoma state law that prohibited the sale of 3.2% beer to males under 21 but allowed its sale to females of the same age group. The case (*Craig v. Boren,* 429 U.S. 190, 1976) was ultimately heard by the U.S. Supreme Court, which ruled that the law was discriminatory.

Laws are allowed to use gender-based differences as long as they "serve important governmental objectives" and "are substantially related to the achievement of these objectives" (Gastwirth, 1988, p. 524). The defense argued that traffic safety was an important governmental objective and that data clearly show that young males are more likely to have alcohol-related accidents than young females.

The Court considered two sets of data. The first set, shown in Table 6.1, consisted of the number of arrests for driving under the influence and for drunkenness for most of the state of Oklahoma, from September 1 to December 31, 1973. The Court also obtained population figures for the age groups in Table 6.1. Based on those figures, they determined that the 1393 young males arrested for one of the two offenses in Table 6.1 represented 2% of the entire male population in the 18–21 age group. In contrast, the 126 young females arrested represented only 0.18% of the

TABLE 6.1	Arrests by Age and Sex in Oklahoma, September–December 1973

	MALES			FEMALES		
	18–21	Over 21	Total	18–21	Over 21	Total
Driving under influence	427	4,973	5,400	24	475	499
Drunkenness	966	13,747	14,713	102	1,176	1,278
Total	1,393	18,720	20,113	126	1,651	1,777

TABLE 6.2	Random Roadside Survey of Driving and Drunkenness in Oklahoma City, August 1972 and August 1973

	MALES			FEMALES		
	Under 21	21 and Over	Total	Under 21	21 and Over	Total
BAC* over .01	55	357	412	13	52	65
Total	481	1926	2407	138	565	703
%BAC over .01	11.4%	18.5%	17.1%	9.4%	9.2%	9.2%

*BAC = Blood alcohol content

young female population. Thus, the arrest rate for young males was about ten times what it was for young females.

The second set of data introduced into the case, partially shown in Table 6.2, came from a "random roadside survey" of cars on the streets and highways around Oklahoma City during August 1972 and August 1973. Surveys like these, despite the name, do not constitute a random sample of drivers. Information is generally collected by stopping some or all of the cars at certain locations, regardless of whether there is a suspicion of wrongdoing

SOURCE: Gastwirth, 1988, pp. 524–528.

Discussion

Suppose you are a justice of the Supreme Court. Based on the evidence presented and the rules regarding gender-based differences, do you think the law should be upheld? Let's go through the seven steps introduced in this chapter with a view toward making the decision the Court was required to make.

STEP 1: Determine if the research was a sample survey, an experiment, an observational study, a combination, or based on anecdotes.

The numbers in Table 6.1 showing arrests throughout the state of Oklahoma for a 4-month period are observational in nature. The figures do represent most of the arrests for those crimes, but the people arrested are obviously only a subset of those who committed the crimes. The data in Table 6.2 constitute a sample survey, based on a convenience sample of drivers passing by certain locations.

STEP 2: Consider the seven critical components in Chapter 2 (p. 16) to familiarize yourself with the details of the research.

A few details are missing, but you should be able to ascertain answers to most of the components. One missing detail is how the "random roadside survey" was conducted.

STEP 3: Based on the answer in step 1, review the "difficulties and disasters" inherent in that type of research and determine if any of them apply.

The arrests in Table 6.1 were used by the defense to show that young males are much more likely to be arrested for incidences related to drinking than are young females. But consider the confounding factors that may be present in the data. For example, perhaps young males are more likely to drive in ways that call attention to themselves, and thus they are more likely to be stopped by the police, whether they have been drinking or not. Thus, young females who were driving while drunk would not be noticed as often. For the data in Table 6.2, because the survey was taken at certain locations, the drivers questioned may not be representative of all drivers. For example, if a sports event had recently ended nearby, there may be more male drivers on the road, and they may have been more likely to have been drinking than normal.

STEP 4: Determine if the information is complete. If necessary, see if you can find the original source of the report or contact the authors for missing information.

The information provided is relatively complete, except for the information on how the random roadside survey was conducted. According to Gastwirth (1994, personal communication), this information was not supplied in the original documentation of the court case.

STEP 5: Ask if the results make sense in the larger scope of things. If they are counter to previously accepted knowledge, see if you can get a possible explanation from the authors.

Nothing is suspicious about the data in either table. Remember that in 1973, when the data were collected, the legal drinking age in the United States had not yet been raised to 21 years of age.

STEP 6: Ask yourself if there is an alternative explanation for the results.

We have discussed one possible source of a confounding variable for the arrest statistics in Table 6.1—namely, that males may be more likely to be stopped for other traffic violations. Let's consider the data in Table 6.2. Notice that almost 80% of the drivers stopped were male. Therefore, at least at that point in time in Oklahoma, males were more likely to be driving than females. That helps explain why ten times more young men than young women had been arrested for alcohol-related reasons. The important point for the law being challenged in this lawsuit was whether young men were more likely to be driving after drinking than young women. Notice from Table 6.2 that of those cars with young males driving, 11.4% had blood alcohol levels over 0.01; of those cars with young females driving, 9.4% had blood alcohol levels over 0.01. These rates are statistically indistinguishable.

STEP 7: Determine if the results are meaningful enough to encourage you to change your lifestyle, attitudes, or beliefs on the basis of the research.

FIGURE 6.1

Original Source: Olds, Henderson, and Tatelbaum, February 1994, pp. 221–227.

News Source: Study: Smoking may lower kids' IQs, 11 February 1994, p. A-10.

STUDY: SMOKING MAY LOWER KIDS' IQS

ROCHESTER, N.Y. (AP)—Second-hand smoke has little impact on the intelligence scores of young children, researchers found. But women who light up while pregnant could be dooming their babies to lower IQs, according to a study released Thursday. Children ages 3 and 4 whose mothers smoked 10 or more cigarettes a day during pregnancy scored about 9 points lower on the intelligence tests than the offspring of nonsmokers, researchers at Cornell University and the University of Rochester reported in this month's *Pediatrics* journal.

That gap narrowed to 4 points against children of nonsmokers when a wide range of interrelated factors were controlled. The study took into account secondhand smoke as well as diet, education, age, drug use, parents' IQ, quality of parental care and duration of breast feeding.

"It is comparable to the effects that moderate levels of lead exposure have on children's IQ scores," said Charles Henderson, senior research associate at Cornell's College of Human Ecology in Ithaca.

In this case study, the important question is whether the Supreme Court justices were convinced that the gender-based difference in the law was reasonable. The Supreme Court overturned the law, concluding that the data in Table 6.2 "provides little support for a gender line among teenagers and actually runs counter to the imposition of drinking restrictions based upon age" (Gastwirth, 1988, p. 527). ■

CASE STUDY 6.4

Smoking During Pregnancy and Child's IQ

Summary

The news article for this case study is shown in Figure 6.1.

Discussion

STEP 1: Determine if the research was a sample survey, an experiment, an observational study, a combination, or based on anecdotes.

This was an observational study because the researchers could not randomly assign mothers to either smoke or not during pregnancy; they could only observe their smoking behavior.

STEP 2: Consider the Seven Critical Components in Chapter 2 (p. 16) to familiarize yourself with the details of the research.

As in the previous case study, the brevity of the news report necessarily meant that some details were omitted. Based on the original report, the seven questions can all be answered. Following is some additional information.

The research was supported by a number of grants from sources such as the Bureau of Community Health Services, the National Center for Nursing Research, and the National Institutes of Health. None of the funders seem to represent special interest groups related to tobacco products.

The researchers described the participants as follows (Olds et al., 1994):

We conducted the study in a semirural county in New York State with a population of about 100,000. Between April 1978 and September 1980, we interviewed 500 primiparous women [those having their first live birth] who registered for prenatal care either through a free antepartum clinic sponsored by the county health department or through the offices of 11 private obstetricians. (All obstetricians in the county participated in the study.) Four hundred women signed informed consent to participate before their 30th week of pregnancy. (p. 221)

The researchers also noted that "eighty-five percent of the mothers were either teenagers (<19 years at registration), unmarried, or poor. Analysis [was] limited to whites who comprised 89% of the sample" (p. 221).

The explanatory variable, smoking behavior, was measured by averaging the reported number of cigarettes smoked at registration and at the 34th week of pregnancy. For the information included in the news report, the only two groups used were mothers who smoked an average of 10 or more cigarettes per day and those who smoked no cigarettes. Those who smoked between 1 and 9 per day were excluded.

The response variable, IQ, was measured at 12 months with the Bayley Mental Development Index, at 24 months with the Cattell Scales, and at 36 and 48 months with the Stanford-Binet IQ test. In addition to those mentioned in the news source (secondhand smoke, diet, education, age, drug use, parents' IQ, quality of parental care, and duration of breast feeding), other potential confounding variables measured were husband/boyfriend support, marital status, alcohol use, maternal depressive symptoms, father's education, gestational age at initiation of prenatal care, and number of prenatal visits. None of those were found to relate to intellectual functioning.

It is not clear if the study was single-blind. In other words, did the researchers who measured the children's IQs know about the mother's smoking status or not?

STEP 3: Based on the answer in step 1, review the "difficulties and disasters" inherent in that type of research and determine if any of them apply.

The study was prospective, so memory is not a problem. However, there are problems with potential confounding variables, and there may be a problem with trying to extend these results to other groups, such as older mothers. The fact that the difference in IQ for the two groups was reduced from 9 points to 4 points with the inclusion of several additional variables may indicate that the difference could be even further reduced by the addition of other variables.

The authors noted both of these as potential problems. They commented that "the particular sample used in this study limits the generalizability of the findings. The sample was at considerable risk from the standpoint of its sociodemographic characteristics, so it is possible that the adverse effects of cigarette smoking may not be as strong for less disadvantaged groups" (Olds et al., 1994, p. 225).

The authors also mentioned two potential confounding variables. First, they noted, "We are concerned about the reliability of maternal report of illegal drug and alcohol use" (Olds et al., 1994, p. 225), and, "in addition, we did not assess fully the child's exposure to side-stream smoke during the first four years after delivery" (Olds et al., 1994, p. 225).

STEP 4: Determine if the information is complete. If necessary, see if you can find the original source of the report or contact the authors for missing information.

The information in the original report is fairly complete, but the news source left out some details that would have been useful, such as the fact that the subjects were young and of lower socioeconomic status than the general population of mothers.

STEP 5: Ask if the results make sense in the larger scope of things. If they are counter to previously accepted knowledge, see if you can get a possible explanation from the authors.

The authors speculate on what the causal relationship might be, if indeed there is one. For example, they speculate that "tobacco smoke could influence the developing fetal nervous system by reducing oxygen and nutrient flow to the fetus" (p. 226). They also speculate that "cigarette smoking may affect maternal/fetal nutrition by increasing iron requirements and decreasing the availability of other nutrients such as vitamins, B12 and C, folate, zinc, and amino acids" (p. 226).

STEP 6: Ask yourself if there is an alternative explanation for the results.

As with most observational studies, there could be confounding factors that were not measured and controlled. Also, if the researchers who measured the children's IQs were aware of the mother's smoking status, that could have led to some experimenter bias. You may be able to think of other potential explanations.

STEP 7: Determine if the results are meaningful enough to encourage you to change your lifestyle, attitudes, or beliefs on the basis of the research.

If you were pregnant and were concerned about allowing your child to have the highest possible IQ, these results may lead you to decide to quit smoking during the pregnancy. A causal connection cannot be ruled out. ■

CASE STUDY 6.5

For Class Discussion: Guns and Homicides at Home

Summary

The news source read as follows:

Challenging the common assumption that guns protect their owners, a multistate study of hundreds of homicides has found that keeping a gun at home nearly triples the likelihood that someone in the household will be slain there.

The study, published in New England Journal of Medicine, *studied the records of three populous counties surrounding Seattle, Washington, Cleveland, Ohio, and Memphis, Tennessee. The counties offered a sample representative of the entire nation because of the mix of urban, suburban, and rural communities.*

Although 1860 homicides occurred during the study period, the team looked only at those that occurred in the homes of the victims—about 400 deaths. The researchers found that members of households with guns were 2.7 times more likely to experience a homicide than those in households without guns.

*In nearly 77 percent of the cases, victims were killed by a relative or someone they knew. In only about 4 percent of the cases were victims killed by a stranger. In most of the remaining cases, the identity of the persons who committed the homicides could not be determined. (*Washington Post, *17–23 October 1993.)*

ORIGINAL SOURCE: Kellerman et al., 7 October 1993, pp. 1084–1091. ■

REFERENCES

The effects of meditation on aging. (Summer 1993). *Noetic Sciences Review,* Science Notes, p. 28.

Gastwirth, Joseph L. (1988). *Statistical reasoning in law and public policy.* Vol. 2. *Tort law, evidence and health.* New York: Academic Press, pp. 524–528.

Glaser, J. L., J. L. Brind, J. H. Vogelman, M. J. Eisner, M. C. Dillbeck, R. K. Wallace, D. Chopra, and N. Orentreich. (1992). Elevated serum dehydroepiandrosterone sulfate levels in practitioners of the Transcendental Meditation (TM) and TM-Sidhi programs. *Journal of Behavioral Medicine* 15, no. 4. pp. 327–341.

Kellerman, A. L., F. R. Rivara, N. B. Rushford, J. G. Banton, D. T. Reay, J. T. Francisco, A. B. Locci, J. Prodzinski, B. B. Hackman, and G. Somes. (7 October 1993). Gun ownership as a risk factor for homicide in the home. *New England Journal of Medicine* 329, no. 15, pp. 1084–1091.

Olds, D. L., C. R. Henderson, Jr., and R. Tatelbaum. (February 1994). Intellectual impairment in children of women who smoke cigarettes during pregnancy. *Pediatrics* 93, no. 2, pp. 221–227.

Rauscher, F. H., G. L. Shaw, and K. N. Ky. (14 October 1993). Music and spatial task performance. *Nature* 365, p. 611.

Study: Smoking may lower kids' IQs. (11 February 1994). *Davis* (CA) *Enterprise,* p. A-10.

Washington Post, weekly edition. (17–23 October 1993). Reprinted in *Chance* (Winter 1994), vol. 7, no. 1, p. 5.

Finding Life in Data

In Part 1, you learned how data should be collected to be meaningful. In Part 2, you will learn some simple things you can do with data after it has been collected. The goal of the material in this part is to increase your awareness of the usefulness of data and to help you interpret and critically evaluate what you read in the press.

First, you will learn how to take a collection of numbers and summarize them in useful ways. For example, you will learn how to find out more about your own pulse rate by taking repeated measurements and drawing a useful picture.

Second, you will learn to critically evaluate presentations of data made by others. From numerous examples of situations where the uneducated consumer could be misled, you will learn how to critically read and evaluate graphs, pictures, and data summaries.

CHAPTER SEVEN

Summarizing and Displaying Measurement Data

THOUGHT QUESTIONS

1. If you were to read the results of a study showing that daily use of a certain exercise machine resulted in an average 10-pound weight loss, what more would you want to know about the numbers in addition to the average? (*Hint:* Do you think everyone who used the machine lost 10 pounds?)

2. Suppose you are comparing two job offers, and one of your considerations is the cost of living in each area. You get the local newspapers and record the price of 50 advertised apartments for each community. What summary measures of the rent values for each community would you need in order to make a useful comparison?

3. A recent newspaper article in California said that the *median* price of single-family homes sold in the past year in the local area was $136,000 and the *average* price was $149,160. How do you think these values are computed? Which do you think is more useful to someone considering the purchase of a home, the median or the average?

4. The Stanford-Binet IQ test is designed to have a mean, or average, for the entire population of 100. It is also said to have a *standard deviation* of 16. What aspect of the population of IQ scores do you think is described by the "standard deviation"?

5. Students in a statistics class at a large state university were given a survey in which one question asked was age (in years); one student was a retired person, and her age was an "outlier." What do you think is meant by an "outlier"? If the students' heights were measured, would this same retired person necessarily have a value that was an "outlier"? Explain.

7.1   TURNING DATA INTO INFORMATION

Looking at a long list of numbers is about as informative as looking at a scrambled set of letters. To get information out of data, the data have to be organized and summarized. As an example, suppose you were told that you received a score of 80 on an examination and that the scores in the class were as follows:

75, 95, 60, 93, 85, 84, 76, 92, 62, 83, 80, 90, 64, 75, 79, 32, 78, 64, 98, 73, 88, 61, 82, 68, 79, 78, 80, 55

How useful would that list of numbers be to you, at first glance? Do you have any idea where you are relative to the rest of the class? The first thought that may occur to you is to put the numbers into increasing order so you could see where your score was relative to the others. Doing that, you find:

32, 55, 60, 61, 62, 64, 64, 68, 73, 75, 75, 76, 78, 78, 79, 79, 80, 80, 82, 83, 84, 85, 88, 90, 92, 93, 95, 98

Now you can see that you are somewhat above the middle, but this list still isn't easy to assimilate into a useful picture. It would help if we could summarize the numbers.

The Mean, the Median, and Outliers

There are three kinds of useful information about a set of data, and each can be measured and expressed in a variety of ways. One useful concept is the idea of the "center" of the data. What's a typical or average value? For the test scores just given, the numerical **average,** or **mean,** is 76.04. As another measure, consider that there were 28 values in that test score set, so the **median,** with half of the scores above and half of the scores below it, is 78.5, halfway between 78 and 79.

You can see that the median is somewhat higher than the mean in this case. That's because a very low score, 32, pulled down the mean. It didn't pull down the median because, as long as that very low score was 78 or less, its effect on the median would be the same. If one or two scores are far removed from the rest of the data, they are called **outliers.** There are no hard and fast rules for determining what qualifies as an outlier, but in this case most people would agree that the score of 32 is an outlier.

Mode

Another measure of "center," called the **mode,** is occasionally useful. The mode is simply the most common value in the list. For the exam scores, no single mode exists because each of the scores 64, 75, 78, 79 and 80 occurs twice. The mode is most useful for discrete or categorical data with a relatively small number of possible values. For example, if you measured the class standing of all the students in your statistics class and coded them with 1 = freshman, 2 = sophomore, and so on, it would probably be more useful to know the mode (most common class standing) than to know the mean or the median.

Variability

The second kind of useful information contained in a set of data is the **variability.** How spread out are the values? Are they all close together? Are most of them together, but a few of them outliers? Knowing that the mean is about 76, your test score of 80 is still hard to interpret. It would obviously have a different connotation for you if the scores ranged from 72 to 80 rather than from 32 to 98.

Range

The simplest measure of variability is the **range,** which is just as it sounds. In this case, the scores went from 32 to 98, for a range of 66 points. We look at a more complicated measure of variability, the standard deviation, later.

Shape

The third kind of useful information is the **shape,** which can be derived from a certain kind of picture of the data. We can answer questions such as: Are most of the values clumped in the middle with values tailing off at each end? Are there two distinct groupings? Are most of the values clumped together at one end with a few very high or low values? You can see that your score of 80 would have different meanings depending on how the other students' scores grouped together. For example, if half of the remaining students had scores of 50 and the other half scores of 100, then even though your score of 80 was "above average," it wouldn't look so good. Next we focus on how to look at the shape of the data.

7.2

THE TOTAL PICTURE: STEMPLOTS AND HISTOGRAMS

About Stemplots

A **stemplot** is a quick and easy way to put a list of numbers into order while getting a picture of their shape. The easiest way to describe a stemplot is to construct one. We could use the test scores we've been using, but instead let's turn to some real data, where each number has an identity.

   Table 7.1 lists per capita income for each of the 50 states and the District of Columbia in 1989. Scanning the list gives us some information, but it would be easier to get the big picture if it were in some sort of numerical order. We could simply list the states by value instead of alphabetically, but that would not give us a picture of the shape.

   Before reading any further, look at the right-most part of Figure 7.1 so you can see what a completed stemplot looks like. Each of the digits extending to the right

TABLE 7.1	**1989 Per Capita Income for the Individual States**

Alabama	13,679	Montana	13,852
Alaska	21,173	Nebraska	15,360
Arizona	15,881	Nevada	18,827
Arkansas	12,984	New Hampshire	20,251
California	19,740	New Jersey	23,764
Colorado	17,494	New Mexico	13,191
Connecticut	24,604	New York	20,540
Delaware	19,116	North Carolina	15,221
D.C.	23,436	North Dakota	13,261
Florida	17,694	Ohio	16,499
Georgia	16,188	Oklahoma	14,151
Hawaii	18,306	Oregon	15,785
Idaho	13,762	Pennsylvania	17,422
Illinois	18,858	Rhode Island	18,061
Indiana	16,005	South Carolina	13,616
Iowa	15,524	South Dakota	13,244
Kansas	16,182	Tennessee	14,765
Kentucky	13,777	Texas	15,483
Louisiana	13,041	Utah	13,027
Maine	16,310	Vermont	16,399
Maryland	21,020	Virginia	18,970
Massachusetts	22,196	Washington	17,640
Michigan	17,745	West Virginia	12,529
Minnesota	17,746	Wisconsin	16,759
Mississippi	11,835	Wyoming	14,135
Missouri	16,431		

Source: *Information please almanac*, 1991, p. 52.

represents one data point. The first thing you see is 11|8. That represents a per capita income of $11,800, which is a truncated version of the per capita income of Mississippi, $11,835, the lowest per capita income in the batch. (To truncate a number, simply drop off the digits at the end. This is slightly different from rounding off.) When more than one state shares the number to the left, a number is given on the right for each of those states. For example, the second item in the stemplot is 12|9 5. This represents two states: Arkansas, $12,984, and West Virginia, $12,529. For one more example, notice the bottom of the picture, 24|6. That represents $24,600, which is the truncated value for the $24,604 per capita income of Connecticut.

FIGURE 7.1

Building a stemplot of 1989 per capita income for the United States

Step 1 Creating the stem	Step 2 Attaching leaves	The finished stemplot
11		11		11	8
12		12	9	12	9 5
13		13	6	13	6 7 7 0 8 1 2 6 2 0
14		14		14	1 7 1
15		15	8	15	8 5 3 2 7 4
16		16		16	1 0 1 3 4 4 3 7
17		17	4 6	17	4 6 7 7 4 6
18		18		18	3 8 8 0 9
19		19	7 1	19	7 1
20		20		20	2 5
21		21	1	21	1 0
22		22		22	1
23		23	4	23	4 7
24		24	6	24	6

Example: 11|8 = $11,800

Creating a Stemplot

Stemplots are sometimes called **stem-and-leaf plots** or **stem-and-leaf diagrams.** Only two steps are needed to create a stemplot.

STEP 1: CREATE THE STEMS

The first step is to divide the range of the data into equal units to be used as the **stems.** In this example, the per capita incomes range from a low of $11,835 (for Mississippi) to a high of $24,604 (for Connecticut). The goal is to use the first few digits in each number as the stem, in such a way that about 6 to 15 numbers are lined up vertically. In this case, we use the first two digits, ranging from 11, representing $11,000, to 24, representing $24,000, resulting in 14 stems.

STEP 2: ATTACH THE LEAVES

The second step is to attach a **leaf** to represent each data point. The next digit in the number is used as the leaf, and the remaining digits are simply dropped. For example, for Alabama, the first number in our list, the per capita income is $13,679. The first two digits have been used in the stem, so we use the third digit, 6, as the leaf. We attach it to the stem value of 13. The middle part of Figure 7.1 shows the picture after leaves have been attached only for the states of Alabama through Florida. The finished picture, on the right, has the leaves attached for the per capita incomes of all 51 states. Sometimes an additional step is taken and the leaves are ordered numerically on each branch.

Attaching Meaning to the Stemplot

Now that we have a picture of the per capita income data, it is easy to make several observations. First, there is a wide range of incomes, with Connecticut, the highest, being more than twice that of Mississippi, the lowest. Second, there appear to be two clusters of values, one in the range of $13,000 and another in the $16,000 to $17,000 range. Third, more states are at the lower end than at the upper end. Although the range from $20,000 and above includes over one-third of the stem values, there are only eight states, or fewer than one-sixth the total number of states, in that range. Finally, we can see that there are no extreme outliers; that is, no states are very far above or below the others.

Further Details for Creating Stemplots

Suppose you wanted to create a picture of what your own pulse rate is when you are relaxed. You collect 25 values over a series of a few days and find that they range from 54 to 78. If you tried to create a stemplot using the first digit as the stem, you would have only three stem values. If you tried to use both digits for the stem, you could have as many as 25 separate values, and the picture would be meaningless.

The solution to this problem is to reuse each of the digits 5, 6, and 7. Because you need to have equally spaced intervals, you could use each of the digits two or five times. If you use them each twice, the first listed would receive leaves from 0 to 4, and the second would receive leaves from 5 to 9. Thus, each stem value would encompass a range of five beats per minute of pulse. If you use each digit five times, each stem value would receive leaves of two possible values. The first stem for each digit would receive 0 and 1, the second would receive 2 and 3, and so on. Notice that if you tried to use the initial pulse digits three or four times each, you could not evenly divide the leaves among them because there are always ten possible values for leaves. Figure 7.2 shows two possible stemplots for the same hypothetical pulse data. Stemplot *a* shows the digits 5, 6, and 7 used twice; stemplot *b* shows them used five times.

Creating a Histogram

Histograms are pictures related to stemplots. For very large data sets, a histogram is more feasible than a stemplot because it doesn't list every data value.

To create a histogram, simply divide the range of the data into intervals in much the same way as we did when creating a stemplot. But instead of listing each individual value, simply count how many fall into each part of the range. Draw a bar whose height is equal to the count for each part of the range. Or, equivalently, make the height equal to the *proportion* of the total count that falls in that interval.

Figure 7.3 shows a histogram for the data from Table 7.1. Notice that the heights of the bars are represented as proportions. For example, there are six values in the per capita income range from $17,000 to $17,999, so that represents $6/51 = 0.12$, or 12% of the data. The heights of all of the bars must sum to 1, or 100%. If you wanted

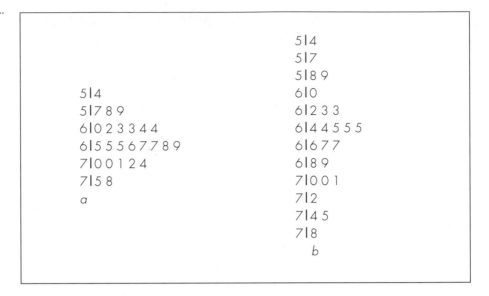

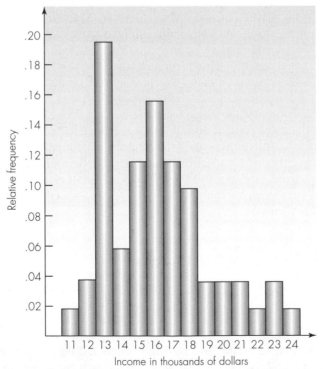

to know what proportion of the data fell into a certain interval or range, you would simply sum the heights of the bars for that range. Also, notice that if you were to turn the histogram on its side, it would look very much like a stemplot except that the labels would differ slightly.

Defining a Common Language about Shape

Symmetric Data Sets

Scientists often talk about the "shape" of data; what they really mean is the shape of the stemplot or histogram resulting from the data. A **symmetric** data set is one in which, if you were to draw a line through the center, the picture on one side would be a mirror image of the picture on the other side. A special case, which will be discussed in detail in Chapter 8, is a **bell-shaped** data set, in which the picture is not only symmetric, but also shaped like a bell. The stemplots in Figure 7.2, displaying pulse rates, are approximately symmetric and bell-shaped.

Unimodal or Bimodal

Recall that the mode is the most common value in a set of data. If there is a single prominent peak in a histogram or stemplot, as in Figure 7.2, the shape is called **unimodal,** meaning "one mode." If there are two prominent peaks, the shape is called **bimodal,** meaning two modes. Figures 7.1 and 7.3 illustrate that the per capita incomes in the United States are bimodal because, as noted earlier, there appear to be two clusters of incomes.

Skewed Data Sets

In common language, something that is skewed is off-center in some way. In statistics, a **skewed** data set is one that is basically unimodal but is substantially off from being bell-shaped. If it is **skewed to the right,** the higher values are more spread out than the lower values, as in the per capita income data. If it is **skewed to the left,** then the lower values are more spread out and the higher ones tend to be clumped. This terminology results from the fact that, before computers were used, shape pictures were always hand drawn using the horizontal orientation in Figure 7.3. Notice that a picture that is skewed to the right, like Figure 7.3, extends further to the right of the highest peak (the tallest bar) than to the left.

Additional Examples of Shapes

Examine Figures 7.4 to 7.6. Each displays a histogram, rotated sideways from the way you have previously seen a histogram. Some computer packages display histograms with this orientation. Try to describe the shapes illustrated in these figures. Are they symmetric? Bell-shaped? Skewed to the left or to the right? Unimodal or bimodal?

Figure 7.4 represents the heights, in millimeters, of a random sample of 199 British men (Marsh, 1988, p. 315; data reproduced in Hand et al., 1994, pp. 179–183). The picture is nearly symmetric, bell-shaped, and, as with all bell-shaped pictures, unimodal.

Figure 7.5 illustrates a histogram of the dollar amounts (in thousands) for collision damage claims paid by an insurance company over a 1-year period for a city in the midwestern United States. The picture is unimodal but clearly not symmetric or

FIGURE 7.4

Heights of British males in millimeters (N = 199)

Source: Data disk from Hand et al., 1994.

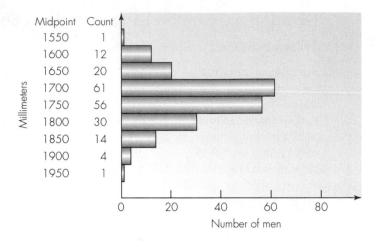

Midpoint	Count
1550	1
1600	12
1650	20
1700	61
1750	56
1800	30
1850	14
1900	4
1950	1

FIGURE 7.5

Insurance claims for collision damage in a midwestern city (N = 187)

Source: Ott and Mendenhall, 1994.

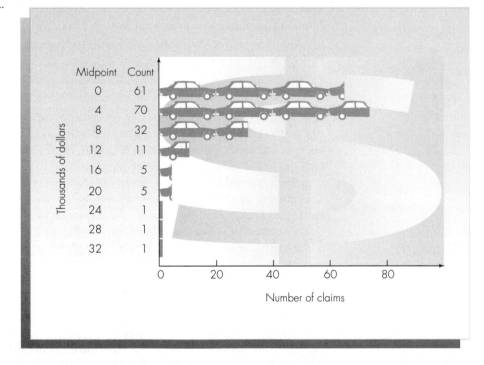

Midpoint	Count
0	61
4	70
8	32
12	11
16	5
20	5
24	1
28	1
32	1

bell-shaped. In fact, as is often the case with monetary data, it is severely skewed to the right.

Figure 7.6 shows a histogram of the times between eruptions of the "Old Faithful" geyser. Notice that the picture is bimodal, with one mode around 50 minutes and another, larger peak around 80 minutes. A picture like this may help scientists figure out what causes the geyser to erupt when it does.

FIGURE 7.6

*Times between
eruptions of "Old
Faithful" geyser
(N = 299)*

Source: Hand et al., 1994.

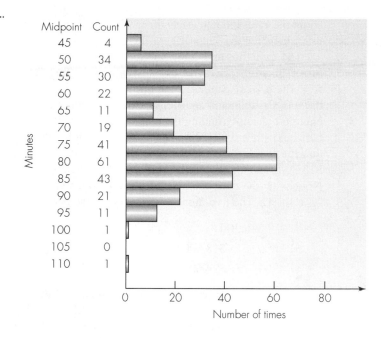

Midpoint	Count
45	4
50	34
55	30
60	22
65	11
70	19
75	41
80	61
85	43
90	21
95	11
100	1
105	0
110	1

Minutes / Number of times

FIGURE 7.7

*The five-number
summary display*

Median

Lower quartile   Upper quartile

Lowest   Highest

7.3   FIVE USEFUL NUMBERS: A SUMMARY

A **five-number summary** is a useful way to summarize a long list of numbers. As the name implies, this is a set of five numbers that provide a good summary of the entire list. Figure 7.7 shows what the five useful numbers are and the order in which they are usually displayed.

The lowest and highest values are self-explanatory. The median, which we discussed earlier, is the number such that half of the values are above it and half are below it. If there are an odd number of values in the data set, the median is simply the middle value in the ordered list. If there is an even number of values, the median is the average of the middle two values. For example, the median of the list 70, 75, 85, 86, 87 is 85 because it is the middle value. If the list had an additional value of 90 in it, the median would be 85.5, the average of the middle two numbers, 85 and 86.

The median can be found quickly from a stemplot, especially if the leaves have been ordered. Using Figure 7.1, convince yourself that the median per capita income is the 26th value (51 = 25 + 1 + 25) from either end, which is $16,300. (Be careful to count the values in numerical order when you get to the stem value 16, and not in the order in which they appear.)

The **quartiles** are simply the medians of the two halves of the ordered list. The **lower quartile**—because it's halfway into the first half—is one quarter of the way from the bottom. Similarly, the **upper quartile** is one quarter of the way down from the top. Complicated algorithms exist for finding exact quartiles. We can get close enough by simply finding the median first, then finding the medians of all the numbers below it and all the numbers above it. For the per capita income data, the lower quartile would be the median of the 25 values below the median of $16,300, or $13,800. The upper quartile would be the median of the upper 25 values, $18,800. Notice that these are the 13th from the bottom and the top of each half because 25 = 12 + 1 + 12. The five-number summary for the per capita income data is thus:

$16,300

$13,800 $18,800

$11,800 $24,600

These five numbers provide a useful summary of the entire set of 51 numbers. We can get some idea of the middle, the spread, and whether or not the values are clumped at one end or the other. Because a much larger gap exists between $18,800 and $24,600 than between $11,800 and $13,800, we know that the values are probably more clumped at the lower end and more spread out at the upper end. In other words, this confirms that the data are skewed to the right.

One final note about using stemplots in this way: Remember that we dropped the last two digits on the per capita incomes when we created the stemplots. If we wanted to find the exact median for the data, we would simply figure out which value corresponded to $16,300 on the stemplot. Scanning the list in Figure 7.1, we see that it is $16,310, the per capita income for Maine. Thus, the exact median is $16,310. Similarly, we could find the exact values for the other four numbers in the five-number summary. However, the truncated values from the stemplot are generally close enough to give us the picture we need.

7.4   BOXPLOTS

A visually appealing and useful way to present a five-number summary is through a *boxplot*, sometimes called a *box and whisker plot*. This simple picture also allows easy comparison of the center and spread of data collected for two or more groups.

EXAMPLE 1   **HOW MUCH DO STATISTICS STUDENTS SLEEP?**

During spring semester, 1998, 190 students in a statistics class at a large university were asked to answer a series of questions in class one day, including how

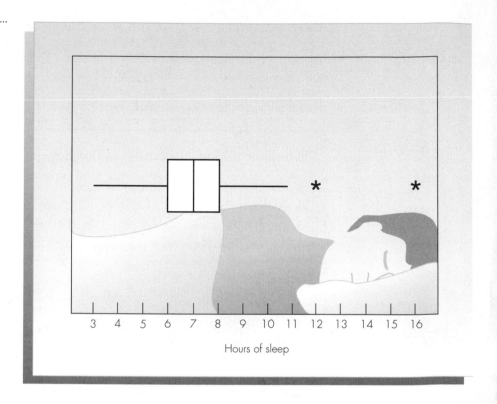

Hours of sleep

many hours they had slept the night before (a Tuesday night). A five-number summary for the reported number of hours of sleep is

$$
\begin{array}{cc}
& 7 \\
6 & 8 \\
3 & 16
\end{array}
$$

Two individuals reported that they slept 16 hours; the maximum for the remaining 188 students was 12 hours. ■

Creating a Boxplot

The boxplot for the hours of sleep is presented in Figure 7.8, and illustrates how a boxplot is constructed. The ends of the "box" represent the lower and upper quartiles. The line in the middle of the box represents the median. The width of the box is the distance between the lower and upper quartiles, so the box has a range that covers half the data values. This distance is called the **interquartile range** because it's the distance between the lower and upper quartiles. The lines or "whiskers" extend from the box until they either reach the minimum (on the left) and maximum (on the right) *or* until they have already extended a specified length, whichever comes first. The specified length is 1.5 times the width of the box. The asterisks indi-cate data values that are beyond that specified length and are considered to be o liers. In Figure 7.8, notice that the box extends from 6 hours to 8 hours (the l and upper quartiles), with the median marked at 7 hours. The interquartile rar

thus 2 hours (from 6 to 8 hours), so the whiskers will extend a maximum of $1.5 \times 2 = 3$ hours beyond the box. The left whisker does exactly that, coincidentally reaching the minimum value of 3 hours of sleep. The right whisker extends to 11 hours $(8 + 3)$, but that leaves outliers at 12 hours and 16 hours, which are marked by asterisks. The asterisk at 16 hours actually represents two equal data values, but there is no way to indicate that on the boxplot.

Summary of How to Construct a Boxplot

1. Draw a box with ends at the lower and upper quartiles.

2. Draw a line in the box at the median.

3. Compute the width of the box; this is the interquartile range.

4. Draw whiskers at each end with length equal to 1.5 times the interquartile range; if the minimum or maximum occurs before the full length is used, stop there.

5. Use an asterisk to indicate any additional data points beyond the range covered by the box and whiskers.

Interpreting Boxplots

Notice that boxplots essentially divide the data into fourths. The lowest fourth of the data values is contained in the range of values below the start of the box, the next fourth is contained in the first part of the box (between the lower quartile and the median), the next fourth is in the upper part of the box, and the final fourth is between the box and the upper end of the picture. Outliers are also easily identified. In the boxplot in Figure 7.8, we can see that one-fourth of the students slept between 3 and 6 hours the previous night, one-fourth slept between 6 and 7 hours, one-fourth slept between 7 and 8 hours, and the final fourth slept between 8 and 16 hours. We can thus immediately see that the data are skewed to the right because the final fourth covers an 8-hour period whereas the lowest fourth covers only a 3-hour period.

As the next example illustrates, boxplots are particularly useful for comparing two or more groups on the same measurement. Although almost the same information is contained in five-number summaries, the visual display makes similarities and differences much more obvious.

....

EXAMPLE 2   **WHO ARE THOSE CRAZY DRIVERS?**

The survey taken in the statistics class in Example 1 also included the question "What's the fastest you have ever driven a car? ____ mph." The boxplots in Figure 7.9 illustrate the comparison of the responses for males and females. Here are the corresponding five-number summaries:

FIGURE 7.9
Boxplots for fastest
ever driven a car

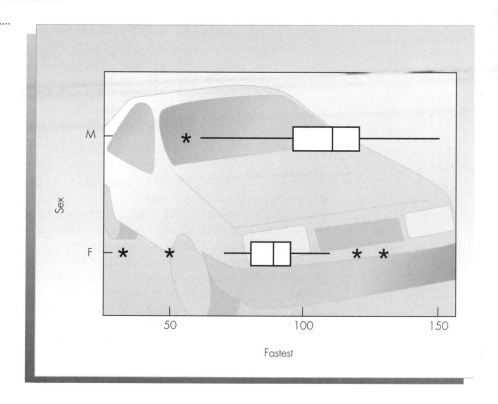

Males (87 Students)		Females (102 Students)	
107		88	
95	120	80	95
55	150	30	130

Some features are more immediately obvious in the boxplots than in the five-number summaries. For instance, the lower quartile for the men is equal to the upper quartile for the women. In other words, 75% of the men have driven 95 mph or faster, but only 25% of the women have done so. Except for a few outliers (120 and 130), all of the women's maximum driving speeds are close to or below the median for the men. Notice how useful the boxplots are for comparing the maximum driving speeds for the sexes. ■

7.5   TRADITIONAL MEASURES: MEAN, VARIANCE, AND STANDARD DEVIATION

The five-number summary has come into use relatively recently. Traditionally, only two numbers have been used to describe a set of numbers: the **mean,** representing the center, and the **standard deviation,** representing the spread or variability in the values. Sometimes the **variance** is given instead of the standard deviation. The standard deviation is simply the square root of the variance, so once you have one you can easily compute the other.

The mean and standard deviation are most useful for symmetric sets of data with no outliers. However, they are very commonly quoted, so it is important to understand what they represent, including their uses and their limitations.

The Mean

As we discussed earlier, the **mean** is the numerical average of a set of numbers. In other words, we add up the values and divide by the number of values. The mean can be distorted by one or more outliers and is thus most useful when there are no extreme values in the data. For example, suppose you are a student taking four classes, and the number of students in each is, respectively, 20, 25, 35, and 200. What is your typical class size? Notice that the median is 30 students. The mean, however, is 280/4 or 70 students. The mean is severely affected by the one large class size of 200 students.

As another example, refer to Figure 7.5, which displays claims paid by an insurance company for collision damages. The majority of claims were for amounts less than $4000, and the median claim is only $3500. But because there were a few very high claims, the mean amount paid is $5178.

Because the mean can be distorted by outliers, data involving incomes or prices are usually summarized by using the median. For example, the median price of a house in a given area, instead of the mean price, is routinely quoted in the newspaper. That's because one house that sold for several million dollars would substantially distort the mean but would have little effect on the median.

Again, the mean is most useful for symmetric data sets with no outliers. As another example, notice that the British male heights in Figure 7.4 fit that description. In such cases, the mean and median should be about equal. For Figure 7.4, the mean height is 1732.5 millimeters (about 68.25 inches) and the median height is 1725 millimeters (about 68 inches).

The Standard Deviation and Variance

It is not easy to compute the **standard deviation** of a set of numbers, but most calculators and computer programs such as *Excel* now handle that task for you. It is more important to know how to interpret the standard deviation, which is a useful measure of how spread out the numbers are.

Consider the following two sets of numbers, both with a mean of 100:

NUMBERS	MEAN	STANDARD DEVIATION
100, 100, 100, 100, 100	100	0
90, 90, 100, 110, 110	100	10

The first set of numbers has no spread or variability to it at all. It has a standard deviation of 0. The second set has some spread to it; on average, the numbers are about 10 points away from the mean, except for the number that is exactly at the mean. That set has a standard deviation of 10.

Computing the Standard Deviation

Here are the steps necessary to compute the standard deviation:

1. Find the mean.
2. Find the deviation of each value from the mean = value – mean.
3. Square the deviations.
4. Sum the squared deviations.
5. Divide the sum by #values (the number of values) –1, resulting in the variance.
6. Take the square root of the variance. The result is the standard deviation.

Let's try this for the set of values 90, 90, 100, 110, 110.

1. The mean is 100.
2. The deviations are –10, –10, 0, 10, 10.
3. The squared deviations are 100, 100, 0, 100, 100.
4. The sum of the squared deviations is 400.
5. The #values –1 = 5 –1 = 4, so the variance is 400/4 = 100.
6. The standard deviation is the square root of 100, or 10.

Although it may seem more logical in step 5 to divide by the number of values, rather than by the number of values minus 1, there is a technical reason for subtracting 1. The reason is beyond the level of this discussion but concerns statistical bias, as discussed in Chapter 3.

The easiest interpretation is to recognize that *the standard deviation is roughly the average distance of the observed values from their mean.* Where the data have a bell shape, the standard deviation is quite useful indeed. For example, the Stanford-Binet IQ test is designed to have a mean of 100 and a standard deviation of 16. If we were to produce a histogram of IQs for a large, representative group, we would find it to be approximately bell-shaped. Its center would be at 100. If we were to determine how far each person's IQ fell from 100, we would find an average distance, on one side or the other, of about 16 points. (In the next chapter, we will see how to use the standard deviation of 16 in a more useful way.) For shapes other than bell shapes, the standard deviation is useful as an intermediate tool for more advanced statistical procedures; it is not very useful on its own, however.

FIGURE 7.10
*Histogram of the
number of matches to
A's answers for each
student*
Source: Data from Boland
and Proschan, Summer 1991,
p. 14.

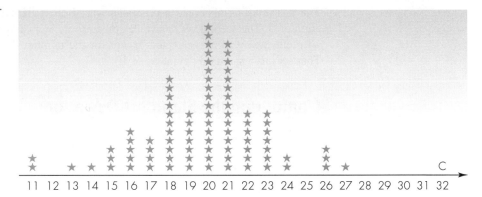

CASE STUDY 7.1   Detecting Exam Cheating with a Histogram

It was summer 1984, and a class of 88 students at a university in Florida was taking a 40-question multiple-choice exam. The proctor happened to notice that one student, whom we will call C, was probably copying answers from a student nearby, whom we will call A. Student C was accused of cheating, and the case was brought before the university's supreme court.

At the trial, evidence was introduced showing that of the 16 questions *missed* by both A and C, both had made the same wrong guess on 13 of them. The prosecution argued that a match that close by chance alone was very unlikely, and student C was found guilty of academic dishonesty.

The case was challenged, however, partly because in calculating the odds of such a strong match, the prosecution had used an unreasonable assumption. They assumed that any of the four wrong answers on a missed question would be equally likely to be chosen. Common sense, as well as data from the rest of the class, made it clear that certain wrong answers were more attractive choices than others.

A second trial was held, and this time the prosecution used a more reasonable statistical approach. The prosecution created a measurement for each student in the class except A (the one from whom C allegedly copied), resulting in 87 data values. For each student, the prosecution simply counted how many of his or her 40 answers matched the answers on A's paper. The results are shown in the histogram in Figure 7.10. Student C is coded as a C, and each asterisk represents one other student. Student C is an obvious outlier in an otherwise bell-shaped picture. You can see that it would be quite unusual for that particular student to match A's answers so well without some explanation other than chance.

Unfortunately, the jury managed to forget that the proctor observed Student C looking at Student A's paper. The defense used this oversight to convince them that, based only on the histogram, A could have been copying from C. The guilty verdict was overturned, despite the compelling statistical picture and evidence.

SOURCE: Boland and Proschan, Summer 1991, pp. 10–14. ∎

FOR THOSE WHO LIKE FORMULAS

The Data

n = number of observations

x_i = the *ith* observation, $i = 1, 2, \ldots, n$

The Mean

$$\bar{x} = \frac{1}{n}(x_1 + x_2 + \cdots + x_n) = \frac{1}{n}\sum_{i=1}^{n} x_i$$

The Variance

$$s^2 = \frac{1}{(n-1)}\sum_{i=1}^{n}(x_i - \bar{x})^2$$

The Computational Formula for the Variance

$$s^2 = \frac{1}{(n-1)}\left(\sum_{i=1}^{n} x_i^2 - \frac{\left(\sum_{i=1}^{n} x_i\right)^2}{n}\right)$$

The Standard Deviation

Use either formula to find s^2; then simply take the square root to get the standard deviation s.

EXERCISES

1. At the beginning of this chapter, the following exam scores were listed: 75, 95, 60, 93, 85, 84, 76, 92, 62, 83, 80, 90, 64, 75, 79, 32, 78, 64, 98, 73, 88, 61, 82, 68, 79, 78, 80, 55.

   a. Create a stemplot for the test scores.

   b. Describe the scores as illustrated by the stemplot, including features such as shape, outliers, and clumping.

2. Refer to the test scores in Exercise 1.

   a. Create a five-number summary.

   b. Create a boxplot.

3. Create a histogram for the test scores in Exercise 1. Comment on the shape.

4. Give an example for which the median would be more useful than the mean as a measure of center.

5. Give an example of a set of five numbers with a standard deviation of 0.

6. Give an example of a set of more than five numbers that has a five-number summary of

40

30   70

10   80

7. All the information contained in the five-number summary for a data set is required for constructing a boxplot. What additional information is required?

8. Find the mean and standard deviation of the following set of numbers: 10, 20, 25, 30, 40.

9. Refer to the pulse rate data displayed in the stemplots in Figure 7.2.

   a. Find the median.

   b. Create a five-number summary.

10. The data on hours of sleep discussed in Example 1 also included whether each student was male or female. Here are the separate five-number summaries for "hours of sleep" for the two sexes:

Males		Females	
7		7	
6	8	6	8
3	16	3	11

   a. Two males reported sleeping 16 hours and one reported sleeping 12 hours. Using this information and the five-number summaries, draw boxplots that allow you to compare the sexes on number of hours slept the previous night. Use a format similar to Figure 7.9.

   b. Based on the boxplots in part a, describe the similarities and differences between the sexes for number of hours slept the previous night.

11. Refer to the data on per capita income in Table 7.1; a five-number summary is given in Section 7.3, page 116).

   a. Construct a boxplot for this data set.

   b. Discuss which picture is more useful for this data set: the boxplot from part a, or the histogram in Figure 7.3.

12. In each of the following cases, which would probably be higher, the mean or the median, or would they be about equal?

   a. Salaries in a company employing 100 factory workers and 2 highly paid executives

   b. Ages at which residents of a suburban city die, including everything from infant deaths to the most elderly

   c. Prices of all new cars sold in 1 month in a large city

   d. Heights of all 7-year-old children in a large city

   e. Shoe sizes of adult women

13. Suppose an advertisement reported that the mean weight loss after using a certain exercise machine for 2 months was 10 pounds. You investigate further and discover that the median weight loss was 3 pounds.

    a. Explain whether it is most likely that the weight losses were skewed to the right, skewed to the left, or symmetric.

    b. As a consumer trying to decide whether to buy this exercise machine, would it have been more useful for the company to give you the mean or the median? Explain.

14. Construct an example and draw a histogram for a measurement that you think would be bell-shaped.

15. Construct an example and draw a histogram for a measurement that you think would be skewed to the right.

16. Construct an example and draw a histogram for a measurement that you think would be bimodal.

17. Give an example of a measurement for which the mode would be more useful than the median or the mean as an indicator of the "typical" value.

18. Explain the following statement in words that someone with no training in statistics would understand: The heights of adult males in the United States are bell-shaped, with a mean of about 70 inches and a standard deviation of about 3 inches.

19. Suppose a set of test scores is approximately bell-shaped, with a mean of 70 and a range of 50. Approximately, what would the minimum and maximum test scores be?

20. Three types of pictures were presented in this chapter: stemplots, histograms, and boxplots. Explain the features of a data set for which:

    a. Stemplots are most useful

    b. Histograms are most useful

    c. Boxplots are most useful

21. Would outliers more heavily influence the range or the quartiles? Explain.

22. What is the variance for the Stanford-Binet IQ test?

23. Give one advantage a stemplot has over a histogram and one advantage a histogram has over a stemplot.

24. Find a set of data of interest to you, such as rents from a newspaper or test scores from a class, with at least 12 numbers. Include the data with your answer.

    a. Create a five-number summary of the data.

    b. Create a boxplot of the data.

    c. Describe the data in a paragraph that would be useful to someone with no training in statistics.

25. Which set of data is more likely to have a bimodal shape: daily New York City temperatures at noon for the summer months or daily New York City temperatures at noon for an entire year? Explain.

26. Suppose you had a choice of two professors to take for a class in which your grade was very important. They both assign scores on a percentage scale (0 to

100). You can have access to three summary measures of the last 200 scores each professor assigned. Of the summary measures discussed in this chapter, which three would you choose? Why?

27. Draw a boxplot illustrating a data set with each of the following features:

   a. Skewed to the right with no outliers

   b. Bell-shaped with the exception of one outlier at the upper end

   c. Values uniformly spread across the range of the data

MINI-PROJECTS

1. Find a set of data that has meaning for you. Some potential sources are the Internet, the sports pages, and the classified ads. Using the methods given in this chapter, summarize and display the data in whatever ways are most useful. Give a written description of interesting features of the data.

2. Measure your pulse rate 25 times over the next few days, but don't take more than one measurement in any 10-minute period. Record any unusual events related to the measurements, such as if one was taken during exercise or one was taken immediately upon awakening. Create a stemplot and a five-number summary of your measurements. Give a written assessment of your pulse rate based on the data.

REFERENCES

Boland, Philip J., and Michael Proschan. (Summer 1991). The use of statistical evidence in allegations of exam cheating. *Chance* 3, no. 3, pp. 10–14.

Hand, D. J., F. Daly, A. D. Lunn, K. J. McConway, and E. Ostrowski. (1994). *A handbook of small data sets.* London: Chapman and Hall.

Information please almanac. (1991). Edited by Otto Johnson. Boston: Houghton Mifflin.

Marsh, C. (1988). *Exploring data.* Cambridge, England: Policy Press.

Ott, R. L., and W. Mendenhall. (1994). *Understanding statistics.* 6th ed. Belmont, CA: Duxbury Press.

CHAPTER EIGHT

Bell-Shaped Curves and Other Shapes

THOUGHT QUESTIONS

1. The heights of adult women in the United States follow, at least approximately, a bell-shaped curve. What do you think this means?

2. What does it mean to say that a man's weight is in the 30th percentile for all adult males?

3. A "standardized score" is simply the number of standard deviations an individual falls above or below the mean for the whole group. (Values above the mean have positive standardized scores, whereas those below the mean have negative ones.) Male heights have a mean of 70 inches and a standard deviation of 3 inches. Female heights have a mean of 65 inches and a standard deviation of 2 1/2 inches. Thus, a man who is 73 inches tall has a standardized score of 1. What is the standardized score corresponding to your own height?

4. Data sets consisting of physical measurements (heights, weights, lengths of bones, and so on) for adults of the same species and sex tend to follow a similar pattern. The pattern is that most individuals are clumped around the average, with numbers decreasing the farther values are from the average in either direction. Describe what shape a histogram of such measurements would have.

FIGURE 8.1
*A normal frequency
curve*

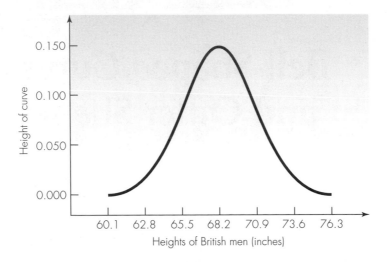

8.1 POPULATIONS, FREQUENCY CURVES, AND PROPORTIONS

In Chapter 7, we learned how to draw a picture of a set of data and how to think about its shape. In this chapter, we learn how to extend those ideas to pictures and shapes for populations of measurements. For example, in Figure 7.4 we illustrated that, based on a sample of 199 men, heights of adult British males are reasonably bell-shaped. Because the men were a representative sample, the picture for all of the millions of British men is probably similar. But even if we could measure them all, it would be difficult to construct a histogram with so much data. What is the best way to represent the shape of a population of measurements?

Frequency Curves

The most common type of picture for a population is a smooth **frequency curve.** Rather than drawing lots of tiny rectangles, the picture is drawn as if the tops of the rectangles were connected with a smooth curve. Figure 8.1 illustrates a frequency curve for the population of British male heights. Notice that the picture is similar to the histogram in Figure 7.4, except that the curve is smooth and the heights have been converted to inches.

Notice that the vertical scale is simply labeled "height of curve." This height is determined by sizing the curve so that the area under the entire curve is 1, for reasons that will become clear in the next few pages. Unlike with a histogram, the height of the curve cannot be interpreted as a proportion or frequency, but is chosen simply to satisfy the rule that the entire area under the curve is 1.

The bell shape illustrated in Figure 8.1 is so common that if a population has this shape, the measurements are said to follow a **normal distribution.** Equivalently,

FIGURE 8.2

A nonnormal frequency curve

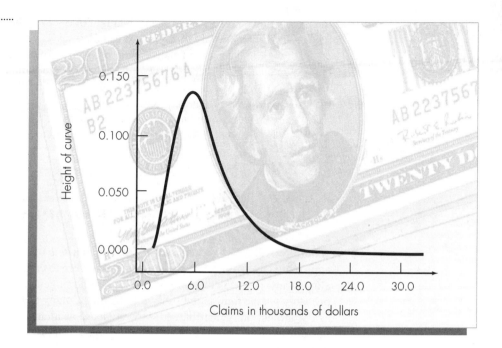

they are said to follow a **bell-shaped curve,** a **normal curve,** or a **Gaussian** curve. This last name comes from the name of Karl Friedrich Gauss (1777–1855), who was one of the first mathematicians to investigate the shape.

Not all frequency curves are bell-shaped. Figure 8.2 shows a likely frequency curve for the population of insurance claims from which the claims in Figure 7.5 were drawn. Notice that the curve is skewed to the right. The majority of claims are below $5,000, but occasionally there will be an extremely high claim. For the remainder of this chapter, we focus on bell-shaped curves.

Proportions

Frequency curves are quite useful for determining what **proportion** or percentage of the population of measurements falls into a certain range. If we wanted to find out what percentage of the data fell into any particular range with a stemplot, we would count the number of leaves that were in that range and divide by the total. If we wanted to find the percentage in a certain range using a histogram, we would simply add up the heights of the rectangles for that range, assuming we had used proportions instead of counts for the heights.

What if we have a frequency curve instead of a stemplot or histogram? Frequency curves are, by definition, drawn to make it easy to represent the proportion of the population falling into a certain range. Recall that they are drawn so the entire area underneath the curve is 1, or 100%. Therefore, to figure out what percentage or proportion of the population falls into a certain range, all you have to do is figure out how much of the area is situated over that range. For example, in Figure 8.1, half of the area is in the range above the mean height of 68.25 inches. In other words, about half of all British men are 68.25 inches or taller.

Although it is easy to visualize what proportion of a population falls into a certain range using a frequency curve, it is not as easy to compute that proportion. For anything but very simple cases, the computation to find the required area involves the use of calculus. However, because bell-shaped curves are so common, tables have been prepared in which the work has already been done (see, for example, Table 8.1 at the end of this chapter), and many calculators and computer applications such as *Excel* will compute these proportions.

8.2   THE PERVASIVENESS OF NORMAL CURVES

Nature provides numerous examples of populations of measurements that, at least approximately, follow a normal curve. If you were to create a picture of the shape of almost any physical measurement within a homogeneous population, you would probably get the familiar bell shape. In addition, many psychological attributes, such as IQ, are normally distributed. Many standard academic tests, such as the Scholastic Assessment Test (SAT), if given to a large group, will result in normally distributed scores.

The fact that so many different kinds of measurements all follow approximately the same shape should not be surprising. The majority of people are somewhere close to average on any attribute, and the farther away you move from the average, either above or below, the fewer people will have those more extreme values for their measurements.

Sometimes a set of data is distorted to make it fit a normal curve. That's what happens when a professor "grades on a bell-shaped curve." Rather than assign the grades students have actually earned, the professor distorts them to make them fit into a normal curve, with a certain percentage of A's, B's, and so on. In other words, grades are assigned *as if* most students were average, with a few good ones at the top and a few bad ones at the bottom. Unfortunately, this procedure has a tendency to artificially spread out clumps of students who are at the top or bottom of the scale, so that students whose original grades were very close together may receive different letter grades.

8.3   PERCENTILES AND STANDARDIZED SCORES

Percentiles

Have you ever wondered what percentage of the population of your sex is taller than you are; or what percentage of the population has a lower IQ than you do? Your **percentile** in a population represents the position of your measurement in comparison with everyone else's. It gives the percentage of the population that falls *below* you. If you are in the 50th percentile, it means that exactly half of the population falls below you. If you are in the 98th percentile, 98% of the population falls below you and only 2% is above you.

Your percentile is easy to find if the population of values has an approximate bell shape and if you have just three pieces of information. All you need to know are your own value and the mean and standard deviation for the population.

Although there are obviously an unlimited number of potential bell-shaped curves, depending on the magnitude of the particular measurements, each one is completely determined once you know its mean and standard deviation. In addition, each one can be "standardized" in such a way that the same table can be used to find percentiles for any of them.

Standardized Scores

Suppose you knew your IQ was 116, as measured by the Stanford-Binet IQ test. Scores from that test have a normal distribution with a mean of 100 and a standard deviation of 16. Therefore, your IQ is exactly 1 standard deviation above the mean of 100. In this case, we would say you have a *standardized score* of 1. In general, a **standardized score** simply represents the number of standard deviations the observed value or score falls from the mean. A positive standardized score indicates an observed value above the mean, whereas a negative standardized score indicates a value below the mean. Someone with an IQ of 84 would have a standardized score of –1 because he or she would be exactly 1 standard deviation below the mean. Sometimes the abbreviated **standard score** is used instead of "standardized score."

Once you know the standardized score for an observed value, all you need to find the percentile is the appropriate table, one that gives percentiles for a normal distribution with a mean of 0 and a standard deviation of 1. A normal curve with a mean of 0 and a standard deviation of 1 is called a **standard normal curve.** It is the curve that results when any normal curve is converted to standardized scores. In other words, the standardized scores resulting from any normal curve will have a mean of 0 and a standard deviation of 1 and will retain the bell shape.

Table 8.1, presented at the end of this chapter, gives percentiles for standardized scores. For example, with an IQ of 116 and a standardized score of +1, you would be at the 84th percentile. In other words, your IQ would be higher than that of 84% of the population.

If we are told the percentile for a score but not the value itself, we can also work backward from the table to find the value. Let's review the steps necessary to find a percentile from an observed value, and vice versa.

> *To find the percentile from an observed value:*
> **1.** Find the standardized score = (observed value – mean) / s.d.,
>   where s.d. = standard deviation. Don't forget to keep the plus or minus sign.
> **2.** Look up the percentile in Table 8.1.

> *To find an observed value from a percentile:*
>
> **1.** Look up the percentile in Table 8.1 and find the corresponding standardized score.
>
> **2.** Compute the observed value = mean + (standardized score)(s.d),
> where s.d. = standard deviation.

EXAMPLE 1

In the Edinburgh newspaper the *Scotsman,* on March 8, 1994, a headline read, "Jury urges mercy for mother who killed baby" (p. 2). The baby had died from improper care. One of the issues in the case was that "the mother . . . had an IQ lower than 98 percent of the population, the jury had heard." From this information, let's compute the mother's IQ. If it was lower than 98% of the population, it was higher than only 2%, so she was in the 2nd percentile. From Table 8.1, we see that her standardized score was –2.05, or 2.05 standard deviations below the mean of 100. We can now compute her IQ:

observed value = mean + (standardized score)(s.d.)

observed value = 100 + (–2.05)(16)

observed value = 100 + (–32.8) = 100 – 32.8

observed value = 67.2

Thus, her IQ was about 67. The jury was convinced that her IQ was, tragically, too low to expect her to be a competent mother. ∎

EXAMPLE 2

The Graduate Record Examination (GRE) is a test taken by college students who intend to pursue a graduate degree in the United States. For all college seniors and graduates who took the exam between October 1, 1989, and September 30, 1992, the mean for the verbal ability portion of the exam was about 497 and the standard deviation was 115 (Educational Testing Service, 1993). If you had received a score of 650 on that GRE exam, what percentile would you be in, assuming the scores were bell-shaped? We can compute your percentile by first computing your standardized score:

standardized score = (observed value – mean)/(s.d.)

standardized score = (650 – 497)/115

standardized score = 153/115 = 1.33

From Table 8.1, we see that a standardized score of 1.33 is between the 90th percentile score of 1.28 and the 91st percentile score of 1.34. In other words, your score was higher than about 90% of the population. Figure 8.3 illustrates the GRE score of 650 for the population of GRE scores and the corresponding stan-

FIGURE 8.3

The 90th percentile for GRE scores and standardized scores

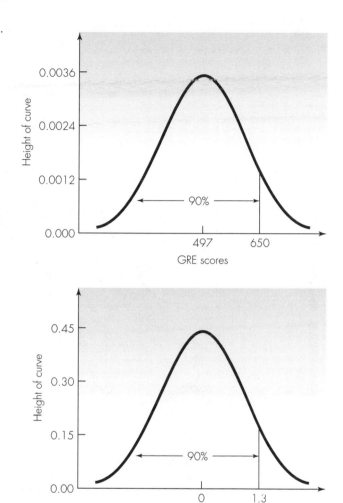

dardized score of 1.3 for the standard normal curve. Notice the similarity of the two pictures. ∎

EXAMPLE 3

Ian Stewart (17 September 1994, p. 14) reported on a problem posed to a statistician by a British company called Molegon, whose business was to remove unwanted moles from gardens. The company kept records indicating that the population of weights of moles in its region was approximately normal, with a mean of 150 grams and standard deviation of 56 grams. The European Union announced that starting in 1995, only moles weighing between 68 grams and 211 grams can be legally caught. Molegon wanted to know what percentage of all moles could be legally caught.

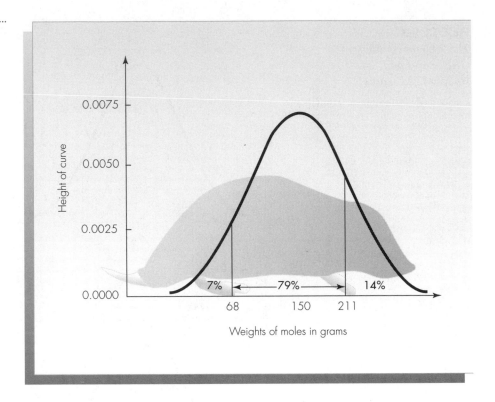

To solve this problem, we need to know what percentage of all moles weigh between 68 grams and 211 grams. We need to find two standardized scores, one for each end of the interval, and then find the percentage of the curve that lies between them:

standardized score for 68 grams = (68 − 150)/56 = −1.46

standardized score for 211 grams = (211 − 150)/56 = 1.09

From Table 8.1, we see that about 86% of all moles weigh 211 grams or less. But we also see that about 7% are below the legal limit of 68 grams. Therefore, about 86% − 7% = 79% are within the legal limits. Of the remaining 21%, 14% are too big to be legal and 7% are too small. Figure 8.4 illustrates this situation. ■

8.4   *z*-SCORES AND FAMILIAR INTERVALS

Any educated consumer of statistics should know a few facts about normal curves. First, a synonym for a standardized score is a **z-score.** Thus, if you are told that your z-score on an exam is 1.5, it means that your score is 1.5 standard deviations above the mean. You can use that information to find your approximate percentile in the class, assuming the scores are approximately bell-shaped.

Second, some easy-to-remember intervals can give you a picture of where values on any normal curve will fall. This information is known as the **Empirical Rule.**

Empirical Rule

For any normal curve, approximately

68% of the values fall within 1 standard deviation of the mean in either direction

95% of the values fall within 2 standard deviations of the mean in either direction

99.7% of the values fall within 3 standard deviations of the mean in either direction

---

A measurement would be an extreme outlier if it fell more than 3 standard deviations above or below the mean. You can see why the standard deviation is such an important measure. If you know that a set of measurements is approximately bell-shaped, and you know the mean and standard deviation, then even without a table like Table 8.1, you can say a fair amount about the magnitude of the values.

For example, because adult women in the United States have a mean height of about 65 inches (5 feet 5 inches) with a standard deviation of about 2.5 inches, and heights are bell-shaped, we know that approximately

- 68% of adult women in the United States are between 62.5 inches and 67.5 inches
- 95% of adult women in the United States are between 60 inches and 70 inches
- 99.7% of adult women in the United States are between 57.5 inches and 72.5 inches

The mean height for adult males in the United States is about 70 inches and the standard deviation is about 3 inches. You can easily compute the ranges into which 68%, 95%, and almost all men's heights should fall.

FOR THOSE WHO LIKE FORMULAS

Notation for a Population

The lowercase Greek letter "mu" = μ represents the **population mean.**

The lowercase Greek letter "sigma" = σ represents the **population standard deviation.**

Therefore, the **population variance** is represented by σ^2.

A **normal distribution** with a mean of μ and variance of σ^2 is denoted by $N(\mu, \sigma^2)$.

For example, the **standard normal distribution** is denoted by $N(0, 1)$.

Standardized Score z for an Observed Value x

$$z = \frac{x - \mu}{\sigma}$$

Observed Value x for a Standardized Score z

$$x = \mu + z\sigma$$

Empirical Rule

If a population of values is $N(\mu, \sigma^2)$, then approximately:

68% of values fall within the interval $\mu \pm \sigma$

95% of values fall within the interval $\mu \pm 2\sigma$

99.7% of values fall within the interval $\mu \pm 3\sigma$

....

EXERCISES

1. Using Table 8.1, a computer, or a calculator, determine the percentage of the population falling *below* each of the following standard scores:

   a. −1.00

   b. 1.96

   c. 0.84

2. Using Table 8.1, a computer, or a calculator, determine the percentage of the population falling *above* each of the following standard scores:

   a. 1.28

   b. −0.25

   c. 2.33

3. Using Table 8.1, a computer, or a calculator, determine the standard score that has the following percentage of the population below it:

   a. 25%

   b. 75%

   c. 45%

   d. 98%

4. Using Table 8.1, a computer, or a calculator, determine the standard score that has the following percentage of the population above it:

   a. 2%

   b. 50%

   c. 75%

   d. 10%

5. Using Table 8.1, a computer, or calculator, determine the percentage of the population falling between the two standard scores given:

TABLE 8.1	**Proportions and Percentiles for Standard Normal Scores**

Standard Score, z	Proportion Below z	Percentile	Standard Score, z	Proportion Below z	Percentile
−3.00	0.0013	0.13	0.03	0.51	51
−2.576	0.005	0.50	0.05	0.52	52
−2.33	0.01	1	0.08	0.53	53
−2.05	0.02	2	0.10	0.54	54
−1.96	0.025	2.5	0.13	0.55	55
−1.88	0.03	3	0.15	0.56	56
−1.75	0.04	4	0.18	0.57	57
−1.64	0.05	5	0.20	0.58	58
−1.55	0.06	6	0.23	0.59	59
−1.48	0.07	7	0.25	0.60	60
−1.41	0.08	8	0.28	0.61	61
−1.34	0.09	9	0.31	0.62	62
−1.28	0.10	10	0.33	0.63	63
−1.23	0.11	11	0.36	0.64	64
−1.17	0.12	12	0.39	0.65	65
−1.13	0.13	13	0.41	0.66	66
−1.08	0.14	14	0.44	0.67	67
−1.04	0.15	15	0.47	0.68	68
−1.00	0.16	16	0.50	0.69	69
−0.95	0.17	17	0.52	0.70	70
−0.92	0.18	18	0.55	0.71	71
−0.88	0.19	19	0.58	0.72	72
−0.84	0.20	20	0.61	0.73	73
−0.81	0.21	21	0.64	0.74	74
−0.77	0.22	22	0.67	0.75	75
−0.74	0.23	23	0.71	0.76	76
−0.71	0.24	24	0.74	0.77	77
−0.67	0.25	25	0.77	0.78	78
−0.64	0.26	26	0.81	0.79	79
−0.61	0.27	27	0.84	0.80	80
−0.58	0.28	28	0.88	0.81	81
−0.55	0.29	29	0.92	0.82	82
−0.52	0.30	30	0.95	0.83	83
−0.50	0.31	31	1.00	0.84	84
−0.47	0.32	32	1.04	0.85	85
−0.44	0.33	33	1.08	0.86	86
−0.41	0.34	34	1.13	0.87	87
−0.39	0.35	35	1.17	0.88	88
−0.36	0.36	36	1.23	0.89	89
−0.33	0.37	37	1.28	0.90	90
−0.31	0.38	38	1.34	0.91	91
−0.28	0.39	39	1.41	0.92	92
−0.25	0.40	40	1.48	0.93	93
−0.23	0.41	41	1.55	0.94	94
−0.20	0.42	42	1.64	0.95	95
−0.18	0.43	43	1.75	0.96	96
−0.15	0.44	44	1.88	0.97	97
−0.13	0.45	45	1.96	0.975	97.5
−0.10	0.46	46	2.05	0.98	98
−0.08	0.47	47	2.33	0.99	99
−0.05	0.48	48	2.576	0.995	99.5
−0.03	0.49	49	3.00	0.9987	99.87
0.00	0.50	50	3.75	0.9999	99.99

    a. −1.00 and 1.00

    b. −1.28 and 1.75

    c. 0.0 and 1.00

6. The 84th percentile for the Stanford-Binet IQ test is 116. (Recall that the mean is 100 and the standard deviation is 16.)

    a. Verify that this is true by computing the standardized score and using Table 8.1.

    b. Draw pictures of the original and standardized scores to illustrate this situation, similar to the pictures in Figure 8.3.

7. Draw a picture of a bell-shaped curve with a mean value of 100 and a standard deviation of 10. Mark the mean and the intervals derived from the Empirical Rule in the appropriate places on the horizontal axis. You do not have to mark the vertical axis.

8. Find the percentile for the observed value in the following situations:

    a. GRE score of 450 (mean = 497, s.d. = 115)

    b. Stanford-Binet IQ score of 92 (mean = 100, s.d. = 16)

    c. Woman's height of 68 inches (mean = 65 inches, s.d. = 2.5 inches)

9. Mensa is an organization that allows people to join only if their IQs are in the top 2% of the population.

    a. What is the lowest Stanford-Binet IQ you could have and still be eligible to join Mensa?

    b. Mensa also allows members to qualify on the basis of certain standard tests. If you were to try to qualify on the basis of the GRE exam, what score would you need on the exam?

10. Every time you have your cholesterol measured, the measurement may be slightly different due to random fluctuations and measurement error. Suppose that for you, the population of possible cholesterol measurements if you are healthy has a mean of 190 and a standard deviation of 10. Further, suppose you know you should get concerned if your measurement ever gets up to the 97th percentile. What level of cholesterol does that represent?

11. Use Table 8.1 to verify that the Empirical Rule is true. You may need to round off the values slightly.

12. Recall from Chapter 7 that the *interquartile range* covers the middle 50% of the data. For a bell-shaped population:

    a. The interquartile range covers what range of standardized scores? In other words, what are the standardized scores for the lower and upper quartiles? (*Hint:* Draw a standard normal curve and locate the 25th and 75th percentiles using Table 8.1.)

    b. How many standard deviations are covered by the interquartile range?

    c. The whiskers on a boxplot can extend a total of 2 interquartile ranges on either side of the median (which for a bell-shaped population is equal to the mean). Beyond that range, data values are considered to be outliers. At what percentiles (at the upper and lower ends) are data values considered to be outliers for bell-shaped populations?

13. Give an example of a population of measurements that you do not think has a normal curve, and draw its frequency curve.

14. A graduate school program in English will admit only students with GRE verbal ability scores in the top 30%. What is the lowest GRE score they will accept? (Recall the mean is 497 and the standard deviation is 115.)

15. Use the Empirical Rule to specify the ranges into which 68%, 95%, and 99.7% of Stanford-Binet IQ scores fall. (Recall the mean is 100 and the standard deviation is 16.)

16. For every 100 births in the United States, the number of boys follows, approximately, a normal curve with mean of 51 boys and standard deviation of 5 boys. If the next 100 births in your local hospital resulted in 36 boys (and thus 64 girls), would that be unusual? Explain.

17. Suppose a candidate for public office is favored by only 48% of the voters. If a sample survey randomly selects 2500 voters, the percentage in the sample who favor the candidate can be thought of as a measurement from a normal curve with a mean of 48% and standard deviation of 1%. Based on this information, how often would such a survey show that 50% or more of the sample favored the candidate?

18. Suppose you record how long it takes you to get to work or school over many months and discover that the times are approximately bell-shaped with a mean of 15 minutes and a standard deviation of 2 minutes. How much time should you allow to get there to make sure you are on time 90% of the time?

19. Assuming heights for each sex are bell-shaped, with means of 70 inches for men and 65 inches for women, and with standard deviations of 3 inches for men and 2.5 inches for women, what proportion of your sex is shorter than you are? (Be sure to mention your sex and height in your answer!)

20. According to *Chance* magazine (vol. 6, no. 3 [1993], p. 5), the mean healthy adult temperature is around 98.2° Fahrenheit, not the previously assumed value of 98.6°. Suppose the standard deviation is 0.6 degree and the population of healthy temperatures is bell-shaped. What proportion of the population have temperatures at or below the presumed norm of 98.6°?

REFERENCES

Educational Testing Service. (1993). *GRE 1993–94 guide.* Princeton, NJ: Educational Testing Service.

Stewart, Ian. (17 September 1994). Statistical modelling. *New Scientist: Inside Science* 74, p. 14.

CHAPTER NINE

Plots, Graphs, and Pictures

THOUGHT QUESTIONS

1. You have seen pie charts and bar graphs and should have some rudimentary idea of how to construct them. Suppose you have been keeping track of your living expenses and find that you spend 50% of your money on rent, 25% on food, and 25% on other expenses. Draw a pie chart and a bar graph to depict this information. Discuss which is more visually appealing and useful.

2. Here is an example of a plot that has some problems. Give two reasons why this is not a good plot.

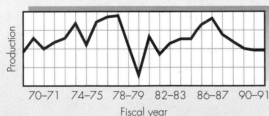

3. Suppose you had a set of data representing two measurement variables—namely, height and weight—for each of 100 people. How could you put that information into a plot, graph, or picture that illustrated the relationship between the two measurements for each person?

4. Suppose you own a company that produces candy bars and you want to display two graphs. One graph is for customers and shows the price of a candy bar for each of the past 10 years. The other graph is for stockholders and shows the

amount the company was worth for each of the past 10 years. You decide to adjust the dollar amounts in one graph for inflation but to use the actual dollar amounts in the other graph. If you were trying to present the most favorable story in each case, which graph would be adjusted for inflation? Explain.

9.1   WELL-DESIGNED STATISTICAL PICTURES

There are many ways to present data in pictures. The most common are plots and graphs, but sometimes a unique picture is used to fit a particular situation. The purpose of a plot, graph, or picture of data is to give you a visual summary that is more informative than simply looking at a collection of numbers. Done well, a picture can convey a message quickly that would take you longer to find if you had to study the data on your own. Done poorly, a picture can mislead all but the most observant of readers. Here are some basic characteristics that all plots, graphs, and pictures should exhibit:

1. The data should stand out clearly from the background.
2. There should be clear labeling that indicates
   a. the title or purpose of the picture.
   b. what each of the axes, bars, pie segments, and so on, denotes.
   c. the scale of each axis, including starting points.
3. A source should be given for the data.
4. There should be as little "chart junk"—that is, extraneous material—in the picture as possible.

9.2   PICTURES OF CATEGORICAL DATA

Categorical data are easy to represent with pictures. The most frequent use of such data is to determine how the whole divides into categories, and pictures are useful in expressing that information. Let's look at three common types of pictures for categorical data and their uses.

Pie Charts

Pie charts are useful when only one categorical variable is measured. Pie charts show what percentage of the whole falls into each category. They are simple to understand, and they convey information about the relative size of groups more readily than a table. Figure 9.1 shows a pie chart that represents the percentage of white American children who have various hair colors.

FIGURE 9.1

Pie chart of hair colors of white American children

Source: Krantz, 1992, p. 188.

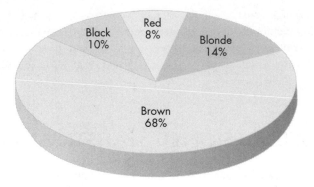

FIGURE 9.2

Percentage of men and women in the labor force

Source: Based on data from U.S. Dept. of Labor, Bureau of Labor Statistics, *Current Population Survey.*

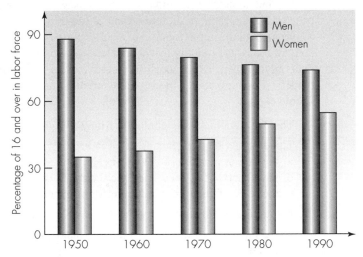

Bar Graphs

Bar graphs also show percentages or frequencies in various categories, but they can be used to represent two or three categorical variables simultaneously. One categorical variable is used to label the horizontal axis. Within each of the categories along that axis, a bar is drawn to represent each category of the second variable. Frequencies or percentages are shown on the vertical axis. A third variable can be included if the graph has only two categories by using percentages on the vertical axis. One category is shown, and the other is implied by the fact that the total must be 100%.

For example, Figure 9.2 illustrates employment trends for men and women across decades. The year in which the information was collected is one categorical variable, represented by the horizontal axis. In each year, people were categorized according to two additional variables: whether they were in the labor force and whether they were male or female. Separate bars are drawn for males and females, and the percentage in the labor force determines the heights of the bars. It is implicit that the remainder were not in the labor force. Respondents were part of the Bureau of Labor Statistics' Current Population Survey, the large monthly survey used to determine unemployment rates.

FIGURE 9.3

*Two pictograms
showing percentages
of Ph.D.s earned by
women*

Source: Alper, 16 April 1993,
p. 409.

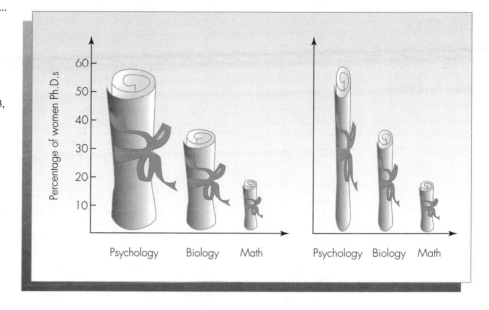

The decision about which variable occupies which position should be made to better convey visually the purpose for the graph. The purpose of the graph in Figure 9.2 is to illustrate that the percentage of women in the labor force has increased since 1950, whereas the percentage of men has decreased slightly. The gap in 1950 was 53 percentage points, but by 1990 it was only 19 percentage points, as is illustrated by the graph.

Bar graphs are not always as visually appealing as pie charts, but they are much more versatile. They can also be used to represent actual frequencies instead of percentages and to represent proportions that are not required to sum to 100%.

Pictograms

A **pictogram** is like a bar graph except that it uses pictures related to the topic of the graph. Figure 9.3 shows a pictogram illustrating the proportion of Ph.D.s earned by women in three fields—psychology (58%), biology (37%), and mathematics (18%)—as reported in *Science* (vol. 260, 16 April, 1993, p. 409). Notice that in place of bars, the graph uses pictures of diplomas.

It is easy to be misled by pictograms. The pictogram on the left shows the diplomas using realistic dimensions. However, it is misleading because the eye tends to focus on the *area* of the diploma rather than just its height. The heights of the three diplomas reach the correct proportions, with heights of 58%, 37%, and 18%, so the height of the one for psychology Ph.D.s is just over three times the height of the one for math Ph.D.s. However, in keeping the proportions realistic, the area of the diploma for psychology is about nine times the area of the one for math, leading the eye to inflate the difference.

The pictogram on the right is drawn by keeping the width of the diplomas the same for each field. The picture is visually more accurate, but it is less appealing

FIGURE 9.4
*Line graph displaying
winning time versus
year for men's
500-meter Olympic
speed skating*
Source: World Almanac and
Book of Facts, 1993, p. 832.

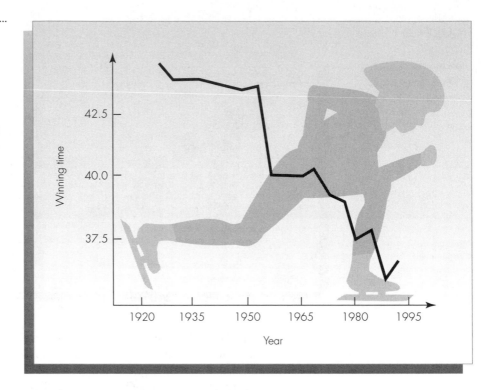

because the diplomas are consequently quite distorted in appearance. When you see a pictogram, be careful to interpret the information correctly and not to let your eye mislead you.

9.3   PICTURES OF MEASUREMENT VARIABLES

Measurement variables can be illustrated with graphs in numerous ways. We saw two ways to illustrate a single measurement variable in Chapter 7—namely, stem-plots and histograms. Graphs are most useful for displaying the relationship between two measurement variables or for displaying how a measurement variable changes over time. Two common types of displays for measurement variables are illustrated in Figures 9.4 and 9.5.

Line Graphs

Figure 9.4 is an example of a **line graph** displayed over time. It shows the winning times for the men's 500-meter speed skating event in the Winter Olympics from

FIGURE 9.5

Scatterplot of grade-point average versus verbal SAT score

Source: Ryan, Joiner, and Ryan, 1985, pp. 309–312.

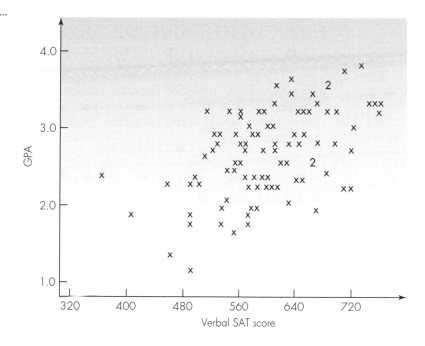

1924 to 1992. Notice the distinct downward trend, with only a few upturns over the years. There was a large drop between 1952 and 1956, followed by a period of relative stability. These patterns are much easier to detect with a picture than they would be by scanning a list of winning times.

Scatterplots

Figure 9.5 is an example of a **scatterplot.** Scatterplots are useful for displaying the relationship between two measurement variables. Each x on the plot represents one individual. In the few cases where a 2 appears on the plot, two individuals had the same data. The plot in Figure 9.5 shows the grade-point averages (GPAs) and verbal scholastic achievement test (SAT) scores for a sample of 100 students at a university in the northeastern United States.

Although a scatterplot can be more difficult to read than a line graph, it displays more information. It shows outliers, as well as the degree of variability that exists for one variable at each location of the other variable. In Figure 9.5, we can see an increasing trend toward higher GPAs with higher SAT scores, but we can also still see substantial variability in GPAs at each level of verbal SAT scores. A scatterplot is definitely more useful than the raw data. Simply looking at a list of the 100 pairs of GPAs and SAT scores, we would find it difficult to detect the trend that is so obvious in the scatterplot.

9.4

DIFFICULTIES AND DISASTERS IN PLOTS, GRAPHS, AND PICTURES

A number of common mistakes appear in plots and graphs that may mislead readers. If you are aware of them and watch for them, you will substantially reduce your chances of misreading a statistical picture.

The most common problems in plots, graphs, and pictures are

1. No labeling on one or more axes
2. Not starting at zero as a way to exaggerate trends
3. Change(s) in labeling on one or more axes
4. Misleading units of measurement
5. Using poor information

No Labeling on One or More Axes

You should always look at the axes in a picture to make sure they are labeled. Figure 9.6a gives an example of a plot for which the units were *not* labeled on the vertical axis. The plot appeared in a newspaper insert titled, "May 1993: Water awareness month." When there is no information about the units used on one of the axes, the plot cannot be interpreted. To see this, consider Figure 9.6b and c, displaying two different scenarios that could have produced the actual graph in Figure 9.6a. In Figure 9.6b, the vertical axis starts at zero for the existing plot. In Figure 9.6c, the vertical axis for the original plot starts at 30 and stops at 40, so what appears to be a large drop in 1979 in the other two graphs is only a minor fluctuation. We do not know which of these scenarios is closer to the truth, yet you can see that the two possibilities represent substantially different situations.

Not Starting at Zero

Often, even when the axes are labeled, the scale of one or both of the axes does not start at zero, and the reader may not notice that fact. A common ploy is to present an increasing or decreasing trend over time on a graph that does not start at zero. As we saw for the example in Figure 9.6, what appears to be a substantial change may actually represent quite a modest change. Always make it a habit to check the numbers on the axes to see where they start.

Figure 9.7 shows what the line graph of winning times for the Olympic speed skating data in Figure 9.4 would have looked like if the vertical axis had started at zero. Notice that the drop in winning times over the years does not look nearly as dramatic as it did in Figure 9.4. Be very careful about this form of potential deception if someone is presenting a graph to display growth in sales of a product, a drop

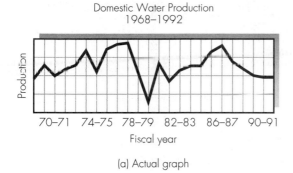

(a) Actual graph

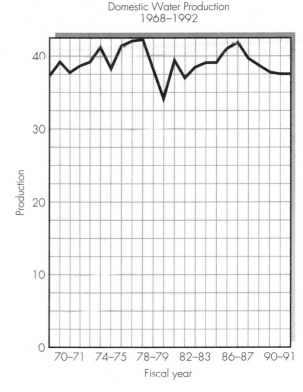

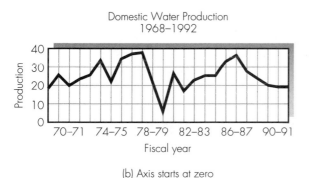

(b) Axis starts at zero

(c) Axis in "actual graph" does not start at zero as a way to exaggerate trends

FIGURE 9.6

Example of a graph with no labeling (a) and possible interpretations (b and c)

Source: Insert in the *California Aggie* (UC Davis), 30 May 1993.

in interest rates, and so on. Be sure to look at the labeling, especially on the vertical axis.

Despite this, be aware that for some graphs it makes sense to start the units on the axes at values different from zero. A good example is the scatterplot of GPAs versus SAT scores in Figure 9.5. It would make no sense to start the horizontal axis (SAT scores) at zero because the range of interest is from about 350 to 800. It is the responsibility of the reader to notice the units. Never assume a graph starts at zero without checking the labeling.

Changes in Labeling on One or More Axes

Figure 9.8 shows an example of a graph where a cursory look would lead one to think the vertical axis starts at zero. However, notice the white horizontal bar just

FIGURE 9.7

An example of the change in perception when axes start at zero

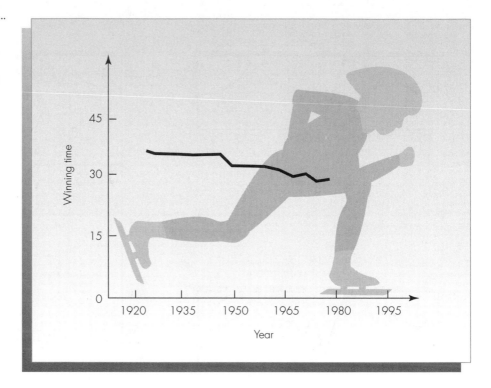

FIGURE 9.8

A bar graph with gap in labeling

Source: Davis (CA) Enterprise, 4 March 1994, p. A-7.

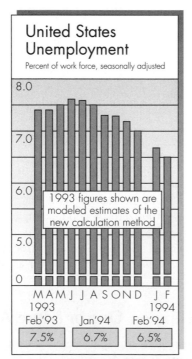

FIGURE 9.9

The distance between successive bars keeps changing.

Source: Washington Post graph reprinted in Wainer, 1984.

above the bottom of the graph, in which the vertical bars are broken. That indicates a gap in the vertical axis. In fact, you can see that the bottom of the graph actually corresponds to about 4.0%. It would have been more informative if the graph had simply been labeled as such, without the break.

Figure 9.9 shows a much more egregious example of changes in labeling. Notice that the horizontal axis does not maintain consistent distances between years and that varying numbers of years are represented by each of the bars. The distance between the first and second bars on the left is 8 years, whereas the 5 bars farthest to the right each represent a single year. This is an extremely misleading graph.

Misleading Units of Measurement

The units shown on a graph can be different from those that the reader would consider important. For example, Figure 9.10 shows a graph with the heading, "Rising Postal Rates." It accurately represents how the cost of a first-class stamp has risen since 1971. However, notice that the fine print at the bottom reads, "In 1971 dollars, the price of a 32-cent stamp in February 1995 would be 8.4 cents." A more truthful picture would show the changing price of a first-class stamp adjusted for inflation. As the footnote implies, such a graph would show little or no rise in postal rates as a function of the worth of a dollar.

FIGURE 9.10

A graph using misleading units

Source: *USA Today,* 7 March 1994, p. 13A.

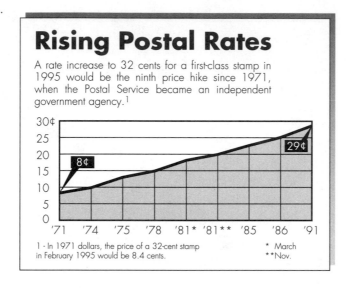

Rising Postal Rates

A rate increase to 32 cents for a first-class stamp in 1995 would be the ninth price hike since 1971, when the Postal Service became an independent government agency.[1]

[graph with y-axis values 30¢, 25, 20, 15, 10, 5, 0 and x-axis values '71, '74, '75, '78, '81*, '81**, '85, '86, '91; data labels 8¢ and 29¢]

1 - In 1971 dollars, the price of a 32-cent stamp in February 1995 would be 8.4 cents.

* March
**Nov.

Using Poor Information

A picture can only be as accurate as the information that was used to design it. All of the cautions about interpreting the collection of information given in Part 1 of this book apply to graphs and plots as well. You should always be told the source of information presented in a picture, and an accompanying article should give you as much information as necessary to determine the worth of that information.

Figure 9.11 shows a graph that appeared in the London newspaper the *Independent on Sunday* on March 13, 1994. The accompanying article was titled, "Sniffers Quit Glue for More Lethal Solvents." The graph appears to show that very few deaths occurred in Britain from solvent abuse before the late 1970s. However, the accompanying article includes the following quote, made by a research fellow at the unit where the statistics are kept: "It's only since we have started collecting accurate data since 1982 that we have begun to discover the real scale of the problem" (p. 5). In other words, the article indicates that the information used to create the graph is not at all accurate until at least 1982. Therefore, the apparent sharp increase in deaths linked to solvent abuse around that time period is likely to have been simply a sharp increase in deaths reported and classified.

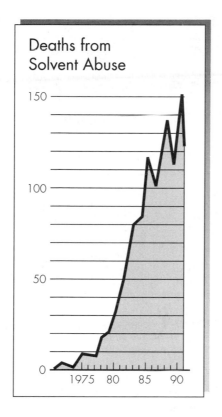

FIGURE 9.11

A graph based on poor information

Source: *The Independent on Sunday* (London), 13 March 1994.

9.5   A CHECKLIST FOR STATISTICAL PICTURES

To summarize, here are ten questions you should ask when you look at a statistical picture—before you even begin to try to interpret the data displayed.

1. Does the message of interest stand out clearly?
2. Is the purpose or title of the picture evident?
3. Is a source given for the data, either with the picture or in an accompanying article?
4. Did the information in the picture come from a reliable, believable source?
5. Is everything clearly labeled, leaving no ambiguity?
6. Do the axes start at zero or not?
7. Do the axes maintain a constant scale?
8. Are there any breaks in the numbers on the axes that may be easy to miss?
9. For financial data, have the numbers been adjusted for inflation?
10. Is there information cluttering the picture or misleading the eye?

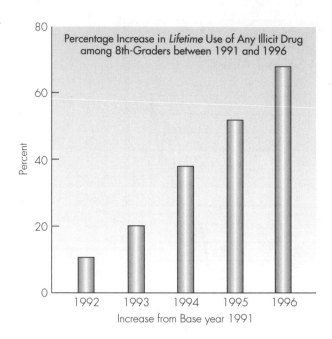

FIGURE 9.12
*Emergency Situation
among Our Youth:
8th-grade drug use*
Source: U.S. Dept. of Justice.

Don't forget that a statistical picture isn't worth much if the data can't be trusted. Once again, you should familiarize yourself to the extent possible with the seven critical components listed in Chapter 2 (p. 16).

| CASE STUDY 9.1 |

Time to Panic about Illicit Drug Use?

The graph illustrated in Figure 9.12 appeared on the website for the U.S. Department of Justice, Drug Enforcement Agency, in spring 1998 (http://www.usdoj.gov/dea/ drugdata/cp-316.htm). The headline over the graph reads "Emergency Situation among Our Youth." Look quickly at the graph, and describe what you see. Did it lead you to believe that almost 80% of 8th-graders used illicit drugs in 1996, compared with only about 10% in 1992? The graph is constructed so that you might easily draw that conclusion. Notice that careful reading indicates otherwise, and crucial information is missing. The graph tells us only that in 1996 the rate of use was 80% higher, or 1.8 times what it was in 1991. The actual rate of use is *not* provided at all in the graph. Only after searching the remainder of the website does that emerge. The rate of illicit drug use among 8th-graders in 1991 was about 11%, and thus, in 1996, it was about 1.8 times that, or about 19.8%. Additional information elsewhere on the website indicates that about 8% of 8th-graders used marijuana in 1991, and thus this was the most common illicit drug used. These are still disturbing statistics, but not as disturbing as the graph would lead you to believe. ■

EXERCISES

1. Give the name of a type of statistical picture that could be used for each of the following kinds of data.

a. One categorical variable

b. One measurement variable

c. Two categorical variables

d. Two measurement variables

2. Suppose a real estate company in your area sold 100 houses last month, whereas their two major competitors sold 50 houses and 25 houses, respectively. The top company wants to display its better record with a pictogram using a simple two-dimensional picture of a house. Draw two pictograms displaying this information, one of which is misleading and one of which is not. (The horizontal axis should list the three companies and the vertical axis should list the number of houses sold.)

3. One method used to compare authors or to determine authorship on unsigned writing is to look at the frequency with which words of different lengths appear in a piece of text. For this exercise, you are going to compare your own writing with that of the author of this book.

a. Using the first full paragraph of this chapter (not the Thought Questions), create a pie chart with three segments, showing the relative frequency of words of 1 to 3 letters, 4 to 5 letters, and 6 or more letters in length.

b. Find a paragraph of your own writing of at least 50 words. Repeat part a of this exercise for your own writing.

c. Display the data in parts a and b of this exercise using a single bar graph that includes the information for both writers.

d. Discuss how your own writing style is similar to or different from that of the author of this book, as evidenced by the pictures in parts a to c.

e. Name one advantage of displaying the information in two pie charts and one advantage of displaying the information in a single bar graph.

4. An article in *Science* (vol. 279, 23 January, 1998, p. 487) reported on a "telephone survey of 2600 parents, students, teachers, employers, and college professors" in which people were asked the question, "Does a high school diploma mean that a student has at least learned the basics?" Results were as follows:

	Professors	Employers	Parents	Teachers	Students
Yes	22%	35%	62%	73%	77%
No	76%	63%	32%	26%	22%

a. The article noted that "there seems to be a disconnect between the producers [parents, teachers, students] and the consumers [professors, employers] of

TABLE 9.1

Kids Live With	1960	1980	1990
Father and mother	80.6%	62.3%	57.7%
Mother only	7.7	18.0	21.6
Father only	1.0	1.7	3.1
Father and stepmother	0.8	1.1	0.9
Mother and stepfather	5.9	8.4	10.4
Neither parent	3.9	5.8	4.3

high school graduates in the United States. Create a bar graph from this study that emphasizes this feature of the data.

b. Create a bar graph that deemphasizes the issue raised in part a.

5. Figure 9.10, which displays rising postal rates, is an example of a graph with misleading units because the prices are not adjusted for inflation. The graph actually has another problem as well. Use the checklist in Section 9.5 to determine the problem; then redraw the graph correctly (but still use the unadjusted prices). Comment on the difference between Figure 9.10 and your new picture.

6. In its February 24–26, 1995, edition (p. 7), *USA Weekend* gave statistics on the changing status of which parent children live with. As noted in the article, the numbers don't total 100% because they are drawn from two sources: the U.S. Census Bureau and *America's Children: Resources from Family, Government, and the Economy* by Donald Hernandez (New York: Russell Sage Foundation, 1995). Using the data shown in Table 9.1, draw a bar graph presenting the information. Be sure to include all the components of a good statistical picture.

7. Figure 10.4 in Chapter 10 displays the success rate for professional golfers when putting at various distances. Discuss the figure in the context of the material in this chapter. Are there ways in which the picture could be improved?

8. Table 9.2 indicates the population (in millions) and the number of violent crimes (in millions) in the United States from 1982 to 1991, as reported in the *World Almanac and Book of Facts* (1993, p. 948).

a. Draw two line graphs representing the trend in violent crime over time. Draw the first graph to try to convince the reader that the trend is quite ominous. Draw the second graph to try to convince the reader that it is not. Make sure all of the other features of your graph meet the criteria for a good picture.

b. Draw a scatterplot of population versus violent crime, making sure it meets all the criteria for a good picture. Comment on the scatterplot. Now explain why drawing a line graph of violent crime versus year, as in part a of this exercise, might be misleading.

TABLE 9.2   U.S. Population and Violent Crime*

Year	1982	1983	1985	1986	1987	1988	1989	1990	1991
U.S. population	231	234	239	241	243	246	248	249	252
Violent crime	1.32	1.26	1.33	1.49	1.48	1.57	1.65	1.82	1.91

* Figures for 1984 were not available in the original.

TABLE 9.3

Blood Type	White Americans		African Americans	
	Rh+	Rh−	Rh+	Rh−
A	38.8%	7.0%	26.0%	2.0%
B	7.0%	1.0%	17.0%	1.5%
AB	3.0%	0.6%	4.0%	0.4%
O	37.0%	6.0%	45.0%	4.0%

c. Rather than using number of violent crimes on the vertical axis, redraw the first line graph (from part a) using a measure that adjusts for the increase in population. Comment on the differences between the two graphs.

9. Find an example of a statistical picture in a newspaper or magazine, or on the Internet. Answer the ten questions in Section 9.5 for the picture. In the process of answering the questions, explain what (if any) features you think should have been added or changed to make it a good picture. Include the picture with your answer.

10. According to the *American Medical Association Family Medical Guide* (1982, p. 422), the distribution of blood types among Americans is as shown in Table 9.3.

a. Draw a pie chart illustrating the blood-type distribution for white Americans, ignoring the RH factor.

b. Draw a statistical picture incorporating all of the information given.

11. Find an example of a statistical picture in a newspaper or magazine that has at least one of the problems listed in Section 9.4, "Difficulties and Disasters in Plots, Graphs, and Pictures." Explain the problem. If you think anything should have been done differently, explain what and why. Include the picture with your answer.

12. Find a graph that does not start at zero. Redraw the picture to start at zero. Discuss the pros and cons of the two versions.

MINI-PROJECTS

1. Collect some categorical data on a topic of interest to you and represent it in a statistical picture. Explain what you have done to make sure the picture is as useful as possible.

2. Collect two measurement variables on each of at least ten individuals. Represent them in a statistical picture. Describe the picture in terms of possible outliers, variability, and relationship between the two variables.

3. Find some data that represent change over time for a topic of interest to you. Present a line graph of the data in the best possible format. Explain what you have done to make sure the picture is as useful as possible.

REFERENCES

Alper, Joe. (16 April 1993). The pipeline is leaking women all the way along. *Science* 260.

American medical association family medical guide. (1982). Edited by Jeffrey R. M. Kunz. New York: Random House.

Krantz, Les. (1992). *What the odds are.* New York: Harper Perennial.

Ryan, B. F., B. L. Joiner, and T. A. Ryan, Jr. (1985). *Minitab handbook.* 2d ed. Boston: PWS Kent.

Wainer, Howard. (1984). How to display data badly. *American Statistician* 38.

World almanac and book of facts. (1993). Edited by Mark S. Hoffman. New York: Pharos Books.

CHAPTER TEN

Relationships Between Measurement Variables

THOUGHT QUESTIONS

1. Judging from the scatterplot in Figure 9.5, there is a *positive correlation* between verbal SAT score and GPA. For used cars, there is a *negative correlation* between the age of the car and the selling price. Explain what it means for two variables to have a positive correlation or a negative correlation.

2. Suppose you were to make a scatterplot of (adult) sons' heights versus fathers' heights by collecting data on both from several of your male friends. You would now like to predict how tall your nephew will be when he grows up, based on his father's height. Could you use your scatterplot to help you make this prediction? Explain.

3. Do you think each of the following pairs of variables would have a positive correlation, a negative correlation, or no correlation?

   a. Calories eaten per day and weight

   b. Calories eaten per day and IQ

   c. Amount of alcohol consumed and accuracy on a manual dexterity test

   d. Number of ministers and number of liquor stores in cities in Pennsylvania

   e. Height of husband and height of wife

4. An article in the *Sacramento Bee* (29 May, 1998, p. A17) noted "Americans are just too fat, researchers say, with 54 percent of all adults heavier than is healthy. If the trend continues, experts say that within a few generations virtually every U.S. adult will be overweight." This prediction is based on "extrapolating," which assumes the current rate of increase will continue indefinitely. Is that a reasonable assumption? Do you agree with the prediction? Explain.

10.1   STATISTICAL RELATIONSHIPS

One of the interesting advances made possible by the use of statistical methods is the quantification and potential confirmation of relationships. In the first part of this book, we discussed relationships between aspirin and heart attacks, meditation and aging, and smoking during pregnancy and child's IQ, to name just a few. In Chapter 9, we saw examples where relationships between two variables were illustrated with pictures, such as the scatterplot of verbal SAT scores and college GPAs.

Although we have examined many relationships up to this point, we have not considered how those relationships could be expressed quantitatively. In this chapter, we discuss **correlation,** which measures the *strength* of a certain type of relationship between two measurement variables, and **regression,** which gives a numerical method for trying to *predict* one measurement variable from another.

Statistical Relationships versus Deterministic Relationships

A **statistical relationship** differs from a **deterministic relationship** in that, in the latter case, if we know the value of one variable, we can determine the value of the other exactly. For example, the relationship between volume and weight of water is deterministic. The old saying, "A pint's a pound the world around," isn't quite true, but the deterministic relationship between volume and weight of water does hold. (A pint is actually closer to 1.04 pounds.) We can express the relationship by a formula, and if we know one value, we can solve for the other (weight in pounds = 1.04 × volume in pints).

Natural Variability

In a statistical relationship, natural variability exists in both measurements. For example, we could describe the average relationship between height and weight for adult females, but very few women would fit that exact formula. If we knew a woman's height, we could predict the average weight for all women with that same height, but we could not predict her weight exactly. Similarly, we can say that, on average, taking aspirin every other day reduces one's chance of having a heart attack, but we cannot predict what will happen to one specific individual.

Statistical relationships are useful for describing what happens to a population, or aggregate. The stronger the relationship, the more useful it is for predicting what will happen for an individual. When researchers make claims about statistical relationships, they are not claiming that the relationship will hold for everyone.

10.2

STRENGTH VERSUS STATISTICAL SIGNIFICANCE

To be convincing, an observed relationship must also be statistically significant. Sometimes this is not the case. To find out if a statistical relationship exists between two variables, researchers must usually rely on measurements from only a sample of individuals. However, for any particular sample, a relationship may exist that does not extend to the population. It may be just the "luck of the draw" that that particular sample exhibited the relationship.

For example, suppose an observational study followed for 5 years a sample of 1000 owners of satellite dishes and a sample of 1000 nonowners and found that 4 of the satellite dish owners developed brain cancer, whereas only 2 of the nonowners did. Could the researcher legitimately claim that the rate of cancer among all satellite dish owners is twice that among nonowners? You would probably not be persuaded that the observed relationship was indicative of a problem in the larger population. The numbers are simply too small to be convincing.

Defining Statistical Significance

To overcome this problem, statisticians try to determine whether an observed relationship in a sample is **statistically significant.** To determine this, we ask what the chances are that a relationship that strong or stronger would have been observed in the sample if there really were nothing going on in the population. If those chances are small, we declare that the relationship is statistically significant and was not just a fluke.

> Most researchers are willing to declare that a relationship is statistically significant if the chances of observing the relationship in the sample when actually nothing is going on in the population are less than 5%. In other words, a relationship is considered to be statistically significant if that relationship is stronger than 95% of the relationships we would expect to see just by chance.

Of course, this reasoning carries with it the implication that of all the relationships that do occur by chance alone, 5% of them will erroneously earn the title of statistical significance. However, this is the price we pay for not being able to measure the entire population—while still being able to determine that statistically significant relationships do exist.

Two Warnings about Statistical Significance

Two important points, which we will study in detail in Part 4, often lead people to misinterpret statistical significance. First, it is easier to rule out chance if the observed relationship is based on very large numbers of observations. Even a minor relationship will achieve "statistical significance" if the sample is very large. However, earning that title does not necessarily imply that there is a *strong* relationship or even one of practical importance.

Second, a very strong relationship won't necessarily achieve "statistical significance" if the sample is very small. If you read about researchers who "failed to find a statistically significant relationship" between two variables, do not be confused into thinking that they have proven that there *isn't* a relationship. It may be that they simply didn't take enough measurements to rule out chance as an explanation.

10.3 MEASURING STRENGTH THROUGH CORRELATION

A Linear Relationship

It is convenient to have a single number to measure the strength of the relationship between two variables and to have that number be independent of the units used to make the measurements. Many types of relationships can occur, but in this chapter, we consider only the most common one. The **correlation** between two measurement variables is an indicator of *how closely their values fall to a straight line.* Sometimes this measure is called the *Pearson product-moment correlation* or the *correlation coefficient* or is simply represented by the letter *r.*

Notice that the statistical definition of *correlation* is more restricted than its common usage. For example, if two variables have a perfect curved relationship of some form, they could still have no statistical correlation. As used in statistics, correlation measures *linear relationships* only; that is, it measures how close the individual points in a scatterplot are to a straight line.

Other Features of Correlations

Here are some other features of correlations:
 1. A correlation of +1 (or 100%) indicates that there is a perfect linear relationship between the two variables; as one increases, so does the other. In other words, all individuals fall on the same straight line, just as when two variables have a deterministic linear relationship.

> **2.** A correlation of –1 also indicates that there is a perfect linear relationship between the two variables; however, as one increases, the other *decreases*.
>
> **3.** A correlation of zero could indicate that there is no linear relationship between the two variables. It could also indicate that the best straight line through the data on a scatterplot is exactly horizontal.
>
> **4.** A *positive correlation* indicates that the variables increase together.
>
> **5.** A *negative correlation* indicates that as one variable increases, the other decreases.
>
> **6.** Correlations are unaffected if the units of measurement are changed. For example, the correlation between weight and height remains the same regardless of whether height is expressed in inches, feet or millimeters.

Examples of Positive and Negative Relationships

Following are some examples of both positive and negative relationships. Notice how the closeness of the points to a straight line determines the *magnitude* of the correlation, whereas whether the line slopes up or down determines if the correlation is positive or negative.

EXAMPLE 1   **VERBAL SAT AND GPA**

In Chapter 9, we saw a scatterplot showing the relationship between the two variables, verbal SAT and GPA, for a sample of college students. The correlation for the data in the scatterplot is .485, indicating a moderate positive relationship. In other words, higher verbal SAT scores tend to indicate higher GPAs as well, but the relationship is nowhere close to being exact. ■

EXAMPLE 2   **HUSBANDS' AND WIVES' AGES AND HEIGHTS**

Marsh (1988, p. 315) and Hand et al. (1994, pp. 179–183) reported data on the ages and heights of a random sample of 200 married couples in Britain, collected in 1980 by the Office of Population Census and Surveys. Figures 10.1 and 10.2 show scatterplots for the ages and the heights, respectively, of the couples. Notice that the ages fall much closer to a straight line than do the heights. In other words, husbands' and wives' ages are likely to be closely related, whereas their heights are less likely to be so. The correlation between husbands' and wives' ages is .94, whereas the correlation between their heights is only .36. Thus, the values for the correlations confirm what we see from looking at the scatterplots. ■

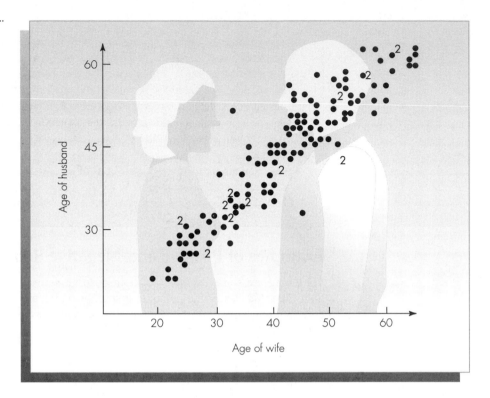

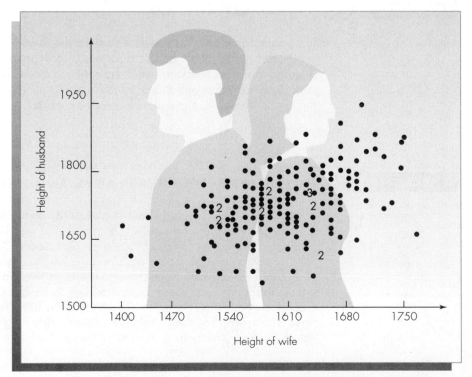

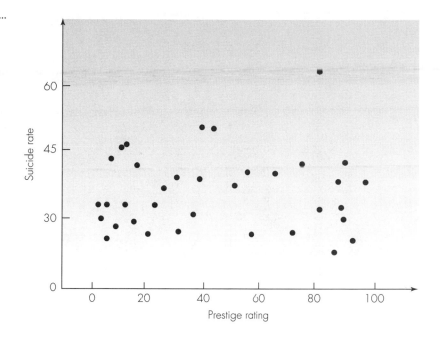

FIGURE 10.3
Plot of suicide rate versus occupational prestige for 36 occupations; correlation = .109
Source: Labovitz, 1970.

EXAMPLE 3

OCCUPATIONAL PRESTIGE AND SUICIDE RATES

Labovitz (1970, Table 1) and Hand et al. (1994, pp. 395–396) listed suicide rates and prestige ratings for 36 occupations in the United States. The suicide rates were for men aged 20 to 64; the prestige ratings were determined by the National Opinion Research Center. Figure 10.3 displays a scatterplot of the data. Notice that there does not appear to be much of a relationship between suicide rates and occupational prestige, and the correlation of .109 confirms that fact. You should also notice the outlier on the plot with a very high suicide rate and a somewhat high prestige rating. That point corresponds to the occupation of "managers, officials, and proprietors—self-employed—manufacturing." The outlier also appears to be responsible for the weak positive correlation. In fact, if that point is removed, the correlation drops to .018, very near zero. Therefore, we can conclude that there is little relationship between occupational prestige and suicide rates. ■

EXAMPLE 4

PROFESSIONAL GOLFERS' PUTTING SUCCESS

Iman (1994, p. 507) reported on a study conducted by *Sports Illustrated* magazine, in which the magazine studied success rates at putting for professional golfers. Using data from 15 tournaments, the researchers determined the percentage of successful putts at distances from 2 feet to 20 feet. We have restricted our attention to the part of the data that follows a linear relationship, which includes

FIGURE 10.4
*Professional golfers'
putting success rates;
correlation = –.94*
Source: Iman, 1994.

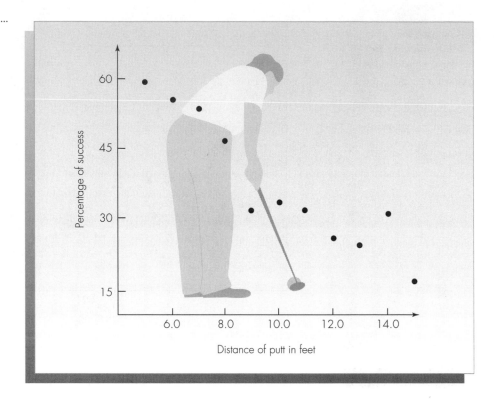

putting distances from 5 feet to 15 feet. Figure 10.4 illustrates this relationship. The correlation between distance and rate of success is –.94. Notice the negative sign, which indicates that as distance goes up, success rate goes down. ■

10.4   SPECIFYING LINEAR RELATIONSHIPS WITH REGRESSION

Sometimes, in addition to knowing the strength of the connection between two variables, we would like a *formula* for the relationship. For example, it might be useful for colleges to have a formula for the connection between verbal SAT score and college GPA. They could use it to predict the potential GPAs of future students. Some colleges do that kind of prediction to decide whom to admit, but they use a collection of variables instead of just one. The simplest kind of relationship between two variables is a straight line, and that's the only type we discuss here. Our goal is to find a straight line that comes as close as possible to the points in a scatterplot.

Defining Regression

We call the procedure we use to find a straight line that comes as close as possible to the points in a scatterplot **regression;** the resulting line, the **regression line;** and the formula that describes the line, the **regression equation.** You may wonder why that word is used. Until now, most of the vocabulary borrowed by statisticians had at least some connection to the common usage of the words. The use of the word *regression* dates back to Francis Galton, who studied heredity in the late 1800s. (See Stigler, 1986 or 1989, for a detailed historical account.) One of Galton's interests was whether a man's height as an adult could be predicted by his parents' heights. He discovered that it could, but that the relationship was such that very tall parents tended to have children who were shorter than they were, and very short parents tended to have children taller than themselves. He initially described this phenomenon by saying there was "reversion to mediocrity" but later changed to the terminology "regression to mediocrity." Henceforth, the technique of determining such relationships has been called *regression.*

How are we to find the best straight line relating two variables? We could just take a ruler and try to fit a line through the scatterplot, but each of us would probably get a different answer. Instead, the most common procedure is to find what is called the **least squares line.** In determining the least squares line, priority is given to how close the points fall to the line for the variable represented by the vertical axis. Those distances are squared and added up for all of the points in the sample. For the least squares line, that sum is smaller than it would be for any other line. The vertical distances are chosen because the equation is often used to predict that variable when the one on the horizontal axis is known. Therefore, we want to minimize how far off the prediction would be in that direction.

The Equation for the Line

All straight lines can be expressed by the same formula. Using standard conventions, we call the variable on the vertical axis y and the variable on the horizontal axis x. We can then write the equation for the line relating them as

$$y = a + bx$$

where for any given situation, a and b would be replaced by numbers. We call the number represented by a the **intercept** and the number represented by b the **slope.** The intercept simply tells us at what particular point the line crosses the vertical axis when the horizontal axis is at zero. The slope tells us how much of an increase there is for one variable (the one on the vertical axis) when the other (on the horizontal axis) increases by one unit. A *negative slope* indicates a decrease in one variable as the other increases, just as a negative correlation does.

For example, Figure 10.5 shows the relationship between y = temperature in Fahrenheit and x = temperature in Celsius. The equation for the relationship is

$$y = 32 + 1.8x$$

The intercept, 32, is the temperature in Fahrenheit when the Celsius temperature is zero. The slope, 1.8, is the amount by which Fahrenheit temperature increases when Celsius temperature increases by one unit.

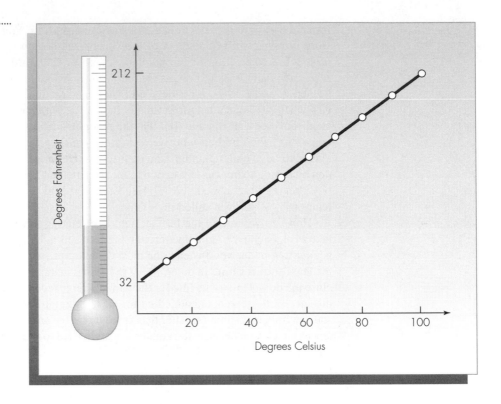

EXAMPLE 5   **HUSBANDS' AND WIVES' AGES, REVISITED**

Figure 10.6 shows the same scatterplot as Figure 10.1, relating ages of husbands and wives, except that now we have added the regression line. This line minimizes the sum of the squared vertical distances between the line and the husbands' actual ages. The regression equation for the line shown in Figure 10.6, relating husbands' and wives' ages, is

$y = 3.6 + .97x$

or, equivalently,

husband's age = 3.6 + (.97)(wife's age)

Let's use this equation to predict husband's age at various wife's ages.

Wife's Age	Predicted Age of Husband
20 years	3.6 + (.97)(20) = 23.0 years
25 years	3.6 + (.97)(25) = 27.9 years
40 years	3.6 + (.97)(40) = 42.4 years
55 years	3.6 + (.97)(55) = 57.0 years

Scatterplot and regression line for British husbands' and wives' ages
Source: Hand et al., 1994.

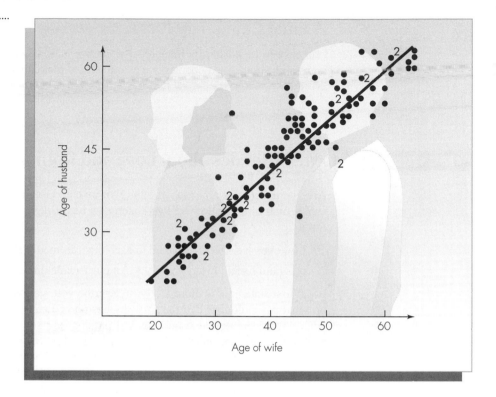

This table shows that for the range of ages in the sample, husbands tend to be 2 to 3 years older than their wives, on average. The older the couple, the smaller the gap in their ages. Remember that with statistical relationships, we are determining what happens to the average and not to any given individual. Thus, although most couples won't fit the pattern given by the regression line exactly, it does show us one way to represent the average relationship for the whole group. ∎

Extrapolation

It is generally not a good idea to use a regression equation to predict values far outside the range where the original data fell. There is no guarantee that the relationship will continue beyond the range for which we have data. For example, using the regression equation illustrated in Figure 10.6, we would predict that women who are 100 years old have husbands whose average age is 100.6 years. It is more likely that if a woman is still married at 100, her husband is younger than she is! In other words, the relationship for much older couples would be affected by differing death rates for men and women, and a different equation would most likely apply. It is typically acceptable to use the equation only for a minor extrapolation beyond the range of the original data.

A Final Cautionary Note

It is easy to be misled by inappropriate interpretations and uses of correlation and regression. In the next chapter, we examine how that can happen, and how you can avoid it.

Are Attitudes about Love and Romance Hereditary?

Are you the jealous type? Do you think of love and relationships as a practical matter? Which of the following two statements better describes how you are likely to fall in love?

My lover and I were attracted to each other immediately after we first met.

It is hard for me to say exactly when our friendship turned into love.

If the first statement is more likely to describe you, you would probably score high on what psychologists call the *Eros* dimension of love, characteristic of those who "place considerable value on love and passion, are self-confident, enjoy intimacy and self-disclosure, and fall in love fairly quickly" (Waller and Shaver, 1994, p. 268). However, if you identify more with the second statement, you would probably score higher on the *Storge* dimension, characteristic of those who "value close friendship, companionship, and reliable affection" (p. 268). Whatever your beliefs about love and romance, do you think they are partially inherited, or are they completely due to social and environmental influences?

Psychologists Niels Waller and Philip Shaver set out to answer the question of whether feelings about love and romance are partially genetic, as are most other personality traits. Waller and Shaver studied the love styles of 890 adult twins and 172 single twins and their spouses from the California Twin Registry. They compared the similarities between the answers given by monozygotic twins (MZ), who share 100% of their genes, to the similarities between those of dizygotic twins (DZ), who share, on average, 50% of their genes. They also studied the similarities between the answers of twins and those of their spouses. If love styles are genetic, rather than determined by environmental and other factors, then the matches between MZ twins should be substantially higher than those between DZ twins.

Waller and Shaver studied 345 pairs of MZ twins, 100 pairs of DZ twins, and 172 spouse pairs (that is, a twin and his or her spouse). Each person filled out a questionnaire called the "Love Attitudes Scale" (LAS), which asked them to read 42 statements like the two given earlier. For each statement, respondents assigned a ranking from 1 to 5, where 1 meant "strongly agree" and 5 meant "strongly disagree." There were seven questions related to each of six love styles, with a score determined for each person on each love style. Therefore, there were six scores for each person.

In addition to the two styles already described (Eros and Storge), scores were generated for the following four:

| TABLE 10.1 | Correlations for Love Styles and for Some Personality Traits | | |

	Correlation		
	Monozygotic Twins	Dizygotic Twins	Spouses
Love Style			
Eros	.16	.14	.36
Ludus	.18	.30	.08
Storge	.18	.12	.22
Pragma	.40	.32	.29
Mania	.35	.27	−.01
Agape	.30	.37	.28
Personality Trait			
Well-being	.38	.13	.04
Achievement	.43	.16	.08
Social closeness	.38	.01	−.04

Source: Waller and Shaver, September 1994.

- *Ludus* characterizes those who "value the fun and excitement of romantic relationships, especially with multiple alternative partners; they generally are not interested in mutual self-disclosure, intimacy or 'getting serious'" (p. 268).

- *Pragma* types are "pragmatic, entering a relationship only if it meets certain practical criteria" (p. 269).

- *Mania* types "are desperate and conflicted about love. They yearn intensely for love but then experience it as a source of pain, a cause of jealousy, a reason for insomnia" (p. 269).

- Those who score high on *Agape* "are oriented more toward what they can give to, rather than receive from, a romantic partner. Agape is a selfless, almost spiritual form of love" (p. 269).

For each type of love style, and for each of the three types of pairs (MZ twins, DZ twins, and spouses) the researchers computed a correlation. The results are shown in Table 10.1. (They first removed effects due to age and gender, so the correlations are not due to a relationship between love styles and age or gender.) Notice that the correlations are not higher for the MZ twins than they are for the DZ twins. This is in contrast to most other personality traits. For comparison purposes, three such traits are also shown in Table 10.1. Notice that for those traits, the correlations are much higher for the MZ twins, indicating a substantial hereditary component. Regarding the findings for love styles, Waller and Shaver conclude:

FIGURE 10.7
Ideal versus actual weight for females

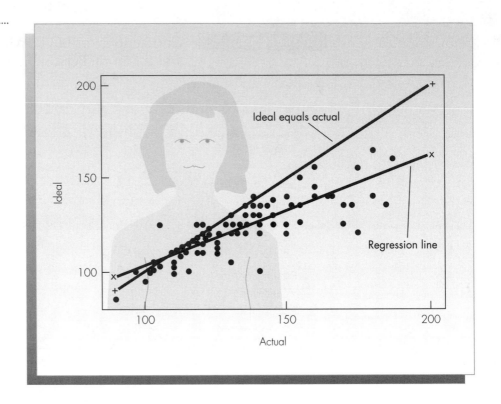

This surprising, and very unusual, finding suggests that genes are not important determinants of attitudes toward romantic love. Rather, the common environment appears to play the cardinal role in shaping familial resemblance on these dimensions. (p. 271)

SOURCE: Waller and Shaver, September 1994. ∎

CASE STUDY 10.2

A Weighty Issue:
Women Want Less, Men Want More

Do you like your weight? Let me guess. . . . If you're male and under about 175 pounds, you probably want to weigh the same or more than you do. If you're female, no matter what you weigh, you probably want to weigh the same or less. Those were the results uncovered in a large statistics class (119 females and 63 males) when students were asked to give their actual and their ideal weights.

Figure 10.7 shows a scatterplot of ideal versus actual weight for the females, and Figure 10.8 is the same plot for the males. Each point represents one student, whose ideal weight can be read on the vertical axis and actual weight can be read on the

FIGURE 10.8
Ideal versus actual
weight for males

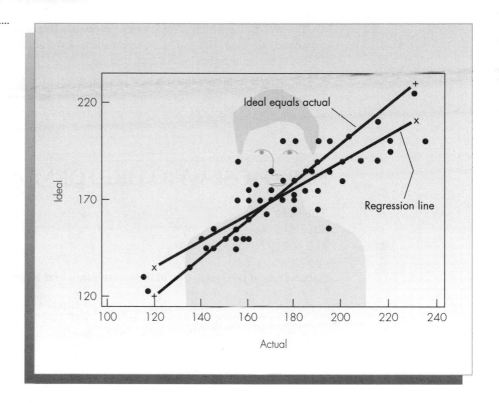

horizontal axis. What is the relationship between ideal and actual weight, on average, for men and for women?

First, notice that if everyone were at their ideal weight, all points would fall on a line with the equation:

Ideal = Actual

That line is drawn in each figure. Most of the women fall below that line, indicating that their ideal weight is below their actual weight. The situation is not as clear for the men, but a pattern is still evident. Most of those weighing under 175 pounds fall on or above the line (would prefer to weigh the same or more than they do), and most of those weighing over 175 pounds fall on or below the line (would prefer to weigh the same or less than they do).

The regression lines are also shown on each scatterplot. The regression equations are:

Women: ideal = 43.9 + 0.6 actual
Men: ideal = 52.5 + 0.7 actual

These equations have several interesting features, which, remember, summarize the relationship between ideal and average weight for the aggregate, not for each individual:

- The weight for which *ideal* = *actual* is about 110 pounds for women and 175 pounds for men. Below those weights, actual weight is less than desired; above them, actual weight is more than desired.

- The slopes represent the increase in ideal weight for each 1 pound increase in actual weight. Thus, every 10 pounds of additional weight indicates an increase of only 6 pounds in ideal weight for women and 7 pounds for men.

FOR THOSE WHO LIKE FORMULAS

The Data

n pairs of observations, (x_i, y_i), $i = 1, 2, \ldots, n$, where x_i is plotted on the horizontal axis and y_i on the vertical axis.

Summaries of the Data, Useful for Correlation and Regression

$$\text{SSX} = \sum_{i=1}^{n} (x_i - \bar{x})^2 = \sum_{i=1}^{n} x_i^2 - \frac{\left(\sum_{i=1}^{n} x_i\right)^2}{n}$$

$$\text{SSY} = \sum_{i=1}^{n} (y_i - \bar{y})^2 = \sum_{i=1}^{n} y_i^2 - \frac{\left(\sum_{i=1}^{n} y_i\right)^2}{n}$$

$$\text{SXY} = \sum_{i=1}^{n} (x_i - \bar{x})(y_i - \bar{y}) = \sum_{i=1}^{n} x_i y_i - \frac{\left(\sum_{i=1}^{n} x_i\right)\left(\sum_{i=1}^{n} y_i\right)}{n}$$

Correlation for a Sample of n Pairs

$$r = \frac{\text{SXY}}{\sqrt{\text{SSX}}\sqrt{\text{SSY}}}$$

The Regression Slope and Intercept

$$\text{slope} = b = \frac{\text{SXY}}{\text{SSX}}$$

$$\text{intercept} = a = \bar{y} - b\bar{x}$$

EXERCISES

1. Suppose 100 different researchers each did a study to see if there was a relationship between coffee consumption and height. Suppose there really is no such relationship in the population. Would you expect any of the researchers to find a statistically significant relationship? If so, approximately how many? Explain your answer.

2. In Figure 10.2, we observed that the correlation between husbands' and wives' heights, measured in millimeters, was .36. Can you determine what the correlation would be if the heights were converted to inches? Explain.

3. A pint of water weighs 1.04 pounds, so 1 pound of water is 0.96 pint. Suppose a merchant sells water in containers weighing 0.5 pound, but customers can fill them to their liking. It is easier to weigh the filled container than to measure the volume of water the customer is purchasing. Define x to be the weight of the container and the water and y to be the volume of the water.

  a. Write the equation the merchant would use to determine the volume y when x is known.

  b. Specify the numerical values of the intercept and the slope, and interpret their physical meanings for this example.

  c. What is the correlation between x and y for this example?

  d. Draw a picture of the relationship between x and y.

4. Are each of the following pairs of variables likely to have a positive correlation or a negative correlation?

  a. Daily temperature at noon in New York City and in Boston

  b. Weight of an automobile and its gas mileage in average miles per gallon

  c. Hours of television watched and grade-point average for college students

  d. Years of education and salary

5. Suppose a weak relationship exists between two variables in a population. Which would be more likely to result in a statistically significant relationship between the two variables: a sample of size 100 or a sample of size 10,000? Explain.

6. The relationship between height and weight is a well-established and obvious fact. Suppose you were to sample heights and weights for a small number of your friends, and you failed to find a statistically significant relationship between the two variables. Would you conclude that the relationship doesn't hold for the population of people like your friends? Explain.

7. Which implies a stronger linear relationship, a correlation of +.4 or a correlation of .6? Explain.

8. Give an example of a pair of variables that are likely to have a positive correlation and a pair of variables that are likely to have a negative correlation.

9. Explain how two variables can have a perfect curved relationship and yet have zero correlation. Draw a picture of a set of data meeting those criteria.

10. The regression line relating verbal SAT scores and GPA for the data exhibited in Figure 9.5 is

  $$GPA = 0.539 + (0.00362)(\text{verbal SAT})$$

  a. Predict the average GPA for those with verbal SAT scores of 500.

  b. Explain what the slope of 0.00362 represents.

  c. The lowest possible SAT score is 200. Does the intercept of 0.539 have any useful meaning for this example? Explain.

11. Refer to Case Study 10.2, in which regression equations are given for males and females relating ideal weight to actual weight. The equations are

   Women: ideal = 43.9 + 0.6 actual

   Men: ideal = 52.5 + 0.7 actual

   a. Predict the ideal weight for a man who weighs 150 pounds and for a woman who weighs 150 pounds. Compare the results.

   b. Does the intercept of 43.9 have a logical physical interpretation in the context of this example? Explain.

   c. Does the slope of 0.6 have a logical interpretation in the context of this example? Explain.

   d. Outliers in scatterplots may be within the range of values for each variable individually but lie outside the general pattern when the variables are examined in combination. A few points in each of Figures 10.7 and 10.8 could be considered as outliers. In the context of this example, explain the characteristics of someone who appears as an outlier.

12. In Chapter 9, we examined a picture of winning time in men's 500-meter speed skating plotted across time. The data represented in the plot started in 1924 and went through 1992. The regression equation relating winning time and year is

   winning time = 255 – (0.1094)(year)

   a. Would the correlation between winning time and year be positive or negative? Explain.

   b. In 1994, the actual winning time for the gold medal was 36.33 seconds. Use the regression equation to predict the winning time for 1994, and compare the prediction to what actually happened.

   c. Explain what the slope of –0.1094 indicates in terms of how winning times change from year to year.

13. Explain why we should not use the regression equation we found in Exercise 12 for speed-skating time versus year to predict the winning time for the 2002 Winter Olympics.

14. The regression equation relating distance (in feet) and success rate (percent) for professional golfers, based on 11 distances ranging from 5 feet to 15 feet, is

   success rate = 76.5 – (3.95)(distance)

   a. What percent success would you expect for these professional golfers if the putting distance is 6.5 feet?

   b. Explain what the slope of –3.95 means in terms of how success changes with distance.

15. The original data for the putting success of professional golfers included values beyond those we used for this example (5 feet to 15 feet), in both directions. At a distance of 2 feet, 93.3% of the putts were successful. At a distance of 20 feet, 15.8% of the putts were successful.

   a. Use the equation in Exercise 14 to predict success rates for those two distances (2 feet and 20 feet). Compare the predictions to the actual success rates.

    b. Use your results from part a to explain why it is not a good idea to use a regression equation to predict information beyond the range of values from which the equation was determined.

    c. Based on the picture in Figure 10.4 and the additional information in this exercise, draw a picture of what you think the relationship between putting distance and success rate would look like for the entire range from 2 feet to 20 feet.

    d. Explain why a regression equation should not be formulated for the entire range from 2 feet to 20 feet.

16. As one of the examples in this chapter, we noticed a very strong relationship between husbands' and wives' ages for a sample of 200 British couples, with a correlation of .94. Coincidentally, the relationship between putting distance and success rate for professional golfers had a correlation of −.94, based on 11 data points. This latter was statistically significant, so we can be pretty sure the observed relationship was not just due to chance. Do you think the observed relationship between husbands' and wives' ages is statistically significant? Explain.

MINI-PROJECTS

1. (*Computer required.*) Measure the heights and weights of ten friends of the same sex. Draw a scatterplot of the data, with weight on the vertical axis and height on the horizontal axis. Using a computer that produces regression equations, find the regression equation for your data. Draw it on your scatter diagram. Use this to predict the average weight for people of that sex who are 67 inches tall.

2. Go to your library and peruse journal articles, looking for examples of scatterplots accompanied by correlations. Find three examples in different journal articles. Present the scatterplots and correlations, and explain in words what you would conclude about the relationship between the two variables in each case.

REFERENCES

Hand, D. J., F. Daly, A. D. Lunn, K. J. McConway, and E. Ostrowski (1994). *A handbook of small data sets.* London: Chapman and Hall.

Iman, R. L. (1994). *A data-based approach to statistics.* Belmont, CA: Wadsworth.

Labovitz, S. (1970). The assignment of numbers to rank order categories. *American Sociological Review* 35, pp. 515–524.

Marsh, C. (1988). *Exploring data.* Cambridge, England: Policy Press.

Stigler, S. M. (1986). *The history of statistics: The measurement of uncertainty before 1900.* Cambridge, MA: Belknap Press.

Stigler, S. M. (1989). Francis Galton's account of the invention of correlation. *Statistical Science* 4, pp. 73–79.

Waller, N. G., and P. R. Shaver (September 1994). The importance of nongenetic influences on romantic love styles: A twin-family study. *Psychological Science* 5, no. 5, pp. 268–274.

CHAPTER ELEVEN

Relationships Can Be Deceiving

THOUGHT QUESTIONS

1. Use the following two pictures to speculate on what influence outliers have on correlation. For each picture, do you think the correlation is higher or lower than it would be without the outlier? (*Hint:* Remember that correlation measures how closely points fall to a straight line.)

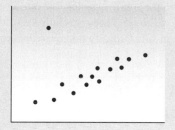 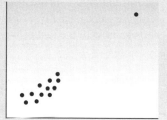

2. A strong correlation has been found in a certain city in the northeastern United States between weekly sales of hot chocolate and weekly sales of facial tissues. Would you interpret that to mean that hot chocolate causes people to need facial tissues? Explain.

3. Researchers have shown that there is a positive correlation between the average fat intake and the breast cancer rate across countries. In other words, countries with higher fat intake tend to have higher breast cancer rates. Does this correlation prove that dietary fat is a contributing cause of breast cancer? Explain.

4. If you were to draw a scatterplot of *number of women in the work force* versus *number of Christmas trees sold* in the United States for each year between 1930 and the present, you would find a very strong correlation. Why do you think this would be true? Does one cause the other?

11.1   ILLEGITIMATE CORRELATIONS

In Chapter 10, we learned that the correlation between two measurement variables provides information about how closely related they are. A strong correlation implies that the two variables are closely associated or related. With a positive correlation, they increase together, and with a negative correlation, one variable tends to increase as the other decreases.

However, as with any numerical summary, correlation does not provide a complete picture. A number of anomalies can cause misleading correlations. Ideally, all reported correlations would be accompanied by a scatterplot. Without a scatterplot, however, you need to ascertain whether any of the problems discussed in this section may be distorting the correlation between two variables.

Watch out for these problems with correlations:

- Outliers can substantially inflate or deflate correlations.
- Groups combined inappropriately may mask relationships.

The Impact Outliers Have on Correlations

In a manner similar to the effect we saw on means, outliers can have a large impact on correlations. This is especially true for small samples. An outlier that is consistent with the trend of the rest of the data will inflate the correlation. An outlier that is not consistent with the rest of the data can substantially decrease the correlation.

EXAMPLE 1   **HIGHWAY DEATHS AND SPEED LIMITS**

The data in Table 11.1 come from the time when the United States still had a maximum speed limit of 55 miles per hour. The correlation between death rate and speed limit is .55, indicating a moderate relationship. Higher death rates tend to be associated with higher speed limits.

A scatterplot of the data is presented in Figure 11.1; the two countries with the highest speed limits are labeled. Notice that Italy has both a much higher speed limit and a much higher death rate than any other country. That fact alone is responsible for the magnitude of the correlation. In fact, if Italy is removed, the correlation drops to .098, a negligible association. Of course, we could now claim that Britain is responsible for the almost zero magnitude of the correlation, and we would be right. If we remove Britain from the plot, the correlation is no longer negligible; it jumps to .70. You can see how much influence outliers have, sometimes inflating correlations and sometimes deflating them. (Of course, the actual rela-

TABLE 11.1	Highway Death Rates and Speed Limits	
COUNTRY	Death Rate (Per 100 Million Vehicle Miles)	Speed Limit (in Miles Per Hour)
Norway	3.0	55
United States	3.3	55
Finland	3.4	55
Britain	3.5	70
Denmark	4.1	55
Canada	4.3	60
Japan	4.7	55
Australia	4.9	60
Netherlands	5.1	60
Italy	6.1	75

Source: Rivkin, 1986.

FIGURE 11.1
An example of how an outlier can inflate correlation
Source: Rivkin, 1986.

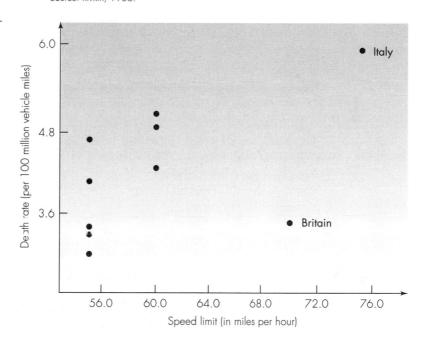

tionship between speed limit and death rate is complicated by many other factors, a point we discuss later in this chapter.) ■

One of the ways in which outliers can occur in a set of data is through erroneous recording of the data. Common wisdom among statisticians is that at least 5% of all

FIGURE 11.2

An example of how an outlier can deflate correlation

Source: Adapted from Figure 10.1.

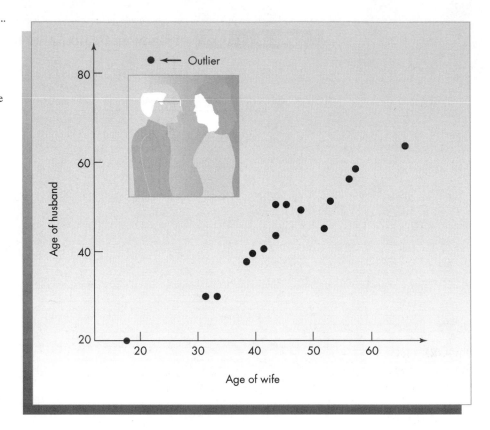

data points are corrupted, either when they are initially recorded or when they are entered into the computer. Good researchers check their data using scatterplots, stemplots, and other methods to ensure that such errors are detected and corrected. However, they do sometimes escape notice, and they can play havoc with numerical measures like correlation.

EXAMPLE 2

AGES OF HUSBANDS AND WIVES, REVISITED

Figure 11.2 shows a subset of the data we examined in Chapter 10, Figure 10.1, relating the ages of husbands and wives in Britain. In addition, an outlier has been added. This outlier could easily have occurred in the data set if someone had erroneously entered one husband's age as 82 when it should have been 28.

The correlation for the picture as shown is .39, indicating a somewhat low correlation between husbands' and wives' ages. However, the low correlation is completely attributable to the outlier. When it is removed, the correlation for the remainder of the points is .964, indicating a very strong relationship. ∎

TABLE 11.2	Major Earthquakes in the Continental United States, 1850–1992		

Date	Location	Deaths	Magnitude
August 31, 1886	Charleston, SC	60	6.6
April 18–19,1906	San Francisco	503	8.3
March 10, 1933	Long Beach, CA	115	6.2
February 9, 1971	San Fernando Valley, CA	65	6.6
October 17, 1989	San Francisco area	62	6.9
June 28, 1992	Yucca Valley, CA	1	7.4

Source: *World Almanac and Book of Facts*, 1993, p. 573.

Legitimate Outliers, Illegitimate Correlation

Outliers can also occur as legitimate data, as we saw in the example for which both Italy and Britain had much higher speed limits than other countries.

However, the theory of correlation was developed with the idea that both measurements were from bell-shaped distributions, so outliers would be unlikely to occur. As we have seen, correlations are quite sensitive to outliers. Be very careful when you are presented with correlations for data in which outliers are likely to occur or when correlations are presented for a small sample, as shown in Example 3. Not all researchers or reporters are aware of the havoc outliers can play with correlation, and they may innocently lead you astray by not giving you the full details.

EXAMPLE 3   **EARTHQUAKES IN THE CONTINENTAL UNITED STATES**

Table 11.2 lists the major earthquakes that occurred in the continental United States between 1850 and 1992. The correlation between deaths and magnitude for these six earthquakes is .732, showing a relatively strong association. This relationship implies that, on average, higher death tolls accompany stronger earthquakes.

However, if you examine the scatterplot of the data shown in Figure 11.3, you will notice that the correlation is entirely due to the famous San Francisco earthquake of 1906. In fact, for the remaining earthquakes, the trend is actually *reversed*. Without the 1906 quake, the correlation for these five earthquakes is actually strongly negative, at –.96. Higher magnitude quakes are associated with fewer deaths.

FIGURE 11.3
A data set for which correlation should not be used
Source: Data from Table 11.2.

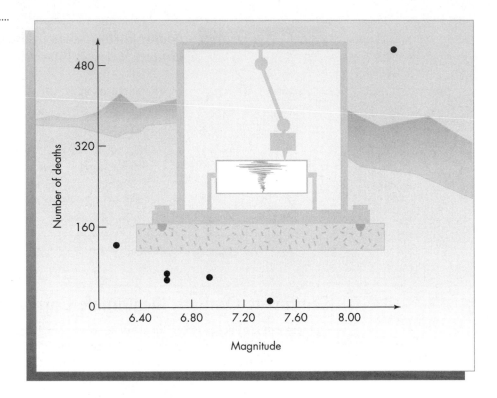

Clearly, trying to interpret the correlation between magnitude and death toll for this small group of earthquakes is a misuse of statistics. The largest earthquake, in 1906, occurred before earthquake building codes were enforced. The next largest quake, with magnitude 7.4, killed only one person but occurred in a very sparsely populated area. ■

The Missing Link: A Third Variable

Another common mistake that can lead to an illegitimate correlation is combining two or more groups when they should be considered separately. The variables for each group may actually fall very close to a straight line, but when the groups are examined together, the individual relationships may be masked. As a result, it will appear that there is very little correlation between the two variables.

This problem is a variation of "Simpson's Paradox" for count data, a phenomenon we will study in the next chapter. However, statisticians do not seem to be as alert to this problem when it occurs with measurement data. When you read that two variables have a very low correlation, ask yourself whether data may have been combined into one correlation when groups should, instead, have been considered separately.

TABLE 11.3		Pages versus Price for the Books on a Professor's Shelf			
Pages	Price	Pages	Price	Pages	Price
104	32.95	342	49.95	436	5.95
188	24.95	378	4.95	458	60.00
220	49.95	385	5.99	466	49.95
264	79.95	417	4.95	469	5.99
336	4.50	417	39.75	585	5.95

EXAMPLE 4

THE FEWER THE PAGES, THE MORE VALUABLE THE BOOK?

If you peruse the bookshelves of a typical college professor, you will find a variety of books ranging from textbooks to esoteric technical publications to paperback novels. To determine whether the price of a book can be determined by the number of pages it contains, a college professor recorded the number of pages and price for 15 books on one shelf. The numbers are shown in Table 11.3. Is there a relationship between number of pages and the price of the book? The correlation for these figures is –.312. The negative correlation indicates that the more pages a book has, the *less* it costs, which is certainly a counterintuitive result.

Figure 11.4 illustrates what has gone wrong. It displays the data in a scatterplot, but it also identifies the books by type. The letter H indicates a hardcover book; the letter S indicates a softcover book. The collection of books on the professor's shelf consisted of softcover novels, which tend to be long but inexpensive, and hardcover technical books, which tend to be shorter but very expensive. If the correlations are calculated within each type, we find the result we would expect. The correlation between number of pages and price is .64 for the softcover books alone, and .35 for the hardcover books alone. Combining the two types of books into one collection not only masked the positive association between length and price, but produced an illogical negative association. ■

11.2   LEGITIMATE CORRELATION DOES <u>NOT</u> IMPLY CAUSATION

Even if two variables are legitimately related or correlated, do not fall into the trap of believing there is a causal connection between them. Although "correlation does not imply causation" is a very well known saying among researchers, relationships

FIGURE 11.4
Combining groups produces misleading correlations (H = hardcover; S = softcover).
Source: Data from Table 11.3.

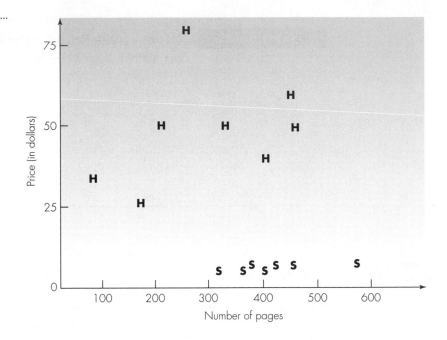

and correlations derived from observational studies are often *reported* as if the connection were causal.

It is easy to construct silly, obvious examples of correlations that do not result from causal connections. For example, a list of weekly tissue sales and weekly hot chocolate sales for a city with extreme seasons would probably exhibit a correlation because both tend to go up in the winter and down in the summer. A list of shoe sizes and vocabulary words mastered by school children would certainly exhibit a correlation because older children tend to have larger feet and to know more words than younger children.

The problem is that sometimes the connections do seem to make sense, and it is tempting to treat the observed association as if there were a causal link. Remember that data from an observational study, in the absence of any other evidence, simply cannot be used to establish causation.

EXAMPLE 5

PROSTATE CANCER AND RED MEAT

The February 1994 issue of the *University of California at Berkeley Wellness Letter* (pp. 2–3) reports the results of a study originally published in the October 1993 issue of the *Journal of the National Cancer Institute*. The study followed 48,000 men who had filled out dietary questionnaires in 1986. By 1990, 300 of the men

TABLE 11.4	Divorce Rates and Prison Rates for Drug Offenses	

Year	Divorce Rate (Per 1000)	Percentage Admitted for Drug Offenses
1960	2.2	4.2
1964	2.4	4.1
1970	3.5	9.8
1974	4.6	12.0
1978	5.2	8.4
1982	5.1	8.1
1986	4.8	16.3

Source: Data for divorce rates are from *Information Please Almanac*, 1991, p. 809. Data for the drug offenses are from *World Almanac and Book of Facts*, 1993, p. 950.

had been diagnosed with prostate cancer and 126 had advanced cases. For the advanced cases, the *Wellness Letter* reported that "men who ate the most red meat had a 164% higher risk than those with the lowest intake. Fats from dairy products, fish, and vegetable oils did not increase the risk" (p. 2).

We may be tempted to believe that this indicates that red meat is a contributing cause of prostate cancer. But perhaps there is no causal connection at all. One possibility is that a third variable both leads men to consume more red meat and increases the risk of prostate cancer. One candidate is the hormone testosterone, which has been implicated in the growth of prostate cancer. ■

EXAMPLE 6   DIVORCE RATES AND DRUG OFFENSES

Table 11.4 shows the divorce rates in the United States for various years from 1960 to 1986, accompanied by the percentage of those admitted to state prisons because of drug offenses.

The correlation between divorce rate and percentage of criminals admitted for drug offenses is quite strong, at .67. Based on this correlation, advocates of traditional family values may argue that increased divorce rates have resulted in more drug offenses. But they both simply reflect a trend across time. The correlation relating year and divorce rate is much higher, at .92. Similarly, the correlation between year and percent admitted for drug offenses is .78. Any two variables that have either both increased or both decreased across time will display this kind of correlation. ■

11.3   SOME REASONS FOR RELATIONSHIPS BETWEEN VARIABLES

We have seen numerous examples of variables that are related but for which there is probably not a causal connection. To help us understand this phenomenon, let's examine some of the reasons two variables could be related, including a causal connection.

Some reasons two variables could be related:

1. The explanatory variable is the direct cause of the response variable.

2. The response variable is causing a change in the explanatory variable.

3. The explanatory variable is a contributing but not sole cause of the response variable.

4. Confounding variables may exist.

5. Both variables may result from a common cause.

6. Both variables are changing over time.

7. The association may be nothing more than coincidence.

Reason 1: The explanatory variable is the direct cause of the response variable. Occasionally, a change in the explanatory variable is the direct cause of a change in the response variable. For example, if we were to measure amount of food consumed in the past hour and level of hunger, we would find a relationship. We would probably agree that the differences in the amount of food consumed were responsible for the difference in levels of hunger.

Unfortunately, even if one variable is the direct cause of another, we may not see a strong association. For example, even though intercourse is the direct cause of pregnancy, the relationship between having intercourse and getting pregnant is not strong; most occurrences of intercourse do not result in pregnancy.

Reason 2: The response variable is causing a change in the explanatory variable. Sometimes the causal connection is the opposite of what might be expected. For example, what do you think you would find if you studied the response variable = hotel occupancy rate and the explanatory variable = advertising sales (in dollars) per room? You would probably expect that higher advertising expenditures would cause higher occupancy rates. Instead, it turns out that the relationship is negative because, when occupancy rates are low, hotels spend more money on advertising to try to raise them. Thus, although we might expect higher advertising dollars to cause higher occupancy rates, if they are measured at the same point in time, we instead find that low occupancy rates cause higher advertising revenues.

Reason 3: The explanatory variable is a contributing but not sole cause of the response variable. The complex kinds of phenomena most often studied by researchers are likely to have multiple causes. Even if there were a causal connection between diet and a type of cancer, for instance, it would be unlikely that the cancer was caused solely by eating that certain type of diet.

It is particularly easy to be misled into thinking you have found a sole cause for a particular outcome, when what you have found is actually a *necessary contributor* to the outcome. For example, scientists generally agree that in order to have AIDS, you must be infected with HIV. In other words, HIV is *necessary* to develop AIDS. But it does not follow that HIV is the *sole* cause of AIDS, and there has been some controversy over whether that is actually the case.

Another possibility, discussed in earlier chapters, is that one variable is a contributory cause of another, but only for a subgroup of the population. If the researchers do not examine separate subgroups, that fact can be masked.

EXAMPLE 7

DELIVERY COMPLICATIONS, REJECTION, AND VIOLENT CRIME

A study summarized in *Science* (Mann, March 1994) and conducted by scientists at the University of Southern California reported a relationship between violent crime and complications during birth. The researchers found that delivery complications at birth were associated with much higher incidence of violent crime later in life. The data came from an observational study of males born in Copenhagen, Denmark, between 1959 and 1961.

However, the connection held only for those men whose mothers rejected them. Rejection meant that the mother had not wanted the pregnancy, had tried to have the fetus aborted, and had sent the baby to an institution for at least a third of his first year of life. Men who were accepted by their mothers did not exhibit this relationship. Men who were rejected by their mothers but for whom there were no complications at birth did not exhibit the relationship either. In other words, it was the interaction of delivery complications and maternal rejection that was associated with higher levels of violent crime.

This example was based on an observational study, so there may not be a causal link at all. However, even if there is a causal connection between delivery complications and subsequent violent crime, the data suggest that it holds only for a particular subset of the population. If the researchers had not measured the additional variable of maternal rejection, the data would have erroneously been interpreted as suggesting that the connection held for all men. ∎

Reason 4: Confounding variables may exist. We defined confounding variables in Chapter 4, but it is worth reviewing the concept here because it is relevant for explaining relationships. Recall that two variables are confounded if their effects on a third variable cannot be separated. Thus, both variables may help cause the change in the third variable, but there is no way to establish how much is due to one and how much is due to the other.

As an example, consider the data from Example 1 in this chapter, relating speed limits and death rates for ten countries. The correlation between the two variables was positive, suggesting that lowering speed limits might cause a drop in vehicular death rates. However, each country in the list has different regulations, different licensing requirements, differing traffic density, and so on. Numerous variables could be confounded with speed limit to produce the observed differences in death rates. We cannot separate the effect of speed limit from the myriad possible confounding effects.

Reason 5: Both variables may result from a common cause. We have seen numerous examples in which a change in one variable was thought to be associated with a change in the other, but for which we speculated that a third variable was responsible. For example, Case Study 6.2 concerned the fact that meditators had levels of an enzyme normally associated with people of a younger age. We could speculate that something in the personality of the meditators caused them to want to meditate and also caused them to have lower enzyme levels than others of the same age.

As another example, recall the scatterplot and correlation between verbal SAT scores and college GPAs, exhibited in Chapters 9 and 10. We would certainly not conclude that higher SAT scores caused higher grades in college, except perhaps for a slight benefit of boosted self-esteem. However, we could probably agree that the causes responsible for one variable being high (or low) are the same as those responsible for the other being high (or low). Those causes would include such factors as intelligence, motivation, and ability to perform well on tests.

Reason 6: Both variables are changing over time. Some of the most nonsensical associations result from correlating two variables that have both changed over time. If they are both changing in a consistent direction, you will indeed see a strong correlation, but it may not have any causal link. For example, you would certainly see a correlation between winning times in two different Olympic events because winning times have all decreased over the years.

Sociological variables are the ones most likely to be manipulated in this way, as demonstrated by Example 6 in this chapter, relating increasing divorce rates and increasing drug offenses. Watch out for reports of a strong association between two such variables, especially when you know that both variables are likely to have had large changes over time.

Reason 7: The association may be nothing more than coincidence. Sometimes an association between two variables is nothing more than coincidence, even though the odds of it happening appear to be very small. For example, suppose a new office building opened and within a year there was an unusually high rate of brain cancer among workers in the building. Suppose someone calculated that the odds of having that many cases in one building were only 1 in 10,000. We might immediately suspect that something wrong in the environment was causing people to develop brain cancer.

The problem with this reasoning is that it focuses on the odds of seeing such a rare event occurring in that particular building in that particular city. It fails to take into account the fact that there are thousands of new office buildings. If the odds really were only 1 in 10,000, we should expect to see this phenomenon just by chance in about 1 of every 10,000 buildings. And that would just be for this particular type of cancer. What about clusters of other types of cancer or other diseases? It would be unusual if we did not occasionally see clusters of diseases as chance occurrences.

We will study this phenomenon in more detail in Part 3. For now, be aware that a connection of this sort should be expected to occur relatively often, even though each individual case has low probability.

11.4   CONFIRMING CAUSATION

Given the number of possible explanations for the relationship between two variables, how do we ever establish that there actually is a causal connection? It isn't easy. Ideally, in establishing a causal connection, we would change nothing in the environment except the suspected causal variable and then measure the result on the suspected outcome variable.

The only legitimate way to try to establish a causal connection statistically is *through the use of designed experiments.* As we have discussed earlier, in designed experiments we try to rule out confounding variables through random assignment. If we have a large sample, and if we use proper randomization, we can assume that the levels of confounding variables will be about equal in the different treatment groups. This reduces the chances that an observed association is due to confounding variables, even those that we have neglected to measure.

Evidence of a possible causal connection:

1. There is a reasonable explanation of cause and effect.

2. The connection happens under varying conditions.

3. Potential confounding variables are ruled out.

Nonstatistical Considerations

If a designed experiment cannot be done, then nonstatistical considerations must be used to determine whether a causal link is reasonable. Following are some features that lend evidence to a causal connection:

1. *There is a reasonable explanation of cause and effect.* A potential causal connection will be more believable if an explanation exists for how the cause and effect occurs. For instance, in Example 4 in this chapter, we established that for hardcover books the number of pages is correlated with the price. We would probably not contend that higher prices result in more pages, but we could reasonably argue that more pages result in higher prices. We can imagine that publishers set the price of a book based on the cost of producing it and that the more pages there are the higher the cost of production. Thus, we have a reasonable explanation for how an increase in the length of a book could cause an increase in the price.

2. *The connection happens under varying conditions.* If many observational studies conducted under different conditions all find the same link between two variables, the evidence for a causal connection is strengthened. This is especially true if the studies are not likely to have the same confounding variables. The evidence is also strengthened if the same type of relationship holds when the explanatory variable falls into different ranges.

For example, numerous observational studies have related cigarette smoking and lung cancer. Further, the studies have shown that the higher the number of cigarettes smoked, the greater the chances of developing lung cancer; similarly, a connection has been established between lung cancer and the age at which smoking began. These facts make it more plausible that smoking actually causes lung cancer.

3. *Potential confounding variables are ruled out.* When a relationship first appears in an observational study, potential confounding variables may immediately come to mind. For example, the researchers in Case Study 6.2, showing the relationship between meditation and aging, did consider that vegetarian diets and low alcohol consumption among many of the meditators may have confounded the results. However, they were able to locate other research that failed to find any connection between these factors and the enzyme they were measuring. The greater the number of confounding factors that can be ruled out, the more convincing the evidence for a causal connection.

A Final Note

As you should realize by now, it is very difficult to establish a causal connection between two variables with the use of anything except designed experiments. Because it is virtually impossible to conduct a flawless experiment, potential problems crop up even with a well-designed experiment. This means that you should look with skepticism on claims of causal connections. Having read this chapter, you should have the tools necessary for making intelligent decisions and for discovering when an erroneous claim is being made.

EXERCISES

1. Explain why a strong correlation would be found between weekly sales of firewood and weekly sales of cough drops over a 1-year period. Would it imply that fires cause coughs?

| TABLE 11.5 | The Eight Greatest Risks | |

Activity or Technology	Experts' Rank	Students' Rank
Motor vehicles	1	5
Smoking	2	3
Alcoholic beverages	3	7
Handguns	4	2
Surgery	5	11
Motorcycles	6	6
X rays	7	17
Pesticides	8	4

Source: Iman, 1994, p. 505.

2. Suppose a study of employees at a large company found a negative correlation between weight and distance walked on an average day. In other words, people who walked more weighed less. Would you conclude that walking causes lower weight? Can you think of another potential explanation?

3. An article in *Science News* (vol. 149, 1 June, 1996, p. 345) claimed that "evidence suggests that regular consumption of milk may reduce a person's risk of stroke, the third leading cause of death in the United States." The claim was based on an observational study of 3150 men, and the article noted that the researchers "report strong evidence that men who eschew milk have more than twice the stroke risk of those who drink 1 pint or more daily." The article concluded by noting that "those who consumed the most milk tended to be the leanest and the most physically active." Go through the list of seven "reasons two variables may be related," and discuss each one in the context of this study.

4. Iman (1994, p. 505) presents data on how college students and experts perceive risks for 30 activities or technologies. Each group ranked the 30 activities. The rankings for the eight greatest risks, as perceived by the experts, are shown in Table 11.5.

   a. Prepare a scatterplot of the data, with students' ranks on the vertical axis and experts' ranks on the horizontal axis.

   b. The correlation between the two sets of ranks is .407. Based on your scatterplot in part a, do you think the correlation would increase or decrease if X rays were deleted? Explain. What if pesticides were deleted instead?

   c. Another technology listed was nuclear power, ranked first by the students and 20th by the experts. If nuclear power was added to the list, do you think the correlation between the two sets of rankings would increase or decrease? Explain.

5. Give an example of two variables that are likely to be correlated because they are both changing over time.

6. Which one of the seven reasons for relationships listed in Section 11.3 is supposed to be ruled out by designed experiments?

7. Refer to Case Study 10.2, in which students reported their ideal and actual weights. When males and females are not separated, the regression equation is

ideal = 8.0 + 0.9 actual

   a. Draw the line for this equation and the line for the equation *ideal = actual* on the same graph. Comment on the graph as compared to those shown in Figures 10.7 and 10.8, in terms of how the regression line differs from the line where ideal and actual weights are the same.

   b. Calculate the ideal weight based on the combined regression equation and the ideal weight based on separate equations, for individuals whose actual weight is 150 pounds. Recall that the separate equations were

   For women: ideal = 43.9 + 0.6 actual

   For men: ideal = 52.5 + 0.7 actual

   c. Comment on the conclusion you would make about individuals weighing 150 pounds if you used the combined equation compared with the conclusion you would make if you used the separate equations.

   d. Explain which of the problems identified in this chapter has been uncovered with this example.

8. Suppose a study measured total beer sales and number of highway deaths for 1 month in various cities. Explain why it would make sense to divide both variables by the population of the city before determining whether a relationship exists between them.

9. Construct an example of a situation where an outlier inflates the correlation between two variables. Draw a scatterplot.

10. Construct an example of a situation where an outlier deflates the correlation between two variables. Draw a scatterplot.

11. According to *The Wellness Encyclopedia* (University of California, 1991, p. 17): "Alcohol consumed to excess increases the risk of cancer of the mouth, pharynx, esophagus, and larynx. These risks increase dramatically when alcohol is used in conjunction with tobacco." It is obviously not possible to conduct a designed experiment on humans to test this claim, so the causal conclusion must be based on observational studies. Explain three potential additional pieces of information that the authors may have used to lead them to make a causal conclusion.

12. Suppose a positive relationship had been found between each of the following sets of variables. In Section 11.3, seven potential reasons for such relationships are given. Explain which of the seven reasons is most likely to account for the relationship in each case. If you think more than one reason might apply, mention them all but elaborate on only the one you think is most likely.

    a. Number of deaths from automobiles and beer sales for each year from 1950 to 1990

    b. Number of ski accidents and average wait time for the ski lift for each day during one winter at a ski resort

    c. Stomach cancer and consumption of barbecued foods, which are known to contain carcinogenic (cancer-causing) substances

    d. Self-reported level of stress and blood pressure

    e. Amount of dietary fat consumed and heart disease

    f. Twice as many cases of leukemia in a new high school, built near a power plant, than at the old high school

13. Explain why it would probably be misleading to use correlation to express the relationship between number of acres burned and number of deaths for major fires in the United States.

14. It is said that a higher proportion of drivers of red cars are given tickets for traffic violations than the drivers of any other color car. Does this mean that if you drove a red car rather than a white car, you would be more likely to receive a ticket for a traffic violation? Explain.

15. Construct an example for which correlation between two variables is masked by grouping over a third variable.

16. An article in the *Davis* (CA) *Enterprise* (5 April, 1994) had the headline, "Study: Fathers key to child's success." The article described a new study as follows: "The research, published in the March issue of the *Journal of Family Psychology,* found that mothers still do a disproportionate share of child care. But surprisingly, it also found that children who gain the 'acceptance' of their fathers make better grades than those not close to their dads." The article implies a causal link, with gaining father's acceptance (the explanatory variable) resulting in better grades (the response variable). Choosing from the remaining six possibilities in Section 11.3 (reasons 2 through 7), give three other potential explanations for the observed connection.

17. Lave (1990) discussed studies that had been done to test the usefulness of seat belts before and after their use became mandatory. One possible method of testing the usefulness of mandatory seat belt laws is to measure the number of fatalities in a particular region for the year before and the year after the law went into effect and to compare them. If such a study were to find substantially reduced fatalities during the year after the law went into effect, could it be claimed that the mandatory seat belt law was completely responsible? Explain. (*Hint:* Consider factors such as weather and the anticipatory effect of the law.)

18. In Case Study 10.1, we learned how psychologists relied on twins to measure the contributions of heredity to various traits. Suppose a study were to find that identical (monozygotic) twins had highly correlated scores on a certain trait but that pairs of adult friends did not. Why would that not be sufficient evidence to conclude that genetic factors were responsible for the trait?

MINI-PROJECTS

1. Find a newspaper or journal article that describes an observational study in which the author's actual goal is to try to establish a causal connection. Read the article, and then discuss how well the author has made a case for a causal connection. Consider the factors discussed in Section 11.4 and discuss whether they have been addressed by the author. Finally, discuss the extent to which the author has convinced you that there is a causal connection.

2. Peruse journal articles and find two examples of scatterplots for which the authors have computed a correlation that you think is misleading. For each case, explain why you think it is misleading.

REFERENCES

Iman, R. L. (1994). *A data-based approach to statistics.* Belmont, CA: Duxbury.

Information please almanac. (1991). Edited by Otto Johnson. Boston: Houghton Mifflin.

Lave, L. B. (1990). Does the surgeon-general need a statistics advisor? *Chance* 3, no. 4, pp. 33–40.

Mann, C. C. (March 1994). War of words continues in violence research. *Science* 263, no. 11, p. 1375.

Rivkin, D. J. (25 November 1986). Fifty-five mph speed limit is no safety guarantee. *New York Times,* letter to the editor, p. 26.

University of California at Berkeley. (1991). *The wellness encyclopedia.* Boston: Houghton Mifflin.

World almanac and book of facts. (1993). Edited by Mark S. Hoffman. New York: Pharos Books.

CHAPTER TWELVE

Relationships Between Categorical Variables

THOUGHT QUESTIONS

1. Students in a statistics class were asked whether they preferred an in-class or a take-home final exam and were then categorized as to whether they had received an A on the midterm. Of the 25 A students, 10 preferred a take-home exam, whereas of the 50 non-A students, 30 preferred a take-home exam. How would you display these data in a table?

2. Suppose a news article claimed that drinking coffee doubled your risk of developing a certain disease. Assume the statistic was based on legitimate, well-conducted research. What additional information would you want about the risk before deciding whether to quit drinking coffee? (*Hint:* Does this statistic provide any information on your actual risk?)

3. A study to be discussed in detail in this chapter classified pregnant women according to whether they smoked and whether they were able to get pregnant during the first cycle in which they tried to do so. What do you think is the question of interest? Attempt to answer it. Here are the results:

	Pregnancy Occurred After		
	First Cycle	Two or More Cycles	Total
Smoker	29	71	100
Nonsmoker	198	288	486
Total	227	359	586

4. A recent study estimated that the "relative risk" of a woman developing lung cancer if she smoked was 27.9. What do you think is meant by the term *relative risk*?

TABLE 12.1	Heart Attack Rates After Taking Aspirin or Placebo		
	Heart Attack	No Heart Attack	Total
Aspirin	104	10,933	11,037
Placebo	189	10,845	11,034
Total	293	21,778	22,071

12.1 DISPLAYING RELATIONSHIPS BETWEEN CATEGORICAL VARIABLES CONTINGENCY TABLES

Summarizing and displaying data resulting from the measurement of two categorical variables is easy to do: Simply count the number of individuals who fall into each combination of categories, and present those counts in a table. Such displays are often called **contingency tables** because they cover all contingencies for combinations of the two variables. Each row and column combination in the table is called a *cell*.

In some cases, one variable can be designated as the explanatory variable and the other as the response variable. In these cases, it is conventional to place the explanatory variables down along the side of the table (as labels for the rows) and the response variables along the top of the table (as labels for the columns). This makes it easier to display the percentages of interest.

EXAMPLE 1   ASPIRIN AND HEART ATTACKS

In Case Study 1.2, we discussed an experiment in which there were two categorical variables:

variable A = explanatory variable = aspirin or placebo
variable B = response variable = heart attack or no heart attack

Table 12.1 illustrates the contingency table for the results of this study. Notice that the explanatory variable (whether the individual took aspirin) is the row variable, whereas the response variable (whether the person had a heart attack) is the col-

| | TABLE 12.2 | Data for Example 1 with Percentage and Rate Added | | | | |

	Heart Attack	No Heart Attack	Total	Heart Attacks (%)	Rate per 1000
Aspirin	104	10,933	11,037	0.94	9.4
Placebo	189	10,845	11,034	1.71	17.1
Total	293	21,778	22,071		

umn variable. There are four cells, one representing each combination of treatment and outcome. ■

Conditional Percentages and Rates

It's difficult to make useful comparisons from a contingency table (unless the number of individuals under each condition is the same) without doing further calculations. Usually, the question of interest is whether the percentages in each category of the response variable change when the explanatory variable changes.

In Example 1, the question of interest is whether the percentage of heart attack sufferers differs, depending on whether the people took aspirin or a placebo. In other words, is the percentage of people who fall into the first column (heart attack) the same for the two rows?

We can calculate the conditional percentages for the response variable by looking separately at each category of the explanatory variable. Thus, in our example, we have two conditional percentages:

Aspirin group. The percentage who had heart attacks was $104/11,037 = 0.0094 = 0.94\%$.

Placebo group: The percentage who had heart attacks was $189/11,034 = 0.0171 = 1.71\%$.

Sometimes, for rare events like these heart attack numbers, percentages are so small that it is easier to interpret a rate. The rate is simply stated as the number of individuals per 1000 or per 10,000 or per 100,000, depending on what's easiest to interpret. Percentage is equivalent to a rate per 100.

Table 12.2 presents the data from Example 1, but also includes the conditional percentages and the rates of heart attacks per 1000 individuals for the two groups. Notice that the rate per 1000 is easier to understand than the percentages.

TABLE 12.3	Results of Roadside Survey for Young Drivers

| | Drank Alcohol in Last 2 Hours? | | | |
	Yes	No	Total	Percentage Who Drank
Males	77	404	481	16.0%
Females	16	122	138	11.6%
Total	93	526	619	15.0%

SOURCE: Gastwirth, 1988, p. 526.

EXAMPLE 2

YOUNG DRIVERS, GENDER, AND DRIVING UNDER THE INFLUENCE OF ALCOHOL

In Case Study 6.3, we learned about a court case challenging an Oklahoma law that differentiated the ages at which young men and women could buy 3.2% beer. The Supreme Court had examined evidence from a "random roadside survey" that measured information on age, gender, and drinking behavior. In addition to the data presented in Case Study 6.3, the roadside survey measured whether the driver had been drinking alcohol in the previous 2 hours. Table 12.3 gives the results for the drivers under 20 years of age.

The Supreme Court concluded that "the showing offered by the appellees does not satisfy us that sex represents a legitimate, accurate proxy for the regulation of drinking and driving" (Gastwirth, 1988, p. 527). Notice the difference in the percentages of young men and women who had been drinking alcohol, with the percentage slightly higher for males. However, later in this chapter, we will see that we cannot rule out chance as a reasonable explanation for this difference. In other words, if there really is no difference among the percentages of young male and female drivers in the population who drink and drive, we still could possibly see a difference as large as the one observed in a sample of this size. Using the language introduced in Chapter 10, this means that the observed difference in percentages is not *statistically significant*. ■

EXAMPLE 3

EASE OF PREGNANCY FOR SMOKERS AND NONSMOKERS

In a retrospective observational study, researchers asked women who were pregnant with planned pregnancies how long it took them to get pregnant (Baird and Wilcox, 1985; see also Weiden and Gladen, 1986). Length of time to pregnancy

| TABLE 12.4 | | Time to Pregnancy for Smokers and Nonsmokers | | | |

	Pregnancy Occurred After			Percentage in First Cycle
	First Cycle	Two or More Cycles	Total	
Smoker	29	71	100	29%
Nonsmoker	198	288	486	41%
Total	227	359	586	

was measured according to the number of cycles between stopping birth control and getting pregnant. Women were also categorized on whether they smoked, with smoking defined as having at least one cigarette per day for at least the first cycle during which they were trying to get pregnant.

For our purposes, we will classify the women on two categorical variables:

variable *A* = explanatory variable = smoker or nonsmoker

variable *B* = response variable = pregnant in first cycle or not

The question of interest is whether the same percentages of smokers and nonsmokers were able to get pregnant during the first cycle. We present the contingency table and the percentages in Table 12.4.

As you can see, a much higher percentage of nonsmokers than smokers were able to get pregnant during the first cycle. Because this is an observational study, we cannot conclude that smoking caused a delay in getting pregnant. We merely notice that there is a relationship between smoking status and time to pregnancy, at least for this sample. It is not difficult to think of potential confounding variables. ■

12.2   ASSESSING THE STATISTICAL SIGNIFICANCE OF A 2×2 TABLE

The percentage responding in a particular way is unlikely to be exactly the same for all categories of an explanatory variable. Therefore, researchers are interested in assessing whether the differences in observed percentages are just chance differences or represent a real difference for the population. Remember that the data observed are usually a sample from a much larger population.

Measuring the Strength of the Relationship

We discussed the concept of statistical significance briefly in Chapter 10. Recall that that term can be applied if an observed relationship is stronger than what would be expected by chance. Specifically, we required such a relationship to be larger than 95% of those that would be observed just by chance. Let's see how that rule can be applied to relationships between categorical variables.

We will consider only the simplest case, that of 2×2 contingency tables. In other words, we will consider only the situation where each variable has two categories. (The same principles and interpretation apply to larger tables, but the details are more cumbersome.)

For 2×2 tables, the strength of the relationship is measured by the *difference* in the percentages of outcomes for the two categories of the explanatory variable. However, before we assess statistical significance, we need to incorporate information about the size of the sample as well. Our examples will illustrate why this feature is necessary.

For Example 1, we note that the difference in percentage of heart attacks between aspirin and placebo takers is only $1.71\% - 0.94\% = 0.77\%$. For Example 2, we notice that the difference between males and females who had been drinking is $16\% - 11.6\% = 4.4\%$. Finally, in Example 3, the difference between the percentage of smokers and nonsmokers who achieved pregnancy during the first cycle in which they tried is $41\% - 29\% = 12\%$.

Are these differences large enough to rule out chance? Can we conclude that the relationships observed for the samples also hold for each population? The difference of only 0.77% between the aspirin and placebo takers seems rather small, and in fact if it had occurred for only a few hundred men, it would probably not be convincing. The difference of 4.4% between male and female drinkers is also rather small and was certainly not convincing to the Supreme Court. The difference of 12% in Example 3 is much larger, but is it large enough to be convincing based on fewer than 600 women? Perhaps another study, on a different 600 women, would yield exactly the opposite result.

Strength of the Relationship versus Size of the Study

At this point, you should be able to see that whether we can rule out chance depends on both the strength of the relationship and on how many people were involved in the study. An observed relationship is much more believable if it is based on 22,000 people (as in Example 1) than if it is based on about 600 people (as in Examples 2 and 3).

The Chi-Squared Statistic

To assess whether a relationship in a 2×2 table achieves statistical significance, we need to know the value of the **chi-squared statistic** for the table. This statistic is a measure that combines the strength of the relationship with information about the size of the sample to give one summary number. If that summary number is larger than 3.84, the relationship in the table is considered to be statistically significant.

The origin of the "magic" number 3.84 is too technical to describe here. It comes from a table of percentiles representing what should happen by chance, similar to the percentile table for z-scores that was included in Chapter 8.

The interpretation of the value 3.84 is straightforward. Of all relationships in 2×2 tables that occur only by chance, 95% of them have a chi-squared statistic smaller than 3.84. Relationships in the sample that reflect a real relationship in the population are likely to produce larger chi squared statistics. Therefore, if we observe a relationship that has a chi-squared statistic larger than 3.84, we can assume that the relationship in the sample did not occur by chance. In that case, we can say that the relationship is statistically significant. Of all relationships that have occurred just by chance, 5% of them will erroneously earn the title of statistically significant. However, if the size of the sample is too small, a real relationship may not be detected. Remember that the chi-squared statistic depends on both the strength of the relationship and the size of the sample.

Computing the Chi-Squared Statistic

The actual computation and assessment of statistical significance is tedious but not difficult. There are different ways to represent the necessary formula, some of them useful only for 2×2 tables. Here we present only one method, but this method can be used for tables with any number of rows and columns. As we list the necessary steps, we will demonstrate the computation using the data from Example 3, shown in Table 12.4.

Determining statistical significance using the chi-squared statistic:

1. Compute the expected numbers.

2. Compare the observed and expected numbers.

3. Compute the chi-squared statistic.

4. Make the decision.

Step 1: Compute the Expected Numbers

Compute the number of individuals that would be expected to fall into each of the cells of the table if there were no relationship. The formula for finding the expected number in any row and column combination is:

expected number = (row total)(column total)/(table total)

The expected numbers in each row and column must sum to the same totals as the observed numbers, so for a 2×2 table we need only compute one of these using the formula. We can obtain the rest by subtraction. It is easy to see why this formula would give the number to be expected if there were no relationship.

Consider the first column. The proportion who fall into the first column overall is (column 1 total)/(table total). If there is no relationship between the two variables, then that proportion should be the same for both rows. Therefore, to find how many of the people in row 1 would be expected to be in column 1, simply take the overall

| TABLE 12.5 | Computing the Expected Numbers for Table 12.4 |

| | Pregnancy Occurred After | | |
	First Cycle	Two or More Cycles	Total
Smoker	38.74	$100 - 38.74 = 61.26$	100
Nonsmoker	$227 - 38.74 = 188.26$	$486 - 188.26 = 297.74$	486
Total	227	359	586

proportion who are in column 1 and multiply it by the number of people who are in row 1. In other words, use (row 1 total)(column 1 total)/(table total).

Step 1 for the Example Let's compute the expected number of smokers achieving pregnancy in the first cycle:

expected number = (100)(227)/586 = 38.74

It's very important that the numbers not be rounded off at this stage. Now that we have the expected number for the first row and column, we can fill in the rest of the expected numbers (see Table 12.5), making sure the row and column totals remain the same as they were in the original table.

Step 2: Compare the Observed and Expected Numbers

For this step, we compute the difference between what we would expect by chance, as shown in step 1, and what we have actually observed. To remove negative signs and to standardize these differences based on the number in each combination, we compute the following for each of the cells of the table:

(observed number − expected number)^2/(expected number)

In a 2×2 table, the numerator will actually be the same for each cell. In contingency tables with more than two rows or columns (or both), this would not be the case.

Step 2 for the Example The numerator for the first cell is

(observed number − expected number)^2 = $(29 - 38.74)^2 = (-9.74)^2 = 94.87$

Convince yourself that this same numerator applies for the other three cells. The contribution for each cell is shown in Table 12.6.

Step 3: Compute the Chi-Squared Statistic

To compute the chi-squared statistic, simply add the numbers in all of the cells from step 2. The result is the chi-squared statistic.

TABLE 12.6	Comparing the Observed and Expected Numbers	

	Pregnancy Occurred After	
	First Cycle	Two or More Cycles
Smoker	94.87/38.74 = 2.45	94.87/61.26 = 1.55
Nonsmoker	94.87/188.26 = 0.50	94.87/297.74 = 0.32

Step 3 for the Example

chi-squared statistic = 2.45 + 1.55 + 0.50 + 0.32 = 4.82

Step 4: Make the Decision

For a 2×2 table, the relationship earns the title of "statistically significant" if the chi-squared statistic is at least 3.84. For larger contingency tables, you would need to look up the appropriate number in a table called "percentiles of the chi-squared distribution," which is found in most statistics books. Many calculators and computer applications such as Excel can also provide these numbers.

Step 4 for the Example The chi-squared statistic computed for the example is 4.82, which is larger than 3.84. Thus, we can say there is a statistically significant relationship between smoking and time to pregnancy. In other words, we can conclude that the difference we observed in time to pregnancy between smokers and nonsmokers in the sample indicates a real difference for the population of all similar women. It was not just the luck of the draw for this sample. This result is based on the assumption that the women studied can be considered to be a random sample from that population.

Computers

Many simple computer programs will compute the chi-squared statistic for you. Figure 12.1 shows how we would use a statistical computing program called Minitab to carry out the example we have just computed by hand. Material in **boldface** was typed into the computer; the rest of the material was the response by the computer. Notice that the computer has done all the work for us and presented it in summary form. The only thing the computer did not supply is the decision.

Practical versus Statistical Significance

You should be aware that "statistical significance" does not mean the two variables have a relationship that you would necessarily consider to be of practical importance. For example, a table based on a very large number of observations will have

```
MTB > READ C1 C2
DATA > 29 71
DATA > 198 288
DATA > END
   2 ROWS READ
MTB > CHISQUARED STATISTIC FOR THE DATA IN C1 AND C2

Expected counts are printed below observed counts

            C1            C2          Total
1           29            71            100
            38.74         61.26

2           198           288           486
            188.26        297.74

Total       227           359           586

ChiSq = 2.448 + 1.548 +
        0.504 + 0.318 = 4.817
```

little trouble achieving statistical significance, even if the relationship between the two variables is only minor. Conversely, an interesting relationship in a population may fail to achieve statistical significance in the sample if there are only a few observations. It is difficult to rule out chance unless you have either a very strong relationship or a sufficiently large sample.

To see this, consider the relationship between taking aspirin instead of a placebo and having a heart attack or not. The chi-squared statistic, based on the result from the 22,071 participants in the study, is 25.01, so the relationship is clearly statistically significant. Now suppose there were only one-tenth as many participants, or 2207—still a fair-sized sample. Further suppose that the heart attack rates remained the same, at 9.4 per thousand for the aspirin group and 17.1 per thousand for the placebo group.

If you look at the method for computing the chi-squared statistic, you will realize that if all numbers in the study are divided by 10, the resulting chi-squared statistic is also divided by 10. Therefore, if the study had only 2207 participants instead of 22,071, the chi-squared statistic would have been only 2.501. It would not have been large enough to conclude that the relationship between heart attacks and aspirin consumption was statistically significant, even though the rates of heart attacks per thousand people were still 9.4 for the aspirin group and 17.1 for the placebo group.

Some researchers report the lack of a statistically significant result erroneously, by implying that a relationship must therefore not exist. When you hear the claim

FIGURE 12.2

Using Minitab to compute the chi-squared statistic for Example 2

```
MTB > read C1 C2
DATA > 77 404
DATA > 16 122
DATA > end
    2 ROWS READ
MTB > chis c1 c2

Expected counts are printed below observed counts

              C1              C2              Total
1             77              404             481
              72.27           408.73

2             16              122             138
              20.73           117.27

Total         93              526             619

ChiSq = 0.310 + 0.055 +
        1.081 + 0.191 = 1.637

df = 1
```

that a study "failed to find a relationship" between two variables, it does not mean that a relationship was not observed. It means that whatever relationship was observed did not achieve statistical significance. When you hear of such a result, always check to make sure the study was not based on a small number of individuals. If it was, remember that with a small sample, it takes a very strong relationship for it to earn the title of "statistical significance."

EXAMPLE 2 REVISITED

Let's examine in more detail the evidence in Example 2 that was presented to the Supreme Court to see if we can rule out chance as an explanation for the higher percentage of male drivers who had been drinking. In Figure 12.2, we present the results of asking the Minitab program to compute the chi-squared statistic. (Boldface is still used to show what we typed, but now we have shown the minimum required rather than spelling out extra verbiage.) Notice that the summary statistic is only 1.637—which is not large enough to find a statistically significant difference in percentages of males and females who had been drinking. You can see why the Supreme Court was reluctant to conclude that the difference in the sample represented sufficient evidence for a real difference in the population. ∎

| TABLE 12.7 | Results of ESP Study | | | |

	Successful ESP Guess?			
	Yes	No	Total	% Success
Static picture	45	119	164	27
Dynamic picture	77	113	190	41
Total	122	232	354	34

SOURCE: Bem and Honorton, 1994.

CASE STUDY 12.1 # Extrasensory Perception Works Best with Movies

Extrasensory perception (ESP) is the apparent ability to obtain information in ways that exclude ordinary sensory channels. Early laboratory research studying ESP focused on having people try to guess at simple targets, such as symbols on cards, to see if the subjects could guess at a better rate than would be expected by chance. In recent years, experimenters have used more interesting targets, such as photographs, outdoor scenes, or short movie segments.

In a study of ESP reported by Bem and Honorton (January 1994), subjects (called "receivers") were asked to describe what another person (the "sender") had just watched on a television screen in another room. The receivers were shown four possible choices and asked to pick which one they thought had been viewed by the sender in the other room. Because the actual target was randomly selected from among the four choices, the guesses should have been successful by chance 25% of the time. Surprisingly, they were actually successful 34% of the time.

For this case study, we are going to examine a categorical variable that was involved and ask whether the results were affected by it. The researchers had hypothesized that moving pictures might be received with better success than ordinary photographs. To test that theory, they had the sender sometimes look at a single, "static" image on the television screen and sometimes look at a "dynamic" short video clip, played repeatedly. The additional three choices shown to the receiver for judging were always of the same type (static or dynamic) as the actual target, to eliminate biases due to a preference for one over the other. The question of interest was whether the success rate changed based on the type of picture. The results are shown in Table 12.7.

Figure 12.3 shows the results from the Minitab program. Notice that the chi-squared statistic is 6.675; this far exceeds the value of 3.84 required to declare the relationship statistically significant. Therefore, it does appear that success in ESP guessing depends on the type of picture used as the target. You can see that guesses for the static pictures were almost at chance (27% compared to 25% expected by chance), whereas the guesses for the dynamic videos far exceeded what was expected by chance (41% compared to 25%). ∎

FIGURE 12.3
*Minitab results for
Case Study 12.1*

```
Expected counts are printed below observed counts

                C1              C2          Total
1               45              119          164
                56.52           107.48

2               77              113          190
                65.48           124.52

Total           122             232          354

ChiSq = 2.348 + 1.235 +
        2.027 + 1.066 = 6.675
```

12.3 RELATIVE RISK, INCREASED RISK, AND ODDS

Various measures are used to report the chances of a particular outcome and how the chances increase or decrease with changes in an explanatory variable. Here are some quotes that use different measures to report chance:

- What they found was that women who smoked had a risk [of getting lung cancer] 27.9 times as great as nonsmoking women; in contrast, the risk for men who smoked regularly was only 9.6 times greater than that for male nonsmokers. (Taubes, 26 November 1993, p. 1375)

- Clinically depressed people are at a 50 percent greater risk of killing themselves. (*Newsweek,* 18 April 1994, p. 48)

- On average, the odds against a high school player playing NCAA football are 25 to 1. But even if he's made his college team, his odds are a slim 30 to 1 against being chosen in the NFL draft. (Krantz, 1992, p. 107)

Risk, Probability, and Odds

There are just two basic ways to express the chances that a randomly selected individual will fall into a particular category for a categorical variable. The first of the two methods involves expressing one category as a **proportion** of the total, the other involves comparing one category to another in the form of relative **odds.**

Suppose a population contains 1000 individuals, of which 400 carry the gene for a disease. The following are all equivalent ways to express this proportion:

Forty *percent* (40%) of all individuals carry the gene.

The *proportion* who carry the gene is 0.40.

The *probability* that someone carries the gene is .40.

The *risk* of carrying the gene is 0.40.

However, to express this in odds requires a different calculation. The equivalent statement represented in odds would be:

The *odds* of carrying the gene are 4 to 6 (or 2 to 3, or 2/3 to 1).

The odds are usually expressed by reducing the numbers with and without the trait to the smallest whole numbers possible. Thus, we would say that the odds are 2 to 3, rather than saying they are 2/3 to 1. Both formulations would be correct.

The general forms of these expressions are as follows:

Percentage with the trait = (number with trait/total) × 100%

Proportion with the trait = number with trait/total

Probability of having the trait = number with trait/total

Risk of having the trait = number with trait/total

Odds of having the trait = (number with trait/number without trait) to 1

Calculating the odds from the proportion and vice versa is a simple operation. If p is the proportion who have the trait, then the odds of having it are $p/(1 - p)$ to 1. If the odds of having the trait are a to b, then the proportion who have it is $a/(a + b)$. For example, if the proportion carrying a certain gene is 0.4, then the odds of having it are (0.4/0.6) to 1, or 2/3 to 1, or 2 to 3. Going in the other direction, if the odds of having it are 2 to 3, then the proportion who have it is $2/(2 + 3) = 2/5 = 4/10 = 0.40$.

Relative Risk

The **relative risk** of an outcome for two categories of an explanatory variable is simply the ratio of the risks for each category. The relative risk is often expressed as a multiple. For example, a relative risk of 3 may be reported by saying that the risk of developing a disease for one group is three times what it is for another group. Notice that a relative risk of 1 would mean that the risk is the same for both categories of the explanatory variable.

....

EXAMPLE 4   **RELATIVE RISK OF DEVELOPING BREAST CANCER**

Pagano and Gauvreau (1993, p. 133) reported data for women participating in the first National Health and Nutrition Examination Survey (Carter, Jones, Schatzkin, and Brinton, January–February 1989). The explanatory variable was whether the age at which a woman gave birth to her first child was 25 or older, and the outcome variable was whether she developed breast cancer (see Table 12.8).

TABLE 12.8	Age at Birth of First Child and Breast Cancer			
First Child at Age 25 or Older?	Breast Cancer	No Breast Cancer	Total	
Yes	31	1597	1628	
No	65	4475	4540	
Total	96	6072	6168	

SOURCE: Pagano and Gauvreau (1993).

To compute the relative risk of developing breast cancer based on whether the age at which a woman had her first child was 25 or older, we first find the risk of breast cancer for each group:

- Risk for women having first child at age 25 or older = 31/1628 = 0.0190
- Risk for women having first child before age 25 = 65/4540 = 0.0143
- Relative risk = 0.0190/0.0143 = 1.33

We can also represent this by saying that the risk of developing breast cancer is 1.33 times greater for women who had their first child at age 25 or older. By the way, the chi-squared statistic for the data in this contingency table is only 1.75. Because it's less than 3.84, we cannot conclude that the increased risk in the sample would hold for the population. The relationship is not statistically significant—it could simply be due to the luck of the draw for this sample. The relative risk in the population may be 1.0, meaning that both groups are at equal risk for developing breast cancer. ■

Notice the direction for which the relative risk was calculated by putting the group with lower risk in the denominator. This is common practice because it's easier for most people to interpret the results in that direction. For the current example, the relative risk in the other direction would be 0.75. In other words, the risk of developing breast cancer is 0.75 as much for women who have had their first child before age 25 as it is for women who have not. You can see that this relative risk statistic is more difficult to read than the relative risk of 1.33 presented in the example.

Increased Risk

Sometimes the increase in risk is presented as a percentage instead of a multiple. The percent increase in risk is calculated as follows:

increased risk = (change in risk/original risk) × 100%

An equivalent way to compute the increased risk is (relative risk − 1.0) × 100%.

INCREASED RISK OF BREAST CANCER

The change in risk of breast cancer for women who have not had their first child before age 25 compared with those who have is (0.0190 – 0.0143) = 0.0047. Because the baseline risk for those who have had a child before age 25 is 0.0143, this change represents an increase of (0.0047/0.0143) = 0.329 = 32.9%, or about 33%. The increased risk would be reported by saying that there is a 33% increase in the chances of breast cancer for women who have not had a child before the age of 25. Notice that this is also (relative risk – 1.0) × 100%. ∎

Odds Ratio

Epidemiologists, who study the causes and progression of diseases and other health risks, often represent comparative risks using the *odds ratio* instead of the relative risk. If the risk of a disease is small, these two measures will be about the same. The relative risk is easier to understand, but the odds ratio is easier to work with statistically. Therefore, you will often find the odds ratio reported in journal articles about health-related issues.

To compute the **odds ratio,** you first compute the odds of getting the disease to not getting the disease for each of the two categories of the explanatory variable. You then take the ratio of those odds. Let's compute it for the example concerning the risk of breast cancer.

- Odds for women having first child at age 25 or older = 31/1597 = 0.0194
- Odds for women having first child before age 25 = 65/4475 = 0.0145
- Odds ratio = 0.0194/0.0145 = 1.34

You can see that the odds ratio of 1.34 is very similar to the relative risk of 1.33. As we have noted, this will be the case as long as the risk of disease in each category is small.

There is a simpler way to compute the odds ratio, but if you were to simply see the formula you might not understand that it was a ratio of the two odds. The formula proceeds as follows:

1. Multiply the two numbers in the upper left and lower right cells of the table.
2. Divide the result by the numbers in the upper right and lower left cells of the table.

For the example we have been studying (Table 12.8), the computation would be as follows:

$$\text{odds ratio} = \frac{31 \times 4475}{1597 \times 65} = 1.34$$

Depending on how your table is constructed, you might have to reverse the numerator and denominator. As with relative risk, it is conventional to construct the odds ratio so that it is greater than 1. The only difference is which category of the explanatory variable gets counted as the numerator of the ratio and which gets counted as the denominator.

12.4   MISLEADING STATISTICS ABOUT RISK

You can be misled in a number of ways by statistics presenting risks. Unfortunately, statistics are often presented in the way that produces the best story rather than in the way that is most informative. Often, you cannot derive the information you need from news reports.

Common ways the media misrepresent statistics about risk:

1. The baseline risk is missing.

2. The time period of the risk is not identified.

3. The reported risk is not necessarily your risk.

Missing Baseline Risk

A study appeared on the front page of the *Sacramento Bee* on March 8, 1984, with the headline, "Evidence of new cancer-beer connection" (p. Al). The article reported that men who drank 500 ounces or more of beer a month (about 16 ounces a day) were *three times more likely* to develop cancer of the rectum than nondrinkers. If you were a beer drinker reading about this study, would it encourage you to reduce your beer consumption?

Although a relative risk of *three times* sounds ominous, it is not much help in making lifestyle decisions without also having information about what the risk is *without* drinking beer. If a threefold risk increase means that your chances of developing this cancer go from 1 in 100,000 to 3 in 100,000, you are much less likely to be concerned than if it means your risk jumps from 1 in 10 to 3 in 10. When a study reports relative risk, it should always give you a baseline risk as well.

In fairness, this article did report an estimate from the American Cancer Society that there are about 40,000 new cases of rectal cancer in the United States each year. Further, it gave enough information so that one could derive the fact that there were about 3600 non-beer drinkers in the study and 20 of them developed this cancer, for a baseline risk of about 0.0056 or about 1 in 180. Therefore, we can surmise that among those who drank more than 500 ounces of beer a month, the risk was about 3 in 180, or 1 in 60. Remember that because these results were based on an observational study, we cannot conclude that drinking beer actually caused the greater observed risk. For instance, there may be confounding dietary factors.

Risk over What Time Period?

"Italian scientists report that a diet rich in animal protein and fat—cheeseburgers, french fries, and ice cream, for example—increases a woman's risk of breast cancer threefold," according to *Prevention Magazine's Giant Book of Health Facts* (1991, p. 122). Couple this with the fact that the American Cancer Society estimates that 1 in 9 women in the United States will get breast cancer. Does that mean that if a woman eats a diet rich in animal protein and fat, her chances of developing breast cancer are 1 in 3?

There are two problems with this line of reasoning. First, the statement attributed to the Italian scientists was woefully incomplete. It did not specify anything about how the study was conducted. It also did not specify the ages of the women studied or what the baseline rate of breast cancer was for the study. Why would we need to know the baseline rate for the study when we already know that 1 in 9 women will develop breast cancer? The answer is that age is a critical factor. The baseline rate of 1 in 9 is a lifetime risk, at least to age 85. As with most diseases, accumulated risk increases with age.

According to the *University of California at Berkeley Wellness Letter* (July 1992, p. 1), the lifetime risk of a woman developing breast cancer by certain ages is

by age 50: 1 in 50

by age 60: 1 in 23

by age 70: 1 in 13

by age 80: 1 in 10

by age 85: 1 in 9

The annual risk of developing breast cancer is only about 1 in 3700 for women in their early 30s but is 1 in 235 for women in their early 70s (Fletcher, Black, Harris, Rimer, and Shapiro, 20 October 1993, p. 1644). If the Italian study had been done on very young women, the threefold increase in risk could represent a small increase. Unfortunately, *Prevention Magazine's Giant Book of Health Facts* did not give even enough information to lead us to the original report of the work. Therefore, it is impossible to intelligently evaluate the claim.

Reported Risk versus Your Risk

The headline was enough to make you want to go out and buy a new car: "Older cars stolen more often than new ones" [*Davis* (CA) *Enterprise,* 15 April 1994, p. C3]. The article reported that "among the 20 most popular auto models stolen [in California] last year, 17 were at least 10 years old."

Suppose you own two cars; one is 15 years old and the other is new. You park them both on the street outside of your home. Are you at greater risk of having the old one stolen? Perhaps, but the information quoted in the article gives you no information about that question.

Numerous factors determine which cars are stolen. We can easily speculate that many of those factors are strongly related to the age of cars as well. Certain neighborhoods are more likely to be targeted than others, and those same neighborhoods are more likely to have older cars parked in them. Cars parked in locked garages are less likely to be stolen and are more likely to be newer cars. Cars with easily opened doors are more likely to be stolen and more likely to be old. Cars that are not locked and/or don't have alarm systems are more likely to be stolen and are more likely to be old. Cars with high value for used parts are more likely to be stolen and are more likely to be old, discontinued models.

You can see that the real question of interest to a consumer is, "If I were to buy a new car, would my chances of having it stolen increase or decrease over those of the car I own now?" That question can't be answered based only on information about which cars have been stolen most often. Simply too many variables are related to both the age of the car and its risk of being stolen.

12.5   SIMPSON'S PARADOX: THE MISSING THIRD VARIABLE

In Chapter 11, we saw an example where omitting a third variable masked the positive correlation between number of pages and price of books. A similar phenomenon can happen with categorical variables, and it goes by the name of **Simpson's Paradox.** It is a paradox because the relationship appears to be in one direction if the third variable is not considered and in the other direction if it is.

EXAMPLE 6   **SIMPSON'S PARADOX**

We illustrate Simpson's Paradox with a hypothetical example of a new treatment for a disease. Suppose two hospitals are willing to participate in an experiment to test the new treatment. Hospital A is a major research facility, famous for its treatment of advanced cases of the disease. Hospital B is a local area hospital in an urban area.

Both hospitals agree to include 1100 patients in the study. Because the researchers conducting the experiment are on the staff of Hospital A, they decide to perform the majority of cases with the new procedure in-house. They randomly assign 1000 patients to the new treatment, with the remaining 100 receiving the standard treatment. Hospital B, which is a bit reluctant to try something new on too many patients, agrees to randomly assign 100 patients to the new treatment, leaving 1000 to receive the standard. The survival rates are shown in Table 12.9.

Table 12.10 shows how the new procedure worked compared with the standard. It looks as though the new treatment is a success. The risk of dying from the standard procedure is higher than that for the new procedure in both hospitals. In fact, the risk of dying when given the standard treatment is an overwhelming ten times higher than it is for the new treatment in Hospital B.

TABLE 12.9   Survival Rates for Standard
and New Treatments at Two Hospitals

	Hospital A			Hospital B		
	Survive	Die	Total	Survive	Die	Total
Standard	5	95	100	500	500	1000
New	100	900	1000	95	5	100
Total	105	995	1100	595	505	1100

TABLE 12.10   Risk Compared for Standard
and New Treatments

	Hospital A	Hospital B
Risk of dying with the standard treatment	95/100 = 0.95	500/1000 = 0.50
Risk of dying with the new treatment	900/1000 = 0.90	5/100 = 0.05
Relative risk	0.95/0.90 = 1.06	0.50/0.05 = 10.0

TABLE 12.11   Estimating the Overall Reduction in Risk

	Survive	Die	Total	Risk of Death
Standard	505	595	1100	595/1100 = 0.54
New	195	905	1100	905/1100 = 0.82
Total	700	1500	2200	

The researchers would now like to estimate the overall reduction in risk for the new treatment, so they combine all of the data (Table 12.11).

What has gone wrong? It now looks as though the standard treatment is superior to the new one. In fact, the relative risk, taken in the same direction as before, is 0.54/0.82 = 0.66. The death rate for the standard treatment is only 66% of what it is for the new treatment.

The problem is that the more serious cases of the disease presumably were treated by the famous research hospital, Hospital A. Because they were more serious cases, they were more likely to die. But because they went to Hospital A, they were also more likely to receive the new treatment. When the results from both hospitals are combined, we lose the information that the patients in Hospital A had both a higher overall death rate *and* a higher likelihood of receiving the new treatment.

Simpson's Paradox makes it clear that it is dangerous to summarize information over groups, especially if patients (or experimental units) were not randomized into the groups. Notice that if patients had been randomly assigned to the two hospitals, this phenomenon probably would not have occurred. It would have been unethical, however, to do such a random assignment. ■

If someone else has already summarized data for you by collapsing three variables into two, you cannot retrieve the information to see whether Simpson's Paradox has occurred. Common sense should help you detect this problem in some cases. When you read about a relationship between two categorical variables, try to find out if the data have been collapsed over a third variable. If so, think about whether separating results for the different categories of the third variable could change the direction of the relationship between the first two. Exercise 14 at the end of this chapter presents an example of this.

CASE STUDY 12.2   # Assessing Discrimination in Hiring and Firing

The term *relative risk* is obviously not applicable to all types of data. It was developed for use with medical data where risk of disease or injury are of concern. An equivalent measure used in discussions of employment is the **selection ratio,** which is the ratio of the proportion of successful applicants for a job from one group (sex, race, and so on) compared with another group. For example, suppose a company hires 10% of the men who apply and 15% of the women. Then the selection ratio for women compared with men is 15/10 = 1.50. Comparing this with our discussion of relative risk, it says that women are 1.5 times as likely to be hired as men. The ratio is often used in the reverse direction when arguing that discrimination has occurred. For instance, in this case it might be argued that men are only 10/15 = 0.67 times as likely to be hired as women.

Gastwirth (1988, p. 209) explains that government agencies in the United States have set a standard for determining whether there is potential discrimination in practices used for hiring. "If the minority pass (hire) rate is *less* than four-fifths (or 0.8) of the majority rate, then the practice is said to have a disparate or disproportionate impact on the minority group, and employers are required to justify its job relevance." In the case where 10% of men and 15% of women who apply are hired, the men would be the minority group. The selection ratio of men to women would be 10/15 = 0.67, so the hiring practice could be examined for potential discrimination.

Unfortunately, as Gastwirth and Greenhouse (1995) argue, this rule may not be as clear as it needs to be. They present a court case contesting the fairness of layoffs by the U.S. Labor Department, in which both sides in the case tried to interpret the rule in their favor. The layoffs were concentrated in Labor Department offices in the Chicago area, and the numbers are shown in Table 12.12.

....

TABLE 12.12   Layoffs by Ethnic Group
for Labor Department Employees

| | Laid Off? | | | |
Ethnic Group	Yes	No	Total	% Laid Off
African American	130	1382	1512	8.6
White	87	2813	2900	3.0
Total	217	4195	4412	

DATA SOURCE: Gastwirth and Greenhouse, 1995.

If we consider the selection ratio based on people who were laid off, it should be clear that the four-fifths rule was clearly violated. The percentage of African Americans who were laid off compared to the percentage of whites who were laid off is 3/8.6 = 0.35, clearly less than the 0.80 required for fairness. However, the defense argued that the selection ratio should have been computed using those who were *retained* rather than those who were laid off. Because 91.4% of African Americans and 97% of whites were retained, the selection ratio is 91.4/97 = 0.94, well above the ratio of 0.80 required to be within acceptable practice. As for which claim was supported by the court, Gastwirth and Greenhouse (1995) report: "The lower court accepted the defendant's claim [using the selection ratio for those retained] but the appellate opinion, by Judge Cudahy, remanded the case for reconsideration." The judge also asked for further statistical information, to rule out chance as an explanation for the difference.

The value of the chi-squared statistic for this case will be given as an exercise (Exercise 18). However, notice that we must be careful in its interpretation. The people in this study are not a random sample from a larger population; they are the only employees of concern. There are other, more appropriate ways to test whether the results could have been due to chance, but the chi-squared statistic is still a useful measure of the strength of the relationship between ethnicity and work status.

Gastwirth and Greenhouse point out that this discrepancy could have been avoided if the *odds ratio* had been used instead of the selection ratio. The **odds ratio** compares the odds of being laid off to the odds of being retained for each group. Therefore, the plaintiffs and defendants could not manipulate the statistics to get two different answers. The odds ratio for this example can be computed using the simple formula

$$\text{odds ratio} = \frac{130 \times 2813}{1382 \times 87} = 3.04$$

This number tells us that the odds of being laid off compared with being retained are three times higher for African Americans than for whites. Equivalently, the odds ratio in the other direction is 1/3.04, or about 0.33. It is this figure that should be assessed using the four-fifths rule. Gastwirth and Greenhouse argue that "should courts accept the odds ratio measure, they might wish to change the 80% rule to about 67% or 70% since some cases that we studied and classified as close, had ORs [odds ratios] in that neighborhood" (1995).

Finally, Gastwirth and Greenhouse argue that the selection ratio may still be appropriate in some cases. For example, some employment practices require applicants to meet certain requirements (such as having a high school diploma) before they can be considered for a job. If 98% of the majority group meets the criteria but only 96% of the minority group does, then the odds of meeting the criteria versus not meeting it would only be about half as high for the minority as for the majority. But the selection ratio would be 96/98 = 98%, which should certainly be legally acceptable. As always, statistics must be combined with common sense to be useful. ■

FOR THOSE WHO LIKE FORMULAS

To represent the *observed numbers* in a 2×2 contingency table, we use the notation:

	Variable 2		
Variable 1	Yes	No	Total
Yes	a	b	$a + b$
No	c	d	$c + d$
Total	$a + c$	$b + d$	n

Therefore, the *expected* numbers are computed as follows:

	Variable 2		
Variable 1	Yes	No	Total
Yes	$(a + b)(a + c)/n$	$(a + b)(b + d)/n$	$a + b$
No	$(c + d)(a + c)/n$	$(c + d)(b + d)/n$	$c + d$
Total	$a + c$	$b + d$	n

Computing the Chi-Squared Statistic, χ^2, for an $r \times c$ Contingency Table

Let O_i = observed number in cell i, and E_i = expected number in cell i. Then:

$$\chi^2 = \sum_{i=1}^{rc} \frac{(O_i - E_i)^2}{E_i}$$

Relative Risk and Odds Ratio

Using the notation for the observed numbers, if variable 1 is the explanatory variable and variable 2 is the response variable, then we can compute

$$\text{relative risk} = \frac{a(c + d)}{c(a + b)}$$

$$\text{odds ratio} = \frac{ad}{bc}$$

EXERCISES

1. Suppose a study on the relationship between gender and political party included 200 men and 200 women and found 180 Democrats and 220 Republicans. Is that information sufficient for you to construct a contingency table for the study? If so, construct the table. If not, explain why not.

2. According to the *World Almanac and Book of Facts* (1995, p. 964), the rate of deaths by drowning in the United States in 1993 was 1.6 per 100,000 population. Express this statistic as a percentage of the population; then explain why it is better expressed as a rate than as a percentage.

3. According to the *University of California at Berkeley Wellness Letter* (February 1994, p. 1), only 40% of all surgical operations require an overnight stay at a hospital. Rewrite this fact as a proportion, as a risk, and as the odds of an overnight stay. In each case, express the result as a full sentence.

4. *Science News* (25 February 1995, p. 124) reported a study of 232 people, aged 55 or over, who had heart surgery. The patients were asked whether their religious beliefs give them feelings of strength and comfort and whether they regularly participate in social activities. Of those who said yes to both, about 1 in 50 died within 6 months after their operation. Of those who said no to both, about 1 in 5 died within 6 months after their operation. What is the relative risk of death (within 6 months) for the two groups? Write your answer in a sentence or two that would be understood by someone with no training in statistics.

5. Raloff (1995) reported on a study conducted by Dimitrios Trichopolous of the Harvard School of Public Health in which researchers "compared the diets of 820 Greek women with breast cancer and 1548 others admitted to Athens-area hospitals for reasons other than cancer." One of the results had to do with consumption of olive oil, a staple in many Greek diets. The article reported that "women who eat olive oil only once a day face a 25 percent higher risk of breast cancer than women who consume it twice or more daily."

   a. The increased risk of breast cancer for those who consume olive oil only once a day is 25%. What is the relative risk of breast cancer for those who consume olive oil only once a day, compared to those who eat it twice or more?

   b. What information is missing from this article that would help individuals assess the importance of the result in their own lives?

6. The headline in an article in the *Sacramento Bee* read "Firing someone? Risk of heart attack doubles"(Haney, 1998). The article explained that "Between 1989 and 1994, doctors interviewed 791 working people who had just undergone heart attacks about what they had done recently. The researchers concluded that firing someone or having a high-stakes deadline doubled the usual risk of a heart attack during the following week. . . . For a healthy 50-year-old man or a healthy 60-year-old woman, the risk of a heart attack in any given hour without any trigger is about 1 in a million."

   a. Refer to Chapter 5. What type of study is this?

TABLE 12.13

	Reportedly Has Seen a Ghost		
	Yes	No	Total
Aged 18 to 29	212	1313	1525
Aged 30 or over	465	3912	4377
Total	677	5225	5902

DATA SOURCE: The Roper Organization, 1992, p. 35.

b. Refer to the reasons for relationships listed in Section 11.3. Which do you think is the most likely explanation for the relationship found between firing someone and having a heart attack? Do you think the headline for this article was appropriate? Explain.

c. Assuming the relationship is indeed as stated in the article, write sentences that could be understood by someone with no training in statistics, giving each of the following for this example:
   i. The odds ratio
   ii. Increased risk
   iii. Relative risk

7. Refer to the ESP experiments described in Case Study 12.1. By chance alone, 25% of the guesses would be expected to be successful. In the experiment, 34% of the guesses were successful. In both parts a and b, express your answer in a full sentence.

a. What are the *odds* of a successful guess by chance alone?

b. What were the *odds* of a successful guess in the experiment?

8. A newspaper story released by the Associated Press noted that "a study by the Bureau of Justice Statistics shows that a motorist has about the same chance of being a carjacking victim as being killed in a traffic accident, 1 in 5000" [*Davis* (CA) *Enterprise,* 3 April 1994, p. A9]. Discuss this statement with regard to your *own chances* of each event happening to you.

9. The Roper Organization (1992) conducted a study as part of a larger survey to ascertain the number of American adults who had experienced phenomena such as seeing a ghost, "feeling as if you left your body," and seeing a UFO. A representative sample of adults (18 and over) in the continental United States were interviewed in their homes during July, August, and September 1991. The results when respondents were asked about seeing a ghost are shown in Table 12.13.

a. Find numbers for each of the following:
   i. The percentage of the younger group who reported seeing a ghost
   ii. The proportion of the older group who reported seeing a ghost
   iii. The risk of reportedly seeing a ghost in the younger group

FIGURE 12.4
*Minitab results for
Exercise 9*

```
Expected counts are printed below observed counts

              Yes              No            Total
1             212             1313            1525
              174.93          1350.07

2             465             3912            4377
              502.07          3874.93

Total         677             5225            5902

ChiSq = 7.857 + 1.018 +
        2.737 + 0.355 = 11.967
```

    iv. The odds of reportedly seeing a ghost to not seeing one in the older group

b. What is the *relative risk* of reportedly seeing a ghost for one group compared to the other? Write your answer in the form of a sentence that could be understood by someone who knows nothing about statistics.

c. Repeat part b using *increased risk* instead of relative risk.

d. Results from asking Minitab to compute the chi-squared statistic are shown in Figure 12.4. What can you conclude about the relationship between age group and reportedly seeing a ghost?

10. Using the terminology of this chapter, what name (for example, odds, risk, relative risk) applies to each of the boldface numbers in the following quotes?

a. "Fontham found increased risks of lung cancer with increasing exposure to secondhand smoke, whether it took place at home, at work, or in a social setting. A spouse's smoking alone produced an overall **30** percent increase in lung-cancer risk" (*Consumer Reports,* January 1995, p. 28).

b. "What they found was that women who smoked had a risk [of getting lung cancer] **27.9** times as great as nonsmoking women; in contrast, the risk for men who smoked regularly was only **9.6** times greater than that for male nonsmokers" (Taubes, 26 November 1993, p. 1375).

c. "**One student in five** reports abandoning safe-sex practices when drunk" (*Newsweek,* 19 December 1994, p. 73).

11. A statement quoted in this chapter was, "Clinically depressed people are at a 50 percent greater risk of killing themselves" (*Newsweek,* 18 April 1994, p. 48). This means that when comparing people who are clinically depressed to those who are not, the former have an *increased risk* of killing themselves of 50%. What is the *relative risk* of suicide for those who are clinically depressed compared with those who are not?

TABLE 12.14

	Program A			Program B		
	Admit	Deny	Total	Admit	Deny	Total
Men	400	250	650	50	300	350
Women	50	25	75	125	300	425
Total	450	275	725	175	600	775

12. According to *Consumer Reports* (1995 January, p. 29), "Among nonsmokers who are exposed to their spouses' smoke, the chance of death from heart disease increases by about 30%." Rewrite this statement in terms of relative risk, using language that would be understood by someone who does not know anything about statistics.

13. Reporting on a study of drinking and drug use among college students in the United States, a *Newsweek* reporter wrote:

Why should college students be so impervious to the lesson of the morning after? Efforts to discourage them from using drugs actually did work. The proportion of college students who smoked marijuana at least once in 30 days went from one in three in 1980 to one in seven last year [1993]; cocaine users dropped from 7 percent to 0.7 percent over the same period. (19 December 1994, p. 72)

a. What was the relative risk of cocaine use for college students in 1980 compared with college students in 1993? Write your answer as a statement that could be understood by someone who does not know anything about statistics.

b. Are the figures given for marijuana use (for example, "one in three") presented as proportions or as odds? Whichever they are, rewrite them as the other.

c. Do you agree with the statement that "efforts to discourage them from using drugs actually did work"? Explain your reasoning.

14. A well-known example of Simpson's Paradox, published by Bickel, Hammel, and O'Connell (1975), examined admission rates for men and women who had applied to graduate programs at the University of California at Berkeley. The actual breakdown of data for specific programs is confidential, but the point can be made with similar, hypothetical numbers. For simplicity, we will assume there are only two graduate programs. The figures for acceptance to each program are shown in Table 12.14.

a. Combine the data for the two programs into one aggregate table. What percentage of all men who applied were admitted? What percentage of all women who applied were admitted? Which sex was more successful?

....

TABLE 12.15

| | Helped Pick Up Pencils? | | |
	Yes	No	Total
Male observer	370	950	1320
Female observer	300	1003	1303
Total	670	1953	2623

DATA SOURCE: Howell, 1992, p. 154, from a study by Latané and Dabbs, 1975.

b. What percentage of the men who applied did Program A admit? What percentage of the women who applied did Program A admit? Repeat the question for Program B. Which sex was more successful in getting admitted to Program A? Program B?

c. Explain how this problem is an example of Simpson's Paradox. Provide a potential explanation for the observed figures by guessing what type of programs A and B might have been.

15. A case-control study in Berlin, reported by Kohlmeier, Arminger, Bartolomeycik, Bellach, Rehm, and Thamm (1992) and by Hand et al. (1994), asked 239 lung cancer patients and 429 controls (matched to the cases by age and sex) whether they had kept a pet bird during adulthood. Of the 239 lung cancer cases, 98 said yes. Of the 429 controls, 101 said yes.

   a. Construct a contingency table for the data.

   b. Compute the risk of lung cancer for bird and nonbird owners for this study.

   c. Can the risks of lung cancer for the two groups, computed in part b, be used as baseline risks for the populations of bird and nonbird owners? Explain.

   d. How much more likely is lung cancer for bird owners than for nonbird owners in this study; that is, what is the increased risk?

   e. What information about risk would you want, in addition to the information on increased risk in part d of this problem, before you made a decision about whether to own a pet bird?

16. Compute the chi-squared statistic and assess the statistical significance for the relationship between bird ownership and lung cancer, based on the data in Exercise 15.

17. Howell (1992, p. 153) reports on a study by Latané and Dabbs (1975) in which a researcher entered an elevator and dropped a handful of pencils, with the appearance that it was an accident. The question was whether the males or females who observed this mishap would be more likely to help pick up the pencils. The results are shown in Table 12.15.

   a. Compute and compare the proportions of males and females who helped pick up the pencils.

TABLE 12.16

	Laid Off?		
Ethnic Group	Yes	No	Total
African American	130	1382	1512
White	87	2813	2900
Total	217	4195	4412

b. Compute the chi-squared statistic and use it to determine whether there is a statistically significant relationship between the two variables in the table. Explain your result in a way that could be understood by someone who knows nothing about statistics.

c. Would the conclusion in part b have been the same if only 262 people had been observed but the pattern of results was the same? Explain how you reached your answer and what it implies about research of this type.

18. The data in Table 12.16 are reproduced from Case Study 12.2 and represent employees laid off by the U.S. Department of Labor.

a. Compute the odds of being retained to being laid off for each ethnic group.

b. Use your results in part a to compute the odds ratio and confirm that it is about 3.0, as computed in Case Study 12.2 (where the shortcut method was used).

c. Minitab computed the chi-squared statistic as 66.595. Explain what this means about the relationship between the two variables. Include an explanation that could be understood by someone with no knowledge of statistics. Make the assumption that these employees are representative of a larger population of employees.

19. Kohler (1994, p. 427) reports data on the approval rates and ethnicity for mortgage applicants in Los Angeles in 1990. Of the 4096 African American applicants, 3117 were approved. Of the 84,947 white applicants, 71,950 were approved.

a. Construct a contingency table for the data.

b. Compute the proportion of each ethnic group that was approved for a mortgage.

c. Compute the ratio of the two proportions you found in part b. Would that ratio be more appropriately called a relative risk or a selection ratio? Explain.

d. Would the data pass the four-fifths rule used in employment and described in Case Study 12.2? Explain.

20. The chi-squared statistic for the data in Exercise 19 is about 220, so the differences observed in the approval rates are clearly statistically significant. Now suppose that a random sample of 890 applicants had been examined, a sample size 100 times smaller than the one reported. Further, suppose the pattern of results had been almost identical, resulting in 40 African American applicants with 30 approved, and 850 white applicants with 720 approved.

a. Construct a contingency table for these numbers.

b. Compute the chi-squared statistic for the table.

c. Make a conclusion based on your result in part b and compare it with the conclusion that would have been made using the full data set. Explain any discrepancies and discuss their implications for this type of problem.

MINI-PROJECTS

1. Carefully collect data cross-classified by two categorical variables for which you are interested in determining whether there is a relationship. Do not get the data from a book or journal; collect it yourself. Be sure to get counts of at least five in each cell and be sure the individuals you use are not related to each other in ways that would influence their data.

a. Create a contingency table for the data.

b. Compute and discuss the risks and relative risks. Are those terms appropriate for your situation? Explain.

c. Determine whether there is a statistically significant relationship between the two variables.

d. Discuss the role of sample size in making the determination in part c.

e. Write a summary of your findings.

2. Find a news story that discusses a study showing increased (or decreased) risk of one variable based on another. Write a report evaluating the information given in the article and discussing what conclusions you would reach based on the information in the article. Discuss whether any of the features in Section 12.4, "Misleading Statistics about Risk," apply to the situation.

REFERENCES

Baird, D. D., and A. J. Wilcox. (1985). Cigarette smoking associated with delayed conception. *Journal of the American Medical Association* 253, pp. 2979–2983.

Bem, D., and C. Honorton. (January 1994). Does psi exist? Replicable evidence for an anomalous process of information transfer. *Psychological Bulletin* 115, no. 1, pp. 4–18.

Bickel, P.J., E. A. Hammel, and J. W. O'Connell. (1975). Sex bias in graduate admissions: Data from Berkeley. *Science* 187, pp. 298–304.

Carter, C. L., D. Y. Jones, A. Schatzkin, and L. A. Brinton. (January–February 1989). A prospective study of reproductive, familial, and socioeconomic risk factors for breast cancer using NHANES I data. *Public Health Reports* 104, pp. 45–49.

Fletcher, S. W., B. Black, R. Harris, B. K. Rimer, and S. Shapiro. (20 October 1993). Report of the international workshop on screening for breast cancer. *Journal of the National Cancer Institute* 85, no. 20, pp. 1644–1656.

Gastwirth, J. L. (1988). *Statistical reasoning in law and public policy.* New York: Academic Press.

Gastwirth, J. L., and S. W. Greenhouse. (1995). Biostatistical concepts and methods in the legal setting. *Statistics in Medicine* 14, no. 15, pp. 1641–1653.

Hand, D. J., F. Daly, A. D. Lunn, K. J. McConway, and E. Ostrowski. (1994). *A handbook of small data sets.* London: Chapman and Hall.

Haney, Daniel Q. (20 March 1998). Firing someone? Risk of heart attack doubles. *Sacramento Bee,* p. E1-2.

Howell, D. C. (1992). *Statistical methods for psychology.* 3d ed. Belmont, CA: Duxbury Press.

Kohler, H. (1994). *Statistics for business and economics.* 3d ed. New York: Harper-Collins College.

Kohlmeier, L., G. Arminger, S. Bartolomeycik, B. Bellach, J. Rehm, and M. Thamm. (1992). Pet birds as an independent risk factor for lung cancer: Case-control study. *British Medical Journal* 305, pp. 986–989.

Krantz, L. (1992). *What the odds are.* New York: HarperPerennial.

Latané, B., and J. M. Dabbs, Jr. (1975). Sex, group size and helping in three cities. *Sociometry* 38, pp. 180–194.

Pagano M., and K. Gauvreau. (1993). *Principles of biostatistics.* Belmont, CA: Duxbury Press.

Prevention magazine's giant book of health facts. (1991). Edited by John Feltman. New York: Wings Books.

Raloff, J. (1995). Obesity, diet linked to deadly cancers. *Science News* 147, no. 3, p. 39.

The Roper Organization. (1992). *Unusual personal experiences: An analysis of the data from three national surveys.* Las Vegas: Bigelow Holding Corp.

Taubes, G. (26 November 1993). Claim of higher risk for women smokers attacked. *Science* 262, p. 1375.

Weiden, C. R., and B. C. Gladen. (1986). The beta-geometric distribution applied to comparative fecundability studies. *Biometrics* 42, pp. 547–560.

World almanac and book of facts. (1995). Edited by Robert Famighetti. Mahwah, NJ: Funk and Wagnalls.

CHAPTER THIRTEEN

Reading the Economic News

THOUGHT QUESTIONS

1. The U.S. government produces a composite index of leading economic indicators as well as one of coincident and lagging economic indicators. These are supposed to "indicate" the status of the economy. What do you think the terms *leading, coincident,* and *lagging* mean in this context?

2. Suppose you wanted to measure the yearly change in the "cost of living" for a college student living in the dorms for the past 4 years. How would you do it?

3. Suppose you were told that the price of a certain product, measured in 1984 dollars, has not risen. What do you think is meant by "measured in 1984 dollars"?

4. How do you think governments determine the "rate of inflation"?

13.1 COST OF LIVING: THE CONSUMER PRICE INDEX

Everyone is affected by inflation. When the costs of goods and services rise, most workers expect their employers to increase their salaries to compensate. But how do employers know what a fair salary adjustment would be?

The most common measure of change in the cost of living in the United States is the Consumer Price Index (CPI), produced by the Bureau of Labor Statistics (BLS). The CPI was initiated during World War I, a time of rapidly increasing prices, to help determine salary adjustments in the shipbuilding industry. As noted by the BLS, the CPI is not a true cost-of-living index:

> Both the CPI and a cost-of-living index would reflect changes in the prices of goods and services, such as food and clothing, that are directly purchased in the marketplace; but a complete cost-of-living index would go beyond this to also take into account changes in other governmental or environmental factors that affect consumers' well-being. (U.S. Dept. of Labor, 1998, p. 3)

Nonetheless, the CPI is the best available measure of changes in the cost of living in the United States.

The CPI measures changes in the cost of a "market basket" of goods and services that a typical consumer would be likely to purchase. The cost of that collection of goods and services is measured during a base period, then again at subsequent time periods. The CPI, at any given time period, is simply a comparison of current cost with cost during the base period. It is supposed to measure the changing cost of maintaining the same standard of living that existed during the base period.

There are actually two different consumer price indexes, but we will focus on the one that is most widely quoted, the CPI-U, for all urban consumers. It is estimated that this CPI covers about 87% of all U.S. consumers (U.S. Dept. of Labor, 1998, p. 3). The CPI-U was introduced in 1978. The other CPI, the CPI-W, is a continuation of the original one from the early 1900s. It is based on a subset of households covered by the CPI-U, for which "more than one-half of the household's income must come from clerical or wage occupations, and at least one of the household's earners must have been employed for at least 37 weeks during the previous 12 months (U.S. Dept. of Labor, 1998, p. 3)." About 32% of the U.S. population is covered by the CPI-W.

To understand how the CPI is calculated, let's first introduce the general concept of *price index numbers*. A price index for a given time period allows you to compare costs with another time period for which you also know the price index.

Price Index Numbers

A **price index number** measures prices (such as the cost of a loaf of bread) at one time period relative to another time period, usually as a percentage. For example, if a loaf of bread cost $1.00 in 1984 and $1.80 in 1998, then the bread price index would be ($1.80/$1.00) × 100 = 180%. In other words, bread in 1998 cost 180% of what it cost in 1984. We could also say the price increased by 80%.

Price index numbers are commonly computed on a collection of products instead of just one. For example, we could compute a price index reflecting the increasing cost of attending college. To define a price index number, decisions about the following three components are necessary:

1. The base year or time period

2. The list of goods and services to be included

3. How to weight the particular goods and services

The general formula for computing a price index number is

price index number = (current cost/base time period cost) × 100

where "cost" is the weighted cost of the listed goods and services. Weights are usually determined by the relative quantities of each item purchased during the base period.

EXAMPLE   **A COLLEGE INDEX NUMBER**

Suppose a senior graduating from college wanted to determine by how much the cost of attending college had increased for each of the four years she was a student. Here is how she might specify the three components:

1. Use her first year as a base.

2. Include yearly tuition and fees, yearly cost of room and board, and yearly average cost of books and related materials.

3. Weight everything equally because the typical student would be required to "buy" one of each category per year.

Here is an example of how the calculation would proceed. Note that we use the formula:

college index number = (current year total/first year total) × 100

Notice that the index for her senior year (listed in Table 13.1) is 121. This means that these components of a college education in her senior year cost 121% of what they cost in her first year. Equivalently, they have increased 21% since she started college. ■

The Components of the Consumer Price Index

The Base Year (or Years)

The base year (or years) for the CPI changes periodically, partly so that the index does not get ridiculously large. If the original base year of 1913 were still used, the

TABLE 13.1 Cost of Attending College

Year	Tuition	Room and Board	Books	Total	College Index
First	$3,000	$4,900	$700	$8,600	100
Sophomore	$3,200	$5,200	$720	$9,120	$(9,120/8,600) \times 100 = 106$
Junior	$3,500	$5,400	$750	$9,650	$(9,650/8,600) \times 100 = 112$
Senior	$4,000	$5,600	$800	$10,400	$(10,400/8,600) \times 100 = 121$

CPI would be well over 1000 and would be difficult to interpret. Since 1988, and continuing as of January 1999, the base period in use was the years 1982–1984. The previous base was the year 1967. Prior to that time, the base period was changed about once a decade. In December 1996, the Bureau of Labor Statistics announced that, beginning with the January 1999 CPI, the base period would change to 1993–1995. Since it made that announcement, however, the BLS has decided that it will retain the 1982–1984 base (U.S. Dept. of Labor, 15 June 1998).

The Goods and Services Included

As in the case with the base year(s), the market basket of goods and services is updated about once every 10 years. Items are added, deleted, and reorganized to represent current buying patterns. In 1998, a major revision occurred that included the addition of a new category called "Education and communication." The market basket now in use consists of over 200 types of goods and services. It was established primarily based on the 1993–1995 Consumer Expenditure Survey, in which a multi-stage sampling plan was used to select families who reported their expenditures. That expenditure information, from about 30,000 individuals and families, was then used to determine the items included in the index.

The market basket includes most things that would be routinely purchased. These are divided into eight major categories, each of which is subdivided into varying numbers of smaller categories. The eight major categories are

1. Food and beverages
2. Housing
3. Apparel
4. Transportation
5. Medical care
6. Recreation
7. Education and communication (the new category added in 1998)
8. Other goods and services

As noted, these categories are broken down into smaller ones. For example, here is the breakdown leading to the item "Ice cream and related products":

Food and beverages → Food at home → Dairy and related products
→ Ice cream and related products

Relative Quantities of Particular Goods and Services

Because consumers spend more on some items than on others, it makes sense to weight those items more heavily in the CPI. The weights assigned to each item are the relative quantities spent, as determined by the Consumer Expenditure Survey. The same weights are used on an ongoing basis and are updated occasionally, just as are the base year and the market basket. Here are the relative weights in effect in December 1997 for the eight categories, rounded to one decimal place:

1. Food and beverages   16.3%
2. Housing   39.6%
3. Apparel   4.9%
4. Transportation   17.6%
5. Medical care   5.6%
6. Recreation   6.1%
7. Education and communication 5.5%
8. Other goods and services   4.3%
                      Total   **100%**

You can see that housing is by far the most heavily weighted category. This makes sense, especially because costs associated with diverse items such as utilities and furnishings are included under the general heading of housing.

Obtaining the Data for the CPI

It is, of course, not possible to actually measure the average price of items paid by all families. The CPI is composed of samples taken from around the United States. Each month the sampling occurs at about 23,000 retail and service establishments in 87 urban areas, and prices are measured on about 80,000 items. Rents are measured from about 50,000 landlords and tenants.

Obviously, determining the Consumer Price Index and trying to keep it current represent a major investment of government time and money. We now examine ways in which the index is used, as well as some of its problems.

13.2   USES OF THE CONSUMER PRICE INDEX

Most Americans, whether they realize it or not, are affected by the Consumer Price Index. It is the most widely used measure of inflation in the United States.

Major Uses of the Consumer Price Index

There are four major uses of the Consumer Price Index:
1. The CPI is used to evaluate and determine economic policy.
2. The CPI is used to compare prices in different years.
3. The CPI is used to adjust other economic data for inflation.
4. The CPI is used to determine salary and price adjustments.

1. The CPI is used to evaluate and determine economic policy. As a measure of inflation, the CPI is of interest to the president, Congress, the Federal Reserve Board, private companies, and individuals. Government officials use it to evaluate how well current economic policies are working. Private companies and individuals also use it to make economic decisions.

2. The CPI is used to compare prices in different years. If you bought a new car in 1983 for $10,000, what would you have paid in 1991 for a similar quality car, using the CPI to adjust for inflation? The CPI in 1983 was very close to 100 (depending on the month), and the average CPI in 1991 was 136.2. Therefore, an equivalent price in 1991 would be about $10,000 × (136.2/100) = $13,620. The general formula for determining the comparable price for two different time periods is

price at time 2 = (price at time 1) × (CPI at time 2)/(CPI at time 1)

For this formula to work, all CPIs must be adjusted to the same base period. When the base period is updated, past CPIs are all adjusted to the new period. Thus, the CPIs of years that precede the current base period are generally less than 100; those of years that follow the current base period are generally over 100. Here are some CPIs since 1950, using the 1982–1984 base period:

Year	1950	1960	1970	1975	1980	1985	1990	1993
CPI	24.1	29.6	38.8	53.8	82.4	107.6	130.7	152.4

Example: If the rent for a particular apartment was $300 a month in 1985, what would the comparable rent be in 1995?

price in 1995 = (price in 1985) × (CPI in 1995)/(CPI in 1985)

price in 1995 = ($300) × (152.4/107.6) = $300 × 1.4164 = $424.91

or about $425 per month.

3. The CPI is used to adjust other economic data for inflation. If you were to plot just about any economic measure over time, you would see an increase simply because—at least historically—the value of a dollar decreases every year. To provide a true picture of changes in conditions over time, most economic data are presented in values adjusted for inflation. You should always check plots of economic data over time to see if they have been adjusted for inflation.

4. The CPI is used to determine salary and price adjustments. According to the Bureau of Labor Statistics:

> *The CPI affects the income of almost 80 million persons, as a result of statutory action: 47.8 million Social Security beneficiaries, about 22.4 million food stamp recipients, and about 4.1 million military and Federal Civil Service retirees and survivors. Changes in the CPI also affect the cost of lunches for 26.7 million children who eat lunch at school, while collective bargaining agreements that tie wages to the CPI cover over 2 million workers. (U.S. Dept. of Labor, 1998, p. 2)*

Because so many government wage and price increases are tied to the CPI, it has recently been the subject of scrutiny by Congress. An Advisory Committee to the U.S. Senate Committee on Finance (U.S. Senate, 1996) has made numerous recommendations for changes, some of which are under consideration. One of these changes, implemented in January 1999, is discussed in the next section.

13.3 CRITICISMS OF THE CONSUMER PRICE INDEX

Although the Consumer Price Index may be the best measure of inflation available, it does have problems. Economists believe that it slightly overestimates increases in the cost of living. A study released in October 1994 by the Congressional Budget Office estimated that "the CPI was overstating inflation by 0.2 percentage point to 0.8 percentage point per year" (Associated Press, 20 October 1994). Further, the CPI may overstate the effect of inflated prices for the items it covers on the average worker's standard of living. The following criticisms of the CPI should help you understand these and other problems with its use.

Some criticisms of the CPI

1. The market basket used in the CPI may not reflect current spending priorities.
2. If the price of one item rises, consumers are likely to substitute another.
3. The CPI does not adjust for changes in quality.
4. The CPI does not take advantage of sale prices.
5. The CPI does not measure prices for rural Americans.

1. The market basket used in the CPI may not reflect current spending priorities. Remember that the market basket of goods and the weights assigned to them are changed infrequently. As of 1998, the CPI was measuring the cost of items typically purchased in 1993–1995. This does not reflect rapid changes in lifestyle and technology. For example, cellular phones and CD players have become much more common in the past decade. You can probably think of many other changes that have occurred since 1995.

2. If the price of one item rises, consumers are likely to substitute another. If the price of beef rises substantially, consumers will buy chicken instead. If the price of fresh vegetables goes up due to poor weather conditions, consumers will use canned or frozen vegetables until prices go back down. When the price of lettuce tripled a few years ago, consumers were likely to buy fresh spinach for salads instead. In the past, the CPI has not taken these substitutions into account. However, starting with the January 1999 CPI, substitutions within a subcategory such as "Ice cream and related products" have been taken into account through the use of a new statistical method for combining data. It is estimated that this change reduces the annual rate of increase in the CPI by approximately 0.2 percentage points (U.S. Dept. of Labor, 16 April 1998).

3. The CPI does not account for changes in quality. The CPI assumes that if you purchase the same items in the current year as you did in the base year, your standard of living will be the same. That may apply to food and clothing but does not apply to many other goods and services. For example, personal computers were not only more expensive in 1982–1984, they were also much less powerful. Owning a new personal computer now would add more to your standard of living than owning one in 1982 would have done.

4. The CPI does not take advantage of sale prices. The outlets used to measure prices for the CPI are chosen by random sampling methods. The outlets consumers choose are more likely to be based on the best price that week. Further, if a supermarket is having a sale on an item you use often, you will probably stock up on the item at the sale price and then not need to purchase it for a while. The CPI does not take this kind of money-saving behavior into account.

5. The CPI does not measure prices for rural Americans. As mentioned earlier, the CPI is relevant for about 87% of the population: those who live in and around urban areas. It does not measure prices for the rural population and we don't know whether it can be extended to that group. The costs of certain goods and services are likely to be relatively consistent. However, if the rise in the CPI in a given time period is mostly due to rising costs particular to urban dwellers, such as the cost of public transportation and apartment rents, then it may not be applicable to rural consumers.

The Bureau of Labor Statistics notes that the CPI is not a cost-of-living index and should not be interpreted as such. It is most useful for comparing prices of similar products in the same geographic area across time. The BLS routinely studies and implements changes in methods that lead to improvements in the CPI. Current information about the CPI can be found on the CPI pages of the BLS website, that address, as of this writing, is www.bls.gov/cpihome.htm.

13.4   ECONOMIC INDICATORS

The Consumer Price Index is only one of many economic indicators produced or used by the U.S. government. The Bureau of Economic Analysis (BEA), part of the Department of Commerce, classifies and monitors a whole host of such indicators. Stratford and Stratford (1992, pp. 36–38) have put together a table listing 103 economic indicators accepted by the BEA as of February 1989.

Most economic indicators are series of data collected across time, like the CPI. Some of them measure financial data, others measure production, and yet others measure consumer behavior. Here is a list of ten series, randomly selected by the author from the previously mentioned table provided by Stratford and Stratford, to give you an idea of the variety. The letters in parentheses will be explained in the following section.

09   Construction contracts awarded for commercial and industrial buildings, floor space (L,C,U)

10   Contracts and orders for plant and equipment in current dollars (L,L,L)

14   Current liabilities of business failure (L,L,L)

25   Changes in manufacturers' unfilled orders, durable goods industries (L,L,L)

27   Manufacturers' new orders in 1982 dollars, nondefense capital goods industries (L,L,L)

39   Percent of consumer installment loans delinquent 30 days or over (L,L,L)

51   Personal income less transfer payments in 1982 dollars (C,C,C)

84   Capacity utilization rate, manufacturing (L,C,U)

110   Funds raised by private nonfinancial borrowers in credit markets (L,L,L)

114   Discount rate on new issues of 91-day Treasury bills (C,LG,LG)

You can see that even this randomly selected subset covers a wide range of information about government, business and consumer behavior, and economic status.

Leading, Coincident, and Lagging Indicators

Most indicators move with the general health of the economy. The BEA classifies economic indicators according to whether their changes precede, coincide with, or lag behind changes in the economy.

A **leading economic indicator** is one in which the highs, lows, and changes tend to *precede* or *lead* similar changes in the economy. (Contrary to what you may have thought, the term does not convey that it is one of the most *important* economic indicators.) A **coincident economic indicator** is one with changes that *coincide* with those in the economy. A **lagging economic indicator** is one whose changes *lag behind* or follow changes in the economy.

To further complicate the situation, some economic indicators have highs that precede or lead the highs in the economy but have lows that are coincident with or

TABLE 13.2   Components of the Index of Leading Economic Indicators

1. Average workweek of production workers in manufacturing
2. Average weekly initial claims for state unemployment insurance
3. New orders for consumer goods and materials, adjusted for inflation
4. Vendor performance (companies receiving slower deliveries from suppliers)
5. Contracts and orders for plant and equipment, adjusted for inflation
6. New building permits issued [private housing units]
7. Change in manufacturers' unfilled orders, durable goods
8. Change in sensitive materials prices
9. Index of stock prices
10. Money supply: M-2, adjusted for inflation
11. Index of consumer expectations

SOURCE: *World Almanac and Book of Facts,* 1993, p. 136.

lag behind the lows in the economy. Therefore, the BEA further classifies the behavior of each series according to how its highs, lows, and changes correspond to similar behavior in the economy. The sample of ten indicators shown in the previous section are classified this way. The letters following each indicator show how the highs, lows, and changes, respectively, are classified for that series.

The code letters are L = Leading, LG = Lagging, C = Coincident, and U = Unclassified. For example, the code letters in indicator 10, "Contracts and orders for plant and equipment in current dollars (L,L,L)," show that this indicator leads the economy in all respects. When this series remains high, remains low, or changes, the economy tends to follow. In contrast, the code letters in indicator 114, "Discount rate on new issues of 91-day Treasury bills (C,LG,LG)," show that this indicator has highs that are coincident with the economy but has lows and changes that tend to lag behind the economy.

Composite Indexes

Rather than require decision makers to follow all of these series separately, the BEA produces composite indexes. The *Index of Leading Economic Indicators* comprises 11 series, listed in Table 13.2. Most, but not all, of the individual component series are collected by the U.S. government. For instance, the Index of Stock Prices is provided by Standard and Poor's Corporation, whereas the Index of Consumer Expectations is provided by the University of Michigan's Survey Research Center.

The *Index of Coincident Economic Indicators* comprises four series, the *Index of Lagging Economic Indicators,* seven series. These indexes are produced monthly, quarterly, and annually.

Behavior of the Index of Leading Economic Indicators is thought to precede that of the general economy by about 6 to 9 months. This is based on observing past performance and not on a causal explanation—that is, it may not hold in the future because there is no obvious cause and effect relationship. In addition, monthly changes can be influenced by external events that may not predict later changes in the economy. Nonetheless, the Index of Leading Economic Indicators is followed closely and used as a predictor of things to come. A news story hints at how this index is influenced by, and influences, events:

> *Index of Leading Indicators shows small decline during February. . . . The government's chief forecasting gauge of future economic activity suffered its first decline in seven months, the Commerce Department reported today. Much of the weakness was blamed on severe winter weather. . . . Today's report provided some assurance to jittery investors who have been dumping stocks and bonds out of fears that the economy was growing so rapidly it would trigger higher inflation.* [Davis (CA) Enterprise, 5 April 1994, p. A10]

Stratford and Stratford (1992, pp. 43–45) discuss other reasons why the Index may be limited as a predictor of the economic future. For instance, they note that about 75% of all jobs in the United States are now in the service sector, yet the Index focuses on manufacturing. Nevertheless, although the Index may not be ideal, it is still the most commonly quoted source of predictions about future economic behavior.

CASE STUDY 13.1   Did Wages Really Go Up in the Reagan–Bush Years?

It was the fall of 1992 and the United States presidential election was imminent. The Republican incumbent, George Bush, had been president for the past 4 years, and vice-president to Ronald Reagan for 8 years before that. One of the major themes of the campaign was the economy. Despite the fact that the federal budget deficit had grown astronomically during those 12 Reagan–Bush years, the Republicans argued that Americans were better off in 1992 than they had been 12 years earlier. One of the measures they used to illustrate their point of view was the average earnings of workers. The average wages of workers in private, nonagricultural production had risen from $235.10 per week in 1980 to $345.35 in 1991 (*World Almanac and Book of Facts,* 1995, p. 150).

Were those workers really better off because they were earning almost 50% more in 1991 than they had been in 1980? Supporters of Democratic challenger Bill Clinton didn't think so. They began to counter the argument with some facts of their own.

Based on the material in this chapter, you can decide for yourself. The Consumer Price Index in 1980, measured with the 1982–1984 baseline, was 82.4. For 1991, it was 136.2. Let's see what average weekly earnings in 1991 should have been, adjusting for inflation, to have remained constant with the 1980 average:

salary at time 2 = (salary at time 1) × (CPI at time 2)/(CPI at time 1)

salary in 1991 = ($235.10) × (136.2)/(82.4) = $388.60

Therefore, the average weekly salary actually dropped during those 11 years, adjusted for inflation, from the equivalent of $388.60 to $345.35. The actual average was only 89% of what it should have been to have kept up with inflation.

There is another reason why the argument made by the Republicans would sound convincing to individual voters. Those voters who had been working in 1980 may very well have been better off in 1991, even adjusting for inflation, than they had been in 1980. That's because those workers would have had an additional 11 years of seniority in the work force, during which their relative positions should have improved. A meaningful comparison, which uses average wages adjusted for inflation, would not apply to an individual worker.

EXERCISES

1. The price of a first-class stamp in 1970 was 8 cents, whereas in 1997 it was 32 cents. The Consumer Price Index for 1970 was 38.8, whereas for 1997 it was 160.5. If the true cost of a first-class stamp did not increase between 1970 and 1997, what should it have cost in 1997? In other words, what would an 8 cent stamp in 1970 cost in 1997, when adjusted for inflation?

2. The CPIs at the start of each decade from 1940 to 1990 were

Year	1940	1950	1960	1970	1980	1990
CPI	14.0	24.1	29.6	38.8	82.4	130.7

   a. Determine the percentage increase in the CPI for each decade.

   b. During which decade was inflation the highest, as measured by the percentage change in the CPI?

   c. During which decade was inflation the lowest, as measured by the percentage change in the CPI?

3. A paperback novel cost $1.50 in 1968, $3.50 in 1981, and $6.99 in 1995. Compute a "paperback novel price index" for 1981 and 1995 using 1968 as the base year. In words that can be understood by someone with no training in statistics, explain what the resulting numbers mean.

4. When the CPI was computed for December 1997, the relative weight for the food and beverages category was 16.3%, whereas for the recreation category it was only 6.1%. Explain why food and beverages received higher weight than recreation.

5. Remember that the CPI is supposed to measure the change in what it costs to maintain the same standard of living that was in effect during the base year(s). Using the material in Section 13.3, explain why it may not do so accurately.

6. Americans spent the following amounts for medical care between 1987 and 1993, in billions of dollars (*World Almanac and Book of Facts,* 1995, p. 128).

Year	1987	1988	1989	1990	1991	1992	1993
Amount Spent	399.0	487.7	536.4	597.8	651.7	704.6	760.5

a. Create a "medical care index" for each of these years, using 1987 as a base.

b. Comment on how the cost of medical care has changed between 1987 and 1993, relative to the change in the Consumer Price Index, which was 113.6 in 1987 and 144.5 in 1993.

7. Suppose that you gave your niece a check for $50 on her 16th birthday in 1997 (when the CPI was 160.5). Your nephew is now about to turn 16. You discover that the CPI is now 180. How much should you give your nephew if you want to give him the same amount you gave your niece, adjusted for inflation?

8. In explaining why it is a costly mistake to have the CPI overestimate inflation, the Associated Press (20 October 1994) reported, "Every 1 percentage point increase in the CPI raises the federal budget deficit by an estimated $6.5 billion." Explain why that would happen.

9. Find out what the tuition and fees were for your school for the previous 4 years and the current year. Using the cost 5 years ago as the base, create a "tuition index" for each year since then. Write a short summary of your results that would be understood by someone who does not know what an index number is.

10. In addition to the overall CPI, the BLS reports the index for the subcategories. The overall CPI in 1993 was 144.5. Following are the values for some of the subcategories, taken from the *World Almanac and Book of Facts* (1995, p. 110):

Dairy products	129.4
Fruits, vegetables	159.0
Alcoholic beverages	149.6
Rent	165.0
House furnishings	109.5
Footwear	125.9
Tobacco products	228.4
Personal, educational expenses	210.7

All of these values are based on using 1982–1984 as the base period, the same period used by the overall CPI. Write a brief report explaining this information that could be understood by someone who does not know what index numbers are.

11. As mentioned in this chapter, both the base year and the relative weights used for the Consumer Price Index are periodically updated.

a. Why is it important to update the relative weights used for the CPI?

b. Explain why the base year is periodically updated.

12. Most newspaper accounts of the Consumer Price Index report the percentage change in the CPI from the previous month rather than the value of the CPI itself. Why do you think that is the case?

13. The Bureau of Labor Statistics reports that one use of the Consumer Price Index is to periodically adjust the federal income tax structure, which sets higher tax rates for higher income brackets. According to the BLS, "These adjustments prevent inflation-induced increases in tax rates, an effect called 'bracket creep'" (U.S. Dept. of Labor, 1998, p. 2). Explain what is meant by "bracket creep" and how you think the CPI is used to prevent it.

14. Many U.S. government payments, such as social security benefits, are increased each year by the percentage change in the CPI. In 1995, the government started discussions about lowering these increases or changing the way the CPI is calculated. According to an article in the *New York Times,* "Most economists who have studied the issue closely say the current system is too generous to Federal beneficiaries . . . the pain of lower COLAs [cost of living adjustments] would be unavoidable but nonetheless appropriate" (Gilpin, 1995, p. D19). Explain in what sense some economists believe the current system is too generous.

15. One of the components of the Index of Leading Economic Indicators is the Index of Consumer Expectations. Why do you think this index would be a leading economic indicator?

16. Examine the 11 series that make up the Index of Leading Economic Indicators, listed in Table 13.2. Choose at least two of the series to support the explanation given by the government in March 1994 that the drop in these indicators in February was partially due to severe winter weather.

17. Two of the economic indicators measured by the U.S. government are "Number of employees on nonagricultural payrolls" and "Average duration of unemployment, in weeks." One of these is designated as a "lagging economic indicator" and the other is a "coincident economic indicator." Explain which you think is which, and why.

MINI-PROJECTS

1. Numerous economic indicators are compiled and reported by the U.S. government and by private companies. Various sources are available at the library and on the Internet to explain these indicators. Write a report on one of the following. Explain how it is calculated, what its uses are, and what limitations it might have.

   a. The Producer Price Index

   b. The Gross Domestic Product

   c. The Dow Jones Industrial Average

2. Find a news article that reports on current values for one of the indexes discussed in this chapter. Discuss the news report in the context of what you have learned in this chapter. For example, does the report contain any information that might be misleading to an uneducated reader? Does it omit any information that you would find useful? Does it provide an accurate picture of the current situation?

REFERENCES

Associated Press. (20 October 1994). U.S. ready to overhaul measure of inflation. *The Press of Atlantic City,* p. A-8.

Gilpin, Kenneth N. (22 February 1995). Changing an inflation gauge is tougher than it sounds. *New York Times,* pp. D1, D19.

Stratford, J. S., and J. Stratford. (1992). *Major U.S. statistical series: Definitions, publications, limitations.* Chicago: American Library Association.

U.S. Department of Labor. Bureau of Labor Statistics. (1998). Understanding the Consumer Price Index: Answers to some questions.

U.S. Department of Labor. Bureau of Labor Statistics. (16 April 1998). Planned changes in the Consumer Price Index formula. News release.

U.S. Department of Labor. Bureau of Labor Statistics. (15 June 1998). Consumer Price Index summary. News release.

U.S. Senate. Committee on Finance. (1996). Final report of the Advisory Commission to Study the Consumer Price Index. Print 104-72, 104 Congress, 2 session. Washington, D.C.: Government Printing Office.

World almanac and book of facts. (1993). Edited by Mark S. Hoffman. New York: Pharos Books.

World almanac and book of facts. (1995). Edited by Robert Famighetti. Mahwah, NJ: Funk and Wagnalls.

CHAPTER FOURTEEN

Understanding and Reporting Trends over Time

THOUGHT QUESTIONS

1. What do you think is meant by the term *time series?*
2. What do you think it means when a monthly economic indicator, such as new housing starts, is reported as having been *seasonally adjusted?*
3. If you were to plot number of ice cream cones sold versus month for 5 years, do you think the plot would show peaks and valleys, or would sales be relatively constant across all months? Explain.
4. If someone is trying to get you to invest in his or her company and shows you a plot of sales or profits over time, what features of the picture do you think you should critically evaluate before you decide to invest?

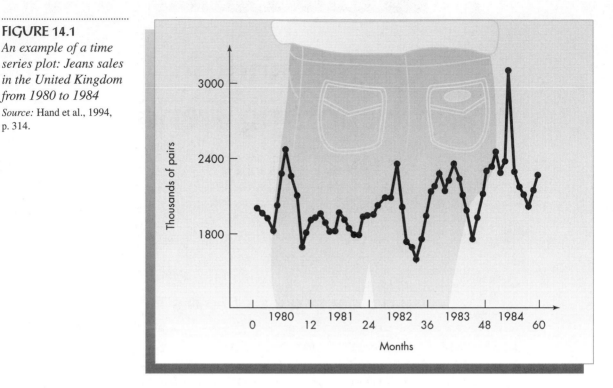

14.1   TIME SERIES

We have already seen examples of *time series* in Chapter 13, although we did not call them by that name. A **time series** is simply a record of a variable across time, usually measured at equally spaced time intervals. Most of the economic indicators discussed in Chapter 13 are time series that are measured monthly.

To understand data presented across time, it is important to know how to recognize the various components that can contribute to the ups and downs in a time series. Otherwise, you could mistake a temporary high in a cycle for a permanent increasing trend and make a very unwise economic decision.

A Time Series Plot

Figure 14.1 illustrates a **time series plot.** The data represent monthly sales of jeans in Britain, for the 5-year period from January 1980 to December 1984. Notice that the data points have been connected to make it easier to follow the ups and downs across time. Data are measured in thousands of pairs sold. Month 1 is January 1980, and month 60 is December 1984.

FIGURE 14.2
*Distortion caused by
displaying only part of
a time series; Jeans
sales for 21 months*

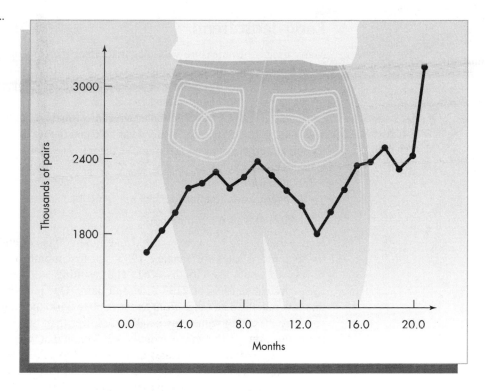

FIGURE 14.2
*Distortion caused by
displaying only part of
a time series; Jeans
sales for 21 months*

Improper Presentation of a Time Series

Before we investigate the components in Figure 14.1 (and other time series), let's look at one way in which you can be fooled by improper presentation of a time series. In Figure 14.2, a subset of the time series is displayed.

Suppose an unscrupulous entrepreneur was anxious to have you invest your hard-earned savings into his blue jeans company. To convince you that sales of jeans can only go up, he presents you with a limited set of data—from October 1982 to June 1984. With only those few months shown, it appears that the basic trend is way up!

A less obvious version of this trick is to present data up to the present time but to start the plot of the series at an advantageous point. Be suspicious of time series showing returns on investments that look too good to be true. They probably are. Notice when the time series begins and compare that with your knowledge of recent economic cycles.

14.2   COMPONENTS OF TIME SERIES

Most time series have the same four basic components: *long-term trend, seasonal components, irregular cycles,* and *random fluctuations.* Let's examine each of these in turn.

Long-Term Trend

Many time series measure variables that either increase or decrease steadily across time. This steady change is called a **trend.** If the trend is even moderately large, it should be obvious by looking at a plot of the series. Figure 14.1 clearly shows an increasing trend for jeans sales.

If the long-term trend is linear, we can estimate it by finding a regression line, with time period as the explanatory variable and the variable in the time series as the response variable. We can then remove the trend to enable us to see what other interesting features exist in the series. When we do that, the result is, aptly enough, called a **detrended time series.**

The regression line for the data in Figure 14.1 is:

sales = 1880 + 6.62 (months)

Notice that month 1 is January 1980 and month 60 is December 1984. If we were to try to forecast sales for January 1985, the first month that is not included in the series, we would use months = 61. The resulting value is 2284 thousand pairs of jeans. Actual sales were 2137 thousand pairs. Our prediction is not far off, given that, overall, the data range from about 1600 to 3100 thousand pairs. Notice one reason why the actual value may be slightly lower than the predicted value; sales tend to be lower during the winter months. We look at that seasonal component next.

The regression line indicates that the trend, on average, shows sales increasing by about 6.62 units per month. Because the units represent thousands of pairs, the actual increase is about 6620 pairs per month. Figure 14.3 presents the time series for jeans sales with the trend removed. Compare Figure 14.3 with Figure 14.1. Notice that the fluctuations remaining in Figure 14.3 are similar in character to those in Figure 14.1, but the upward trend is gone.

Let's look at what we would have estimated as the trend if we had been fooled by the picture in Figure 14.2. We would have predicted a much higher increase per month. The regression line for the data in Figure 14.2 is:

sales = 1832 + 32.1 (months)

In other words, the trend is estimated to show an increase of 32,100 pairs a month, compared with 6620 pairs computed from the full time series.

Seasonal Components

Most time series involving economic data or data related to people's behavior also have **seasonal components.** In other words, they tend to be high in certain months or seasons and low in others every year. For example, new housing starts are much higher in warmer months. Sales of toys and other standard gifts are much higher just before Christmas. U.S. unemployment rates tend to rise in January, when outdoor jobs are minimal and the Christmas season is over, and again in June, when a new graduating class enters the job market.

Most of the economic indicators discussed in Chapter 13 are subject to seasonal fluctuations. As we shall see in Section 14.3, they are usually reported after they have been *seasonally adjusted.*

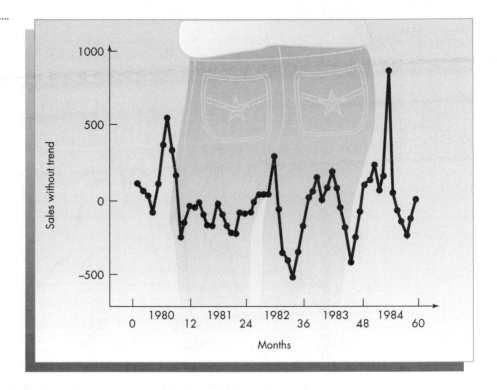

Notice that there is indeed a seasonal component to the time series of sales of jeans. It is evident in both Figure 14.1 and Figure 14.3. Sales appear to peak during June and July and reach a low in October every year. Manufacturers need to know that information. Otherwise, they might mistake increased sales during June, for example, as a general trend and overproduce their product.

Economists have sophisticated methods for seasonally adjusting time series. They use data from the same month or season in prior years to construct a *seasonal factor*, which is a number either greater than or less than one by which the current figure is multiplied. According to the U.S. Department of Labor (1992, p. 243), "The standard practice at BLS for current seasonal adjustment of data, as it is initially released, is to use projected seasonal factors which are published ahead of time." In other words, when figures such as the Consumer Price Index become available for a given month, the BLS already knows the amount by which the figures should be adjusted up or down to account for the seasonal component.

Irregular Cycles and Random Fluctuations

There are two remaining components of time series: the *irregular* (but smooth) *cycles* that economic systems tend to follow and unexplainable *random fluctuations*. It is often hard to distinguish between these two components, especially if the cycles are not regular.

FIGURE 14.4

An example of a time series with irregular cycles: Adjusted January unemployment rates, 1950–1982

Source: Based on data from Miller, 1988, p. 284.

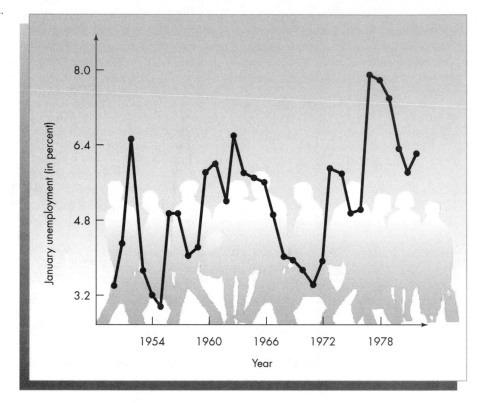

Figure 14.4 shows the U.S. unemployment rate, seasonally adjusted, for each January from 1950 to 1982. Notice the definite **irregular cycles** during which unemployment rates rise and fall over a number of years. Some of these can be at least partially explained by social and political factors. For example, the Vietnam War era spanned the years from the mid-1960s to the early 1970s. The mandatory draft ended in 1973, freeing many young men to enter the job market. You can see a decreasing cycle during the Vietnam War years that ends in 1972.

The **random fluctuations** in a time series are defined as what's left over when the other three components have been removed. They are part of the natural variability present in all measurements. Notice from Figures 14.1 and 14.3 that, even if you were to account for the trend, seasonal components, and smooth irregular cycles, you would still not be able to perfectly explain the jeans sales each month. The remaining, unexplainable components are labeled random fluctuations.

14.3   SEASONAL ADJUSTMENTS: REPORTING THE CONSUMER PRICE INDEX

It is unusual to see the Consumer Price Index itself reported in the news. More commonly, what you see reported is the *change* from the previous month, which is generally reported in the middle of each month. Following is an example of how it was reported in the *New York Times:*

> Consumer Prices Rose 0.3% in June   *Washington, July 13—Consumer prices climbed three-tenths of 1 percent in June, as increases for cars, gasoline, air fares and clothing more than offset moderation in housing, the Labor Department reported today. (Hershey, 19 July 1994, p. C1)*

Most news reports never tell you the actual value of the CPI. In this report, the CPI itself was finally given at the end of a long article, in the following paragraph:

> *The index now stands at 148.0, meaning that an array of goods and services that cost $10 in the 1982–84 reference period now costs $14.80. The value of the 1982–84 dollar is now 67.6 cents. (Hershey, 19 July 1994, p. C17)*

One piece of information is blatantly missing from the article itself. You find it only when you read the accompanying graph, which shows the change in the CPI for the previous 12 months. The heading on the graph reads: "Consumer prices—percent change, month to month, *seasonally adjusted*" (italics added).

In other words, the change of 0.3% does not represent an absolute change in the Consumer Price Index; rather it represents a change *after* seasonal adjustments have been made. Adjustments have already been made for the fact that certain items are expected to cost more during certain months of the year. According to the *BLS Handbook of Methods:*

> *An economic time series may be affected by regular intrayearly (seasonal) movements which result from climatic conditions, model changeovers, vacation practices, holidays, and similar factors. Often such effects are large enough to mask the short-term, underlying movement of the series. If the effect of such intrayearly repetitive movements can be isolated and removed, the evaluation of a series may be made more perceptive. (U.S. Dept. of Labor, 1992, p. 243)*

The *BLS Handbook* is thus recognizing what you should admit as common sense. It is important that economic indicators be reported with seasonal adjustments; otherwise, it would be impossible to determine the direction of the real trend. For example, it would probably always appear as if new housing starts dipped in February and jumped in May. Therefore, it is prudent reporting to include a seasonal adjustment.

Why Are Changes in the CPI Big News?

You may wonder why changes in the CPI are reported as the big news. The reason is that financial markets are extremely sensitive to changes in the rate of inflation. The same *New York Times* article quoted earlier reported:

> *Unlike Tuesday's surprisingly favorable report that prices at the producer level were unchanged last month, the C.P.I. data provided little comfort to the majority of analysts, who say that inflation—higher in June than in either April or*

> *May—has begun a gradual upswing, and that the Federal Reserve will need to raise short-term interest rates again by mid-August. (Hershey, 19 July 1994, p. C1)*

Like anything else in the world, it is the changes that attract concern and attention, not the continuation of the status quo.

14.4   CAUTIONS AND CHECKLIST

Some time series data are not adjusted for inflation or for seasonal components. When you read a report of an economic indicator, you should check to see if it has been adjusted for inflation and/or seasonally adjusted. You can then compensate for those factors in your interpretation.

EXAMPLE   **THE DOW JONES INDUSTRIAL AVERAGE**

The Dow Jones Industrial Average (DJIA) is a weighted average of the price of 30 major stocks on the New York Stock Exchange. It reached an all-time high of $9337.97 on July 17, 1998. In fact, it reaches an all-time high almost every year. But the DJIA is not adjusted for inflation; it is simply reported in current dollars. Thus, to compare the high in one year with that in another, we need to adjust it using the CPI.

For example, in 1970, the high for the DJIA was $842.00, occurring on December 29. The high in 1993 was $3794.33, also occurring on December 29. The CPI in 1970 was 38.8, whereas in 1993 it was 144.5. Did the DJIA rise faster than inflation? To determine if it did, let's calculate what the 1970 high would have been in 1993 dollars:

$$\text{value in 1993} = (\text{value in 1970}) \times (\text{CPI in 1993})/(\text{CPI in 1970})$$
$$= (\$842.00) \times (144.5)/(38.8) = \$3135.80$$

Therefore, the high of $3794.33 cannot be completely explained by inflation. If we take the ratio of the two numbers, we find $3794.33/$3135.80 = 1.21. In other words, the increase in the DJIA highs from 1970 to 1993 is 21% after adjusting for inflation using the Consumer Price Index. ∎

Checklist for Reading Time-Series Data

When you see a plot of a variable across time or when you read about a monthly change in a time series, keep in mind that you could be misled in various ways.

> *You should ask the following questions when reading time series data:*
> 1. Are the time periods equally spaced?
> 2. Is the series adjusted for inflation?
> 3. Are the values seasonally adjusted?
> 4. Does the series cover enough of a time span to represent typical long-term behavior?
> 5. Is there an upward or downward trend?
> 6. Are there other seasonal components that have not been removed?
> 7. Are there smooth cycles?

Based on your answers to those questions, you may need to calculate or approximate an adjustment to the data reported.

CASE STUDY 14.1

If You're Looking for a Job, Try May and October

SOURCE: Miller, 1988, pp. 283–285.

How much do you think unemployment rates fluctuate from season to season? Do you think they reach a yearly high and low during the same month each year? In Figure 14.4, we saw that unemployment rates tend to follow cycles over time. In this case study, we look at a 5-year period only, to see the effect of monthly components.

Figure 14.5 shows the monthly unemployment rates from January 1977 to December 1981. Each month has been coded with a letter: A = January, B = February, and so on. Notice that there are definite monthly components. For each of the 5 years, a sharp increase occurs between December (L) and January (A) and another between May (E) and June (F). The yearly lows occur in the spring, particularly in May and in the fall.

Figure 14.6 shows the official, seasonally adjusted unemployment rates for the same time period. Notice that the extremes have been removed by the process of seasonal adjustment. One month no longer dominates each year as the high or the low. In fact, the series in Figure 14.6 shows much less variability than the one in Figure 14.5. Much of the variability in Figure 14.5 was not due to random fluctuations but to explainable monthly components. The remaining variability apparent in Figure 14.6, after adjusting for the obvious trend, can be attributed to random monthly fluctuations.

As a final note, compare Figure 14.6 with Figure 14.4, in which January unemployment rates from 1950 to 1982 were presented. Notice that the downward trend in Figure 14.6 is simply part of the longer cyclical behavior of unemployment rates. It would not be projected to continue indefinitely. In fact, by 1983, the yearly unemployment rate had risen to 9.7%, as the cyclical behavior evident in Figure 14.4

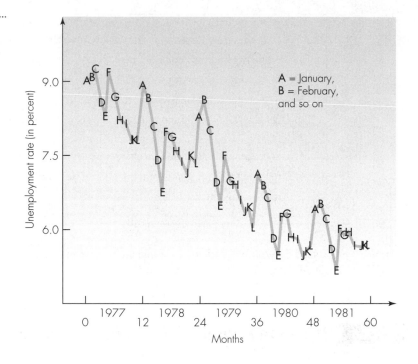

FIGURE 14.5
Unemployment rates for 1977–1981 before being seasonally adjusted
Source: Miller, 1988.

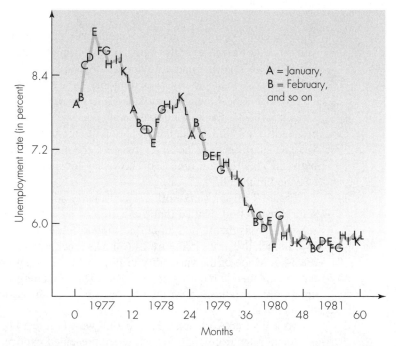

FIGURE 14.6
Unemployment rates for 1977–1981 after being seasonally adjusted
Source: Miller, 1988.

continued. This example illustrates that what appears as a trend in a short time series may actually be part of a cycle in a longer time series. For that reason, it is not wise to forecast a trend very far into the future. ■

EXERCISES

1. For each of the following time series, do you think the long-term trend would be positive, negative, or nonexistent?

   a. The cost of a loaf of bread measured monthly from 1960 to 1998

   b. The temperature in Boston measured at noon on the first day of each month from 1960 to 1998

   c. The price of a basic computer, adjusted for inflation, measured monthly from 1970 to 1998

   d. The number of personal computers sold in the United States measured monthly from 1970 to 1998

2. For each of the time series in Exercise 1, explain whether there is likely to be a seasonal component.

3. Global warming is a major concern because it implies that temperatures around the world are going up on a permanent basis. Suppose you were to examine a plot of monthly temperatures in one location for the past 50 years. Explain the role that the three time series components (trend, seasonal, cycles) would play in trying to determine whether global warming was taking place.

4. If you were to present a time series of the yearly cost of tuition at your local college for the past 30 years, would it be better to first adjust the costs for inflation? Explain.

5. If you wanted to present a time series of the yearly cost of tuition at your local college for the past 30 years, adjusted for inflation, how would you do the adjustment?

6. The population of the United States rose from about 179 million people in 1960 to about 249 million people in 1990. Suppose you wanted to examine a time series to see if homicides had become an increasing problem over that time period. Would you simply plot the number of homicides versus time, or is there a better measure to plot against time? Explain.

7. Many statistics related to births, deaths, divorces, and so on across time are reported as rates per 100,000 of population rather than as actual numbers. Explain why those rates may be more meaningful as a measure of change across time than the actual numbers of those events.

8. Suppose a time series across 60 months has a long-term positive trend. Would you expect to find a correlation between the values in the series and the months 1 to 60? If so, can you tell from the information given whether it would be positive or negative?

9. Explain which one of the components of an economic time series would be most likely to be influenced by a major war. (See Section 14.2.)

10. Discuss which of the three components of a time series (trend, seasonal, and cycles) are likely to be present in each of the following series, reported monthly for the past 10 years:

    a. Unemployment rates

    b. Hours per day the average child spends watching television

    c. Interest rates paid on a savings account at a local bank

11. Which of the three nonrandom components of time series (trend, seasonal, or cycles) is likely to contribute the most to the unadjusted Consumer Price Index? Explain.

12. Draw an example of a time series that has:

    a. Trend, cycles, and random fluctuations, but not seasonal components

    b. Seasonal components and random fluctuations, but not trend or cycles

13. Explain why it is important for economic time series to be seasonally adjusted before they are reported.

14. Suppose you have been hired as a salesperson, selling computers and software. In January, after 6 months on the job, your sales suddenly plummet. They had been high from August to December. Your boss, who is also new to the position, chastises you for this drop. What would you say to your boss to protect your job?

15. The CPI in July 1977 was 60.9; in July 1994, it was 148.4.

    a. The salary of the governor of California in July 1977 was $49,100; in July 1994, it was $120,000. Compute what the July 1977 salary would be in July 1994, adjusted for inflation, and compare it with the actual salary in July 1994.

    b. The salary of the president of the United States in July 1977 was $200,000. In July 1994, it was still $200,000. Compute what the July 1977 salary would be in July 1994, adjusted for inflation, and compare it with the actual salary.

16. The Dow Jones Industrial Average reached a high of $7801.63 on December 29, 1997. Recall from Section 14.4 that it reached a high of $842.00 on December 29, 1970. The Consumer Price Index averaged 38.8 for 1970; for 1997, it averaged 160.5. By what percentage did the high in the DJIA increase from December 29, 1970, to December 29, 1997, after adjusting for inflation?

17. Explain why it is important to examine a time series for many years before making conclusions about the contribution of each of the three nonrandom components.

18. According to the *World Almanac and Book of Facts* (1995, p. 380), the population of Austin, Texas (reported in thousands), has grown as follows:

Year	1950	1960	1970	1980	1990
Population	132.5	186.5	253.5	345.5	465.6

a. Of the three nonrandom components of time series (trends, seasonal, and cycles), which do you think would be most likely to explain the data if you were to see the population of Austin, Texas, by month, from 1950 to 1990? Explain.

b. The regression equation relating the last two digits of each year (50, 60, and so on) to the population for Austin, Texas, is:

$$\text{population} = -301 + 8.25 \text{ (year)}$$

Use this equation to predict the population of Austin for the year 2000.

c. Discuss the method you used for the prediction in part b. Do you think it is likely to be accurate? Would the same method continue to give accurate predictions for the years 2010, 2020, and so on?

....

MINI-PROJECTS

1. Plot your own resting pulse rate taken at regular intervals for 5 days. Comment on which of the components of time series are present in your plot. Discuss what you have learned about your own pulse from this exercise.

2. Find an example of a time series plot presented in a newspaper, magazine, journal, or website. Discuss the plot based on the information given in this chapter. Comment on what you can learn from the plot.

3. In addition to the Dow Jones Industrial Average, there are other indicators of fluctuation in stock prices. Two examples are the New York Stock Exchange Composite Index and the Standard and Poor's 500. Choose a stock index (other than the Dow Jones) and write a report about it. Include whether it is adjusted for inflation, seasonally adjusted, or both. Give information about its recent performance, and compare it with performance a few decades ago. Make a conclusion about whether the stock market has gone up or down in that time period, based on the index you are using, adjusted for inflation.

....

REFERENCES

Hand, D. J., F. Daly, A. D. Lunn, K. J. McConway, and E. Ostrowski. (1994). *A handbook of small data sets.* London: Chapman and Hall.

Hershey, Robert D., Jr. (19 July 1994). Consumer prices rose 0.3% in June. *New York Times,* pp. C1, C17.

Miller, Robert B. (1988). *Minitab handbook for business and economics.* Boston: PWS Kent.

U.S. Department of Labor. Bureau of Labor Statistics. (September 1992). *BLS handbook of methods.* Bulletin 2414.

World almanac and book of facts. (1995). Edited by Robert Famighetti. Mahwah, NJ: Funk and Wagnalls.

Understanding Uncertainty in Life

In Parts 1 and 2 of this book, you learned how data should be collected and summarized. Some simple ideas about chance were introduced in the context of whether chance could be ruled out as an explanation for a relationship observed in a sample.

The purpose of the material in Part 3 is to acquaint you with some simple ideas about probability in ways that can be applied to your daily life. In Chapter 15, you will learn how to determine and interpret probabilities for simple events. You will also see that it is sometimes possible to make long-term predictions, even when specific events can't be predicted well. In Chapters 16 and 17, you will learn how psychological factors can influence judgments involving uncertainty. As a consequence, you will learn some hints that will help you make better decisions in your own life.

CHAPTER FIFTEEN

Understanding Probability and Long-Term Expectations

THOUGHT QUESTIONS

1. Here are two very different queries about probability:

   a. If you flip a coin and do it fairly, what is the probability that it will land heads up?

   b. What is the probability that you will eventually own a home; that is, how likely do you think it is? (If you already own a home, what is the probability that you will own a different home within the next 5 years?)

   For which question was it easier to provide a precise answer? Why?

2. Explain what it means for someone to say that the probability of his or her eventually owning a home is 70%.

3. Explain what's wrong with the following statement, given by a student as a partial answer to Thought Question 1: "The probability that I will eventually own a home, or of any other particular event happening, is 1/2 because either it will happen or it won't."

4. Why do you think insurance companies charge young men more than they do older men for automobile insurance, but charge older men more for life insurance?

5. How much would you be willing to pay for a ticket to a contest in which there was a 1% chance that you would win $500 and a 99% chance that you would win nothing? Explain your answer.

15.1   PROBABILITY

The word *probability* is so common that in all probability you will run across it today in everyday language. But we rarely stop to think about what the word means. For instance, when we speak of the probability of winning a lottery based on buying a single ticket, are we using the word in the same way as when we speak of the probability that we will eventually buy a home? In the first case, we can quantify the chances exactly. In the second case, we are basing our assessment on personal beliefs about how life will evolve for us. The conceptual difference illustrated by these two examples leads to two distinct interpretations of what is meant by the term *probability*.

15.2   THE RELATIVE-FREQUENCY INTERPRETATION

The **relative-frequency interpretation** of probability applies to situations in which we can envision observing results over and over again. For example, it is easy to envision flipping a coin over and over again and observing whether it lands heads or tails. It then makes sense to discuss the probability that the coin lands heads up. It is simply the relative frequency, over the long run, with which the coin lands heads up.

Here are some more interesting situations to which this interpretation of probability can be applied:

- Buying a weekly lottery ticket and observing whether it is a winner
- Commuting to work daily and observing whether a certain traffic signal is red when we encounter it
- Testing individuals in a population and observing whether they carry a gene for a certain disease
- Observing births and noting if the baby is male or female.

The Idea of Long-Run Relative Frequency

If we have a situation such as those just described, we can define the **probability** of any specific outcome as *the proportion of time it occurs over the long run*. This is also called the **relative frequency** of that particular outcome.

Notice the emphasis on what happens *in the long run*. We cannot assess the probability of a particular outcome by observing it only a few times. For example, consider a family with five children, in which only one child is a boy. We would not take that as evidence that the probability of having a boy is only 1/5. However, if we noticed that out of thousands of births only one in five of the babies were boys, then it would be reasonable to conclude that the probability of having a boy is only 1/5.

TABLE 15.1	Relative Frequency of Male Births					
Weeks of Watching	1	4	12	24	36	52
Number of boys	12	47	160	310	450	618
Number of babies	30	100	300	590	880	1200
Proportion of boys	.400	.470	.533	.525	.511	.515

According to the *Information Please Almanac* (1991, p. 815), the long-run relative frequency of males born in the United States is about .512. In other words, over the long run, 512 male babies are born to every 488 female babies.

Suppose we were to record births in a certain city for the next year. Table 15.1 shows what we might observe.

Notice how the proportion, or relative frequency, of male births jumps around at first but starts to settle down to something just above .51 in the long run. If we had tried to determine the true proportion after just 1 week, we would have been seriously misled.

Determining the Probability of an Outcome

Method 1: Make an Assumption about the Physical World

Two methods for determining the probability of a particular outcome fit the relative-frequency interpretation. The first method is to *make an assumption about the physical world* and use it to determine the probability of an outcome. For example, we generally assume that coins are manufactured in such a way that they are equally likely to land with heads up or tails up when flipped. Therefore, we conclude that the probability of a flipped coin showing heads up is 1/2.

As a second example, we can determine the probability of winning the lottery by assuming that the physical mechanism used to draw the winning numbers gives each number an equal chance. For instance, many state-run lotteries in the United States have participants choose three digits, each from the set 0 to 9. If the winning set is drawn fairly, each of the 1000 possible combinations should be equally likely. (The 1000 possibilities are 000, 001, 002, . . . , 999.) Therefore, each time you play your probability of winning is 1/1000. You win only on those rare occasions when the set of numbers you chose is actually drawn. In the long run, that should happen about 1 out of 1000 times. Notice that this does not mean it will happen exactly once in every thousand draws.

Method 2: Observe the Relative Frequency

The other way to determine the probability of a particular outcome is by *observing the relative frequency over many, many repetitions of the situation*. We used that method when we observed the relative frequency of male births in a given city over the course of a year. By using this method, we can get a very accurate figure for the probability that a birth will be a male. As mentioned, the relative frequency of male births in the United States has been consistently close to .512 (*Information Please Almanac,* 1991, p. 815). For example, in 1987 there were a total of 3,809,394 live births in the United States, of which 1,951,153 were males. Therefore, the probability that a live birth will result in a male is 1,951,153/3,809,394 = .5122.

Sometimes relative-frequency probabilities are reported on the basis of sample surveys. In such cases, a margin of error should be included but often is not. For example, the *World Almanac and Book of Facts* (1993, p. 38), reported that "on any given day, 71 percent of Americans read a newspaper . . . according to a 1991 Gallup Poll."

Summary of the Relative-Frequency Interpretation of Probability

- The relative-frequency interpretation of probability can be applied when a situation can be repeated numerous times, at least conceptually, and the outcome can be observed each time.

- In scenarios for which this interpretation applies, the relative frequency with which a particular outcome occurs should settle down to a constant value over the long run. That value is then defined to be the probability of that outcome.

- The interpretation does not apply to situations where the outcome one time is influenced by or influences the outcome the next time because the probability would not remain the same from one time to the next. We cannot determine a number that is always changing.

- Probability cannot be used to determine whether the outcome will occur on a single occasion but can be used to predict the long-term proportion of the times the outcome will occur.

Relative-frequency probabilities are quite useful in making daily decisions. For example, suppose you have a choice between two flights to reach your destination. All other factors are equivalent, but your travel agent tells you that one has a probability of .90 of being on time, whereas the other has only a probability of .70 of being on time. Even though you can't predict the outcome for your particular flight, you would be likely to choose the one that has the better performance in the long run.

15.3   THE PERSONAL-PROBABILITY INTERPRETATION

The relative-frequency interpretation of probability is clearly limited to repeatable conditions. Yet, uncertainty is a characteristic of most events, whether they are repeatable under similar conditions or not. We need an interpretation of probability that can be applied to situations even if they will never happen again.

Will you fare better by taking calculus than by taking statistics? If you decide to drive downtown this Saturday afternoon, will you be able to find a good parking space? Should a movie studio release a potential new hit movie before Christmas, when many others are released, or wait until January, when it might have a better chance of being the top box-office attraction? Would a trade alliance with a new country cause problems in relations with a third country?

These are unique situations, not likely to be repeated. They require people to make decisions based on an assessment of how the future will evolve. We could each assign a personal probability to these events, based on our own knowledge and experiences, and we could use that probability to help us with our decisions. We may not agree on what the probabilities of differing outcomes are, but none of us would be considered wrong.

Defining Personal Probability

We define the **personal probability** of an event to be the degree to which a given individual believes the event will happen. There are very few restrictions on personal probabilities. They must fall between 0 and 1 (or, if expressed as a percentage, between 0 and 100%). They must also fit together in certain ways if they are to be *coherent*. By being **coherent,** we mean that your personal probability of one event doesn't contradict your personal probability of another. For example, if you thought that the probability of finding a parking space downtown Saturday afternoon was .20, then to be coherent, you must also believe that the probability of not finding one is .80. We explore some of these logical rules later in this chapter.

How We Use Personal Probabilities

People routinely base decisions on personal probabilities. This is why committee decisions are often so difficult. For example, suppose a committee is trying to decide which candidate to hire for a job. Each member of the committee has a different assessment of the candidates, and each may disagree with the others on the probability that a particular candidate would fit the job best. We are all familiar with the problem juries sometimes have when trying to agree on someone's guilt or innocence. Each member of the jury has his or her own personal probability of guilt and innocence. One of the benefits of committee or jury deliberations is that such deliberations may help members reach some consensus in their personal probabilities.

Personal probabilities often take relative frequencies of similar events into account. For example, the late astronomer Carl Sagan believed that the probability of a major asteroid hitting the Earth soon is high enough to be of concern. "The probability that the Earth will be hit by a civilization-threatening small world in the next century is a little less than one in a thousand" (Arraf, 14 December, 1994, p. 4). To arrive at that probability, Sagan obviously could not use the long-run frequency definition of probability. He would have to use his own knowledge of astronomy, combined with past asteroid behavior.

15.4   APPLYING SOME SIMPLE PROBABILITY RULES

Situations often arise where we already know probabilities associated with simple events, such as the probability that a birth will result in a girl, and we would like to find probabilities of more complicated events, such as the probability that we will eventually have at least one girl if we ultimately have four children. Some simple, logical rules about probability allow us to do this.

These rules apply naturally to relative-frequency probabilities, and they must apply to personal probabilities if they are to be coherent. For example, we can never have a probability below 0 or above 1. An impossible event has a probability of 0 and a sure thing has a probability of 1. Here are four additional useful rules:

> *Rule 1:* If there are only two possible outcomes in an uncertain situation, then their probabilities must add to 1.

EXAMPLE 1

If the probability of a single birth resulting in a boy is .51, then the probability of it resulting in a girl is .49. ■

EXAMPLE 2

If you estimate the chances that you will eventually own a home to be 70%, in order to be coherent (consistent with yourself) you are also estimating that there is a 30% chance that you will never own one. ■

EXAMPLE 3

According to Krantz (1992), the probability that a piece of checked luggage will be temporarily lost on a flight with a U.S. airline is 1/176. Thankfully, that means the probability of finding the luggage waiting at the end of a trip is 175/176. ■

> *Rule 2:* If two outcomes cannot happen simultaneously, they are said to be **mutually exclusive.** The probability of one or the other of two mutually exclusive outcomes happening is the sum of their individual probabilities.

EXAMPLE 4

The two most common primary causes of death in the United States are heart attacks, which kill about one-third (0.33) of the Americans who die each year, and various cancers, which kill about one-fifth (0.20). Therefore, if this year is like the past, the probability that a randomly selected American who dies will die of either a heart attack or cancer is the sum of these two probabilities, or about 0.53 (53%). Notice that this is based on 1994 death rates, and could well change long before you have to worry about it. This calculation also assumes that one cannot die simultaneously of both causes—in other words, the two causes of death are mutually exclusive. Given the way deaths are recorded, this fact is actually guaranteed because only one primary cause of death may be entered on a death certificate. ∎

EXAMPLE 5

If you estimate your chances of getting an A in your statistics class to be 50% and your chances of getting a B to be 30%, then you are estimating your chances of getting either an A or a B to be 80%. Notice that you are therefore estimating your chances of getting a C or less to be 20% by Rule 1. ∎

EXAMPLE 6

If you estimate your chances of getting an A in your statistics class to be 50% and your chances of getting an A in your history class to be 60%, are you estimating your chances of getting one or the other, or both, to be 110%? Obviously not, because probabilities cannot exceed 100%. The problem here is that Rule 2 stated explicitly that the events under consideration couldn't happen simultaneously. Because it is possible for you to get an A in both courses simultaneously, Rule 2 does not apply here. In case you are curious, Rule 2 could be modified to apply. You would have to subtract the probability that both events happen, which would require you to estimate that probability as well. We see one way to do that using Rule 3. ∎

> *Rule 3:* If two events do not influence each other, and if knowledge about one doesn't help with knowledge of the probability of the other, the events are said to be **independent** of each other. If two events are independent, the probability that they both happen is found by multiplying their individual probabilities.

EXAMPLE 7   Suppose a woman has two children. Assume that the outcome of the second birth is independent of what happened the first time and that the probability that each birth results in a boy is .51, as observed earlier. Then the probability that she has a boy *followed by* a girl is $(.51) \times (.49) = .2499$. In other words, there is about a 25% chance that a woman having two children will have a boy and then a girl. ■

EXAMPLE 8   From Example 6, suppose you continue to believe that your probability of getting an A in statistics is .5 and an A in history is .6. Further, suppose you believe that the grade you receive in one is independent of the grade you receive in the other. Then you must also believe that the probability that you will receive an A in both is $(.5) \times (.6) = .3$. Notice that we can now complete the calculation we started at the end of Example 6. The probability that you will receive at least one A is found by taking $.5 + .6 - .3 = .8$, or 80%. Note that by Rule 1 you must also believe that the probability of not receiving an A in either class is 20%. ■

Rule 3 is sometimes difficult for people to understand, but if you think of it in terms of the relative-frequency interpretation of probability, it's really quite simple. Consider women who have had two children. If about 51% of the women had a boy for their first child, and only 49 percent of those women had a girl the second time around, it makes sense that we are left with only about 25% of the women.

EXAMPLE 9   Let's try one more example of Rule 3, using the logic just outlined. Suppose you encounter a red light on 30% of your commutes and get behind a bus on half of your commutes. The two are unrelated because whether you have the bad luck to get behind a bus presumably has nothing to do with the red light. The probability of having a really bad day and having both happen is 15%. This is logical because you get behind a bus half of the time. Therefore, you get behind a bus half of the 30% of the time you encounter the red light, resulting in total misery only 15% of the time. Using Rule 3 directly, this is equivalent to $(.30) \times (.50) = .15$, or 15% of the time both events happen. ■

One more rule is such common sense that it almost doesn't warrant writing down. However, as we will see in Chapter 16, in certain situations this rule will actually seem counterintuitive. Here is the rule:

> *Rule 4:* If the ways in which one event can occur are a subset of those in which another event can occur, then the probability of the first event *cannot* be higher than the probability of the one for which it is a subset.

TABLE 15.2	Calculating Probabilities

Try on Which the Outcome First Happens	Probability
1	p
2	$(1 - p)p$
3	$(1 - p)(1 - p)p = (1 - p)^2 p$
4	$(1 - p)(1 - p)(1 - p)p = (1 - p)^3 p$
5	$(1 - p)(1 - p)(1 - p)(1 - p)p = (1 - p)^4 p$

EXAMPLE 10

Suppose you are 18 years old and speculating about your future. You decide that the probability that you will eventually get married and have children is 75%. By Rule 4, you must then assume that the probability that you will eventually get married is *at least* 75%. The possible futures in which you get married *and* have children are a subset of the possible futures in which you just get married. ∎

15.5   WHEN WILL IT HAPPEN?

Often, we would like an event to occur and will keep trying to make it happen until it does, such as when a couple keeps having children until they have one of the desired sex. Also, we often gamble that something won't go wrong, even though we know it could, such as when people have unprotected sex and hope they won't get infected with HIV, the virus that is suspected to cause AIDS.

A simple application of our probability rules allows us to determine the chances of waiting one, two, three, or any given number of repetitions for such events to occur. Suppose (1) we know the probability of each possible outcome on any given occasion, (2) those probabilities remain the same for each occasion, and (3) the outcome each time is *independent* of the outcome all of the other times.

Let's use some shorthand. Define the probability that the outcome of interest will occur on any given occasion to be p so that the probability that it will not occur is $(1 - p)$ by Rule 1. For instance, if we are interested in giving birth to a girl, p is .49 and $(1 - p)$ is .51.

We already know the probability that the outcome occurs on the first try is p. By Rule 3, the probability that it *doesn't* occur on the first try but *does* occur on the *second* try is found by multiplying two probabilities. Namely, it doesn't happen at first $(1 - p)$ and then it does happen (p). Thus, the probability that it happens for the *first* time on the second try is $(1 - p)p$. We can continue this logic. We multiply $(1 - p)$ for each time it *doesn't* happen, followed by p for when it finally *does* happen. We can represent these probabilities as shown in Table 15.2, and you can see the emerging pattern.

TABLE 15.3	Probability of a Birth Resulting in a First Girl

Number of Births to First Girl	Probability
1	.49
2	(.51)(.49) = .2499
3	(.51)(.51)(.49) = .1274
5	(.51)(.51)(.51)(.51)(.49) = .0331
7	(.51)(.51)(.51)(.51)(.51)(.51)(.49) = .0086

EXAMPLE 11

NUMBER OF BIRTHS TO FIRST GIRLS

The probability of a birth resulting in a boy is about .51, and the probability of a birth resulting in a girl is about .49. Suppose a couple would like to continue having children until they have a girl. Assuming the outcomes of births are independent of each other, the probabilities of having the first girl on the first, second, third, fifth, and seventh tries are shown in Table 15.3. ∎

Accumulated Probability

We are often more interested in the *cumulative probability* of something happening by a certain time than just the specific occasion on which it will occur. For example, we would probably be more interested in knowing the probability that we would have had the first girl *by the time of* the fifth child, rather than the probability that it would happen at that specific birth. It is easy to use the probability rules to find this accumulated probability. Notice that the probability of the first occurrence *not happening* by occasion n is $(1 - p)^n$ Therefore, the probability that the first occurrence *has happened* by occasion n is $[1 - (1 - p)^n]$ from Rule 1.

EXAMPLE 12

GETTING INFECTED WITH HIV

According to Krantz (1992, p. 13), the probability of getting infected with HIV from a single heterosexual encounter without a condom, with a partner whose risk status you don't know, is between 1/500 and 1/500,000. For the sake of this example, let's assume it is the higher figure of 1/500 = .002. Therefore, on such a single encounter, the probability of not getting infected is 499/500 = .998. However, the risk of getting infected goes up with multiple encounters, and, using the strategy we have just outlined, we can calculate the probabilities associated

The Probability of Getting Infected with HIV from Unprotected Sex

Number of Encounters	Probability of First Infection	Accumulated Probability of HIV
1	.002	.002
2	$(.998)(.002) = .001996$	.003996
4	$(.998)^3(.002) = .001988$	.007976
10	$(.998)^9(.002) = .001964$	.019821

with the number of encounters it would take to become infected. Of course, the real interest is in whether infection will have occurred after a certain number of encounters, and not just on the exact encounter during which it occurs. In Table 15.4, we show this accumulated probability as well. It is found by adding the probabilities to that point, using Rule 2. Equivalently, we could use the general form we found just prior to this example. For instance, the probability of HIV infection *by the second encounter* is $[1 - (1 - .002)^2]$ or $[1 - .998^2]$ or .003996.

Table 15.4 tells us that, although the risk after a single encounter is only 1 in 500, after ten encounters the accumulated risk has risen to almost .02, or almost 1 in 50. This means that out of all those people who have ten encounters, about 1 in 50 of them is likely to get infected with HIV. Also, according to Krantz (1992, p. 13), the probability of infection with a partner whose status is unknown if a condom is used is between 1/5000 and 1 in 5 million. Assuming the higher figure of 1/5000, the probability of infection after ten encounters is only 1/5009, or about .0002.

The base rate, or probability of infection, may have changed by the time you read this. But the method for calculating the risk remains the same, and you can reevaluate all of the numbers if you know the current risk for a single encounter. ■

WINNING THE LOTTERY

To play the California Super Lotto game, a player picks six numbers from the choices 1 to 51. Six winning numbers are selected. If the player has matched at least three of the winning numbers, the ticket is a winner. Matching three numbers results in a prize of $5.00; matching four or more results in a prize determined by the number of other successful entries. Information on the back of the ticket places the probability of winning anything at all at 1/60. How many times would you have to play before winning anything? See Table 15.5.

If you do win, your most likely prize is only $5.00. Notice that even after purchasing five tickets, which cost one dollar each, your probability of having won

TABLE 15.5	Probabilities of Winning Super Lotto	
Number of Plays	Probability of First Win	Accumulated Probability of Win
1	$1/60 = .0167$	$1/60 = .0167$
2	$(59/60)(1/60) = .0164$	.0331
5	$(59/60)^4(1/60) = .0156$	.0806
10	$(59/60)^9(1/60) = .0143$	.1547
20	$(59/60)^{19}(1/60) = .0121$	.2855

anything is still under 10%; in fact, it is just 8%. After 20 tries, your probability of having won anything is less than 30%. In the next section, we learn how to determine the average expected payoff from playing games like this. ■

15.6   LONG-TERM GAINS, LOSSES, AND EXPECTATIONS

The concept of the long-run relative frequency of various outcomes can be used to predict long-term gains and losses. Although it is impossible to predict the result of one random happening, we can be remarkably successful in predicting aggregate or long-term results. For example, we noted that the probability of winning anything at all in the California Super Lotto game with a single ticket is 1 in 60. Among the millions of people who play that game regularly, some will be winners and some will be losers. We cannot predict who will win and who will lose. However, we can predict that in the long run, about 1 in every 60 tickets sold will be a winner.

Long-Term Outcomes Can Be Predicted

It is because aggregate or long-term outcomes can be accurately predicted that lottery agencies, casinos, and insurance companies are able to stay in business. Because they can closely predict the amount they will have to pay out over the long run, they can determine how much to charge and still make a profit.

EXAMPLE 14   **INSURANCE POLICIES**

Suppose an insurance company has thousands of customers, and each customer is charged $500 a year. The company knows that about 10% of them will submit a claim in any given year and that claims will always be for $1500. How much can the company expect to make per customer?

Notice that there are two possibilities. With probability .90 (or for about 90% of the customers), the amount gained by the company is $500, the cost of the policy. With probability .10 (or for about 10% of the customers), the "amount gained" by the company is the $500 cost of the policy minus the $1500 payoff, for a loss of $1000. We represent the loss by saying that the "amount gained" is −$1000, or negative one thousand dollars. Here are the possible amounts "gained" and their probabilities:

Claim Paid?	Probability	Amount Gained
Yes	.10	−$1000
No	.90	+$ 500

What is the average amount gained, per customer, by the company? Because the company gains $500 from 90% of its customers and loses $1000 from the remaining 10%, its average "gain" per customer is:

average gain = .90 ($500) − .10 ($1000) = $350

In other words, the company makes an average of $350 per customer. Of course, to succeed this way, it must have a large volume of business. If it had only a few customers, the company could easily lose money in any given year. As we have seen, long-run frequencies apply only to a large aggregate. For example, if the company only had two customers, we could use Rule 3 to find that the probability of the company's having to pay both of them during a given year is .1 × .1 = .01 = 1/100. This calculation assumes that the probability of paying for one individual is independent of that for the other individual, which is a reasonable assumption unless the customers are somehow related to each other. ∎

Expected Value

Statisticians use the phrase **expected value (EV)** to represent the average value of any measurement over the long run. The average gain of $350 per customer for our hypothetical insurance company is called the *expected value* for the amount the company earns per customer. Notice that the expected value does not have to be one of the possible values. For our insurance company, the two possible values were $500 and − $1000. Thus, the expected value of $350 was not even a possible value for any one customer. In that sense, "expected value" is a real misnomer. It doesn't have to be a value that's ever expected in a single outcome.

To compute the expected value for any situation, we need only be able to specify the possible *amounts*—call them $A_1, A_2, A_3, \ldots, A_k$—and the associated *probabilities*, which can be denoted by $p_1, p_2, p_3, \ldots, p_k$. Then the expected value can be found by multiplying each possible amount by its probability and adding them up. Remember, the expected value is the average value per measurement over the long run and not necessarily a typical value for any one occasion or person.

Computing the expected value

$$EV = \text{expected value} = A_1 p_1 + A_2 p_2 + A_3 p_3 + \ldots + A_k p_k$$

TABLE 15.6   Probability of Winning the Decco Game

Number of Matches	Prize	Net Gain	Probability
4	$5000	$4999	1/28,561 = .000035
3	$50	$49	1/595 = .00168
2	$5	$4	1/33 = .0303
1	Free ticket	0	.2420
0	None	–$1	.7260

EXAMPLE 15   **CALIFORNIA DECCO LOTTERY GAME**

The California lottery has offered a number of games over the years. One such game is Decco, in which players chose one card from each of the four suits in a regular deck of playing cards. For example, the player might choose the 4 of hearts, 3 of clubs, 10 of diamonds, and jack of spades. A winning card is then drawn from each suit. If even one of the choices matches the winning cards drawn, a prize is awarded. It costs one dollar for each play, so the net gain for any prize is one dollar less than the prize. Table 15.6 shows the prizes and the probability of winning each prize, taken from the back of a game card.

We can thus compute the expected value for this Decco game:

$$EV = (\$4999 \times 1/28,561) + (\$49 \times 1/595) + (\$4 \times .0303) + (-\$1 \times .726)$$
$$= -\$0.35$$

Notice that we count the free ticket as an even trade because it is worth $1, the same amount it cost to play the game. This result tells us that over many repetitions of the game, you will *lose* an *average* of 35 cents each time you play. From the perspective of the Lottery Commission, about 65 cents is paid out for each one dollar ticket sold for this game. ■

Expected Value as Mean Number

If the measurement in question is one taken over a large group of individuals, rather than across time, the expected value can be interpreted as the mean value per individual. As a simple example, suppose we had a population in which 40% of the people smoked a pack of cigarettes a day (20 cigarettes) and the remaining 60% smoked none. Then the expected value for the number of cigarettes smoked per day would be:

$$EV = (.40 \times 20 \text{ cigarettes}) + (.60 \times 0 \text{ cigarettes}) = 8 \text{ cigarettes}$$

In other words, on average, eight cigarettes are smoked per person per day. If we were to measure each person in the population by asking them how many cigarettes they smoked per day (and they answered truthfully), then the arithmetic average would be eight. This example further illustrates the fact that the expected value is not a value we actually expect to measure on any one individual.

CASE STUDY 15.1   Birthdays and Death Days—Is There a Connection?

SOURCE: Phillips, Van Voorhies, and Ruth, 1992.

Is the timing of death random or does it depend on significant events in one's life? That's the question U.C. San Diego sociologist David Phillips and his colleagues attempted to answer. Previous research had shown a possible connection between the timing of death and holidays and other special occasions. This study focused on the connection between birthday and death day.

The researchers studied death certificates of all Californians who had died between 1969 and 1990. Because of incomplete information before 1978, we report only on the part of their study that included the years 1979 to 1990. They limited their study to adults (over 18) who had died of natural causes. They eliminated anyone for whom surgery had been a contributing factor to death because there is some choice as to when to schedule surgery. They also omitted those born on February 29 because there was no way to know on which date these people celebrated their birthday in non–leap years.

Because there is a seasonal component to birthdays and death days, the researchers adjusted the numbers to account for those as well. They determined the number of deaths that would be expected on each day of the year if date of birth and date of death were independent of each other. Each death was then classified as to how many weeks after the birthday it occurred. For example, someone who died from 0 to 6 days after his or her birthday was classified as dying in "Week 0," whereas someone who died from 7 to 13 days after the birthday was classified in "Week 1," and so on. Thus, people who died in Week 51 died within a few days before their birthdays.

Finally, the researchers compared the actual numbers of deaths during each week with what would be expected based on the seasonally adjusted data. Here is what they found. For women, the biggest peak was in Week 0. For men, the biggest peak was in Week 51. In other words, the week during which the highest number of women died was the week *after* their birthdays. The week during which the highest number of men died was the week *before* their birthdays.

Perhaps this observation is due only to chance. Each of the 52 weeks is equally likely to show the biggest peak. What is the probability that the biggest peak for the women would be Week 0 *and* the biggest peak for the men would be Week 51? Using Rule 3, the probability of both events occurring is $(1/52) \times (1/52) = 1/2704 = .0004$.

As we will learn in Chapter 17, unusual events often do happen just by chance. Many facts given in the original report, however, add credence to the idea that this is not a chance result. For example, the peak for women in Week 0 remained even when the deaths were separated by age group, by race, and by cause of death. It was also present in the sample of deaths from 1969 to 1977. Further, earlier studies from various cultures have shown that people tend to die just after holidays important to that culture. ■

FOR THOSE WHO LIKE FORMULAS

Notation
Denote "events" or "outcomes" with capital letters A, B, C, and so on.
If A is one outcome, all other possible outcomes are part of "A complement" $= A^C$.

$P(A)$ is the probability that the event or outcome A occurs. For any event A,
$0 \leq P(A) \leq 1$.

Rule 1
$$P(A) + P(A^C) = 1$$

A useful formula that results from this is

$$P(A^C) = 1 - P(A)$$

Rule 2
If events A and B are *mutually exclusive*, then

$$P(A \text{ or } B) = P(A) + P(B)$$

Rule 3
If events A and B are *independent*, then

$$P(A \text{ and } B) = P(A)\,P(B)$$

Rule 4
If the ways in which an event B can occur are a subset of those for event A, then

$$P(B) \leq P(A)$$

EXERCISES

1. Recall that there are two interpretations of probability: relative frequency and personal probability.

    a. Which interpretation applies to this statement: "The probability that I will get the flu this winter is 30%"? Explain.

    b. Which interpretation applies to this statement: "The probability that a randomly selected adult in America will get the flu this winter is 30%"? Explain. (Assume it is known that the proportion of adults who get the flu each winter remains at about 30%.)

2. Use the probability rules in this chapter to solve each of the following:

    a. The probability that a randomly selected Caucasian-American child will have blonde or red hair is 23%. The probability of having blonde hair is 14%. What is the probability of having red hair?

    b. According to Blackenhorn (24–26 February 1995), in 1990 the probability that a randomly selected child was living with his or her mother as the sole parent was .216 and with his or her father as the sole parent was .031. What was the probability that a child was living with just one parent?

    c. In 1991, the probability that a birth would result in twins was .0231, and the probability that a birth would result in triplets or more was .0008 (*USA Weekend*, 25–27 November 1994, p. 15). What was the probability that a birth in 1991 resulted in a single child only?

3. There is something wrong in each of the following statements. Explain what is wrong.

    a. The probability a randomly selected driver will be wearing a seat belt is .75, whereas the probability that he or she will not be wearing one is .30.

    b. The probability that a randomly selected car is red is 1.20.

    c. The probability that a randomly selected car is red is .20, whereas the probability that a randomly selected car is a red sports car is .25.

4. According to Krantz (1992, p. 111), the probability of being born on a Friday the 13th is about 1/214.

    a. What is the probability of not being born on a Friday the 13th?

    b. In any particular year, Friday the 13th can occur once, twice, or three times. Is the probability of being born on Friday the 13th the same every year? Explain.

    c. Explain what it means to say that the probability of being born on Friday the 13th is 1/214.

5. Explain which of the following more closely describes what it means to say that the probability of a tossed coin landing with heads up is 1/2:

Explanation 1: After more and more tosses, the fraction of heads will get closer and closer to 1/2.

Explanation 2: The number of heads will always be about half the number of tosses.

6. Explain why probabilities cannot always be interpreted using the relative-frequency interpretation. Give an example of where that interpretation would not apply.

7. Suppose you wanted to test your ESP using an ordinary deck of 52 cards, which has 26 red and 26 black cards. You have a friend shuffle the deck and draw cards at random, replacing the card and reshuffling after each guess. You attempt to guess the color of each card.

    a. What is the probability that you guess the color correctly by chance?

    b. Is the answer in part a based on the relative-frequency interpretation of probability or is it a personal probability?

    c. Suppose another friend has never tried the experiment but believes he has ESP and can guess correctly with probability .60. Is the value of .60 a relative-frequency probability or a personal probability? Explain.

    d. Suppose another friend guessed the color of 1000 cards and got 600 correct. The friend claims she has ESP and has a .60 probability of guessing correctly. Is the value of .60 a relative-frequency probability or a personal probability? Explain.

8. Suppose you wanted to determine the probability that someone randomly selected from the phone book in your town or city has the same first name as you.

    a. Assuming you had the time and energy to do it, how would you go about determining that probability? (Assume all names listed are spelled out.)

    b. Using the method you described in part a, would your result be a relative-frequency probability or a personal probability? Explain.

9. A small business performs a service and then bills its customers. From past experience, 90% of the customers pay their bills within a week.

    a. What is the probability that a randomly selected customer will not pay within a week?

    b. The business has billed two customers this week. What is the probability that neither of them will pay within a week? What assumption did you make to compute that probability? Is it a reasonable assumption?

10. Suppose the probability that you get an interesting piece of mail on any given weekday is 1/10. Is the probability that you get at least one interesting piece of mail during the week (Monday to Friday) equal to 5/10? Why or why not?

11. The probability that a randomly selected American adult belongs to the American Automobile Association (AAA) is .10 (10%), and the probability that that person belongs to the American Association of Retired Persons (AARP) is .11 (11%) (Krantz, 1992, p. 175). What assumption would we have to make in order to use Rule 3 to conclude that the probability that a person belongs to both is $(.10) \times (.11) = .011$? Do you think that assumption holds in this case? Explain.

12. A study by Kahneman and Tversky (1982, p. 496) asked people the following question: "Linda is 31 years old, single, outspoken, and very bright. She majored in philosophy. As a student, she was deeply concerned with issues of discrimination and social justice, and also participated in antinuclear demonstrations. Please check off the most likely alternative:

A. Linda is a bank teller.

B. Linda is a bank teller and is active in the feminist movement.

Nearly 90% of the 86 respondents chose alternative B. Explain why alternative B *cannot* have a higher probability than alternative A.

13. Example 3 in this chapter states that "the probability that a piece of checked luggage will be temporarily lost on a flight with a U.S. airline is 1/176." Interpret that statement, using the appropriate interpretation of probability.

14. In Section 15.2, you learned two ways in which relative-frequency probabilities can be determined. Explain which method you think was used to determine each of the following probabilities:

    a. The probability that a particular flight from New York to San Francisco will be on time is .78.

    b. On any given day, the probability that a randomly selected American adult will read a book for pleasure is .33.

    c. The probability that a five-card poker hand contains "four of a kind" is .00024.

15. People are surprised to find that it is not all that uncommon for two people in a group of 20 to 30 people to have the same birthday. We will learn how to find that probability in a later chapter. For now, consider the probability of finding two people who have birthdays in the same month. Make the simplifying assumption that the probability that a randomly selected person will have a birthday in any given month is 1/12. Suppose there are three people in a room and you consecutively ask them their birthdays. Your goal, following parts a–d (below), is to determine the probability that at least two of them were born in the same calendar month.

    a. What is the probability that the second person you ask will not have the same birth month as the first person? (*Hint:* Use Rule 1.)

    b. Assuming the first and second persons have different birth months, what is the probability that the third person will have yet a different birth month? (*Hint:* Suppose January and February have been taken. What proportion of all people will have birth months from March to December?)

    c. Explain what it would mean about overlap among the three birth months if the outcomes in part a and part b both happened. What is the probability that the outcomes in part a and part b will both happen?

    d. Explain what it would mean about overlap among the three birth months if the outcomes in part a and part b did *not* both happen. What is the probability of that occurring?

16. Use your own particular expertise to assign a personal probability to something, such as the probability that a certain sports team will win next week. Now assign a personal probability to another *related* event. Explain how you determined each probability, and explain how your assignments are *coherent*.

17. Read the definition of "independent events" given in Rule 3. Explain whether each of the following pairs of events is likely to be independent:

a. A married couple goes to the voting booth. Event A is that the husband votes for the Republican candidate; Event B is that the wife votes for the Republican candidate.

b. Event A is that it snows tomorrow; Event B is that the high temperature tomorrow is at least 60 degrees Fahrenheit.

c. You buy a lottery ticket, betting the same numbers two weeks in a row. Event A is that you win in the first week; Event B is that you win in the second week.

d. Event A is that a major earthquake will occur somewhere in the world in the next month; Event B is that the Dow Jones Industrial Average will be higher in one month than it is now.

18. Suppose you routinely check coin-return slots in vending machines to see if they have any money in them. You have found that about 10% of the time you find money.

a. What is the probability that you do not find money the next time you check?

b. What is the probability that the next time you will find money is on the third try?

c. What is the probability that you will have found money by the third try?

19. Lyme disease is a disease carried by ticks, which can be transmitted to humans by tick bites. Suppose the probability of contracting the disease is 1/100 for each tick bite.

a. What is the probability that you will not get the disease when bitten once?

b. What is the probability that you will not get the disease from your first tick bite and will get it from your second tick bite?

20. According to Krantz (1992, p. 161), the probability of being injured by lightning in any given year is 1/685,000. Assume that the probability remains the same from year to year and that avoiding a strike in one year doesn't change your probability in the next year.

a. What is the probability that someone who lives 70 years will never be struck by lightning? You do not need to compute the answer, but write down how it would be computed

b. According to Krantz, the probability of being injured by lightning over the average lifetime is 1/9100. Show how that probability should relate to your answer in part a, assuming that average lifetime is about 70 years.

c. Do the probabilities given in this exercise apply specifically to you? Explain.

d. About 260 million people live in the United States. In a typical year, assuming Krantz's figure is accurate, about how many would be expected to be struck by lightning?

21. Suppose you have to cross a train track on your commute. The probability that you will have to wait for a train is 1/5, or .20. If you don't have to wait, the commute takes 15 minutes, but if you have to wait, it takes 20 minutes.

a. What is the expected value of the time it takes you to commute?

....

TABLE 15.7

Number of Matches	Amount Won	Probability
3	$20	.012
2	$2	.137
0 or 1	$0	.851

b. Is the expected value ever the actual commute time? Explain.

22. Remember that the probability that a birth results in a boy is about .51. You offer a bet to an unsuspecting friend. Each day you will call the local hospital and find out how many boys and how many girls were born the previous day. For each girl, you will give your friend $1 and for each boy your friend will give you $1.

   a. Suppose that on a given day there are 3 births. What is the probability that you lose $3 on that day? What is the probability that your friend loses $3?

   b. Notice that your net profit is $1 if a boy is born and –$1 if a girl is born. What is the expected value of your profit for each birth?

   c. Using your answer in part b, how much can you expect to make after 1000 births?

23. In the "3 Spot" version of the California Keno lottery game, the player picks three numbers from 1 to 40. Ten possible winning numbers are then randomly selected. It costs $1 to play. Table 15.7 shows the possible outcomes. Compute the expected value for this game. Interpret what it means.

24. Suppose the probability that you get an A in any class you take is .3, and the probability that you get a B is .7. To construct a grade-point average, an A is worth 4.0 and a B is worth 3.0. What is the expected value for your grade-point average? Would you expect to have this grade-point average separately for each quarter or semester? Explain.

25. In 1991, 72% of children in the United States were living with both parents, 22% were living with mother only, and 3% were living with father only and 3% were not living with either parent (*World Almanac and Book of Facts,* 1993, p. 945). What is the expected value for the number of parents a randomly selected child was living with? Does the concept of expected value have a meaningful interpretation for this example? Explain.

26. We have seen many examples for which the term *expected value* seems to be a misnomer. Construct an example of a situation where the term *expected value* would *not* seem to be a misnomer for what it represents.

27. Find out your yearly car insurance cost. If you don't have a car, find out the yearly cost for a friend or relative. Now assume you will either have an accident or not, and if you do, it will cost the insurance company $5000 more than the premium you pay. Calculate what yearly accident probability would result in a "break-even" expected value for you and the insurance company. Comment on whether you think your answer is an accurate representation of your yearly probability of having an accident.

MINI-PROJECTS

1. Refer to Exercise 12. Present the question to ten people, and note the proportion who answer with alternative B. Explain to the participants why it cannot be the right answer, and report on their reactions.

2. Flip a coin 100 times. Stop each time you have done 10 flips (that is, stop after 10 flips, 20 flips, 30 flips, and so on) and compute the proportion of heads using *all* of the flips up to that point. Plot that proportion versus the number of flips. Comment on how the plot relates to the relative-frequency interpretation of probability.

3. Pick an event that will result in the same outcome for everyone, such as whether it will rain next Saturday. Ask ten people to assess the probability of that event, and note the variability in their responses. (Don't let them hear each other's answers, and make sure you don't pick something that would have 0 or 1 as a common response.) At the same time, ask them the probability of getting a heart when a card is randomly chosen from a fair deck of cards. Compare the variability in responses for the two questions and explain why one is more variable than the other.

4. Find two lottery or casino games that have fixed payoffs and for which the probabilities of each payoff are available. (Some lottery tickets list them on the back of the ticket. Some books about gambling give the payoffs and probabilities for various casino games.)

   a. Compute the expected value for each game. Discuss what they mean.

   b. Using both the expected values and the list of payoffs and probabilities, explain which game you would rather play and why.

REFERENCES

Arraf, Jane. (14 December 1994). Leave Earth or perish: Sagan. *China Post* (Taiwan), p. 4.

Blackenhorn, David. (24–26 February 1995). Life without father. *USA Weekend,* pp. 6–7.

Information please almanac. (1991). Edited by Otto Johnson. Boston: Houghton Mifflin.

Kahneman, D., and A. Tversky. (1982). On the study of statistical intuitions. In D. Kahneman, P. Slovic, and A. Tversky (eds.), *Judgment under uncertainty: Heuristics and biases* (Chapter 34). Cambridge, England: Cambridge University Press.

Krantz, Les. (1992). *What the odds are.* New York: Harper Perennial.

Phillips, D. P., C. A. Van Voorhies, and T. E. Ruth. (1992). The birthday: Lifeline or deadline? *Psychosomatic Medicine* 54, pp. 532–542.

World almanac and book of facts. (1993). Edited by Mark S. Hoffman. New York: Pharos Books.

CHAPTER SIXTEEN

Psychological Influences on Personal Probability

THOUGHT QUESTIONS

1. Plous (1993) presented readers with the following test.

   Place a check mark beside the alternative that seems most likely to occur within the next 10 years:

   - An all-out nuclear war between the United States and Russia
   - An all-out nuclear war between the United States and Russia in which neither country intends to use nuclear weapons, but both sides are drawn into the conflict by the actions of a country such as Iraq, Libya, Israel, or Pakistan.

   Using your intuition, pick the more likely statement. Now consider the probability rules discussed in Chapter 15 to try to determine which statement is more likely.

2. Which is a more likely cause of death in the United States, homicide or diabetes? How did you arrive at your answer?

3. Do you think people are more likely to pay to reduce their risk of an undesirable event from 95% to 90% or to reduce it from 5% to zero? Explain whether there should be a preferred choice, based on the material from Chapter 15.

4. A fraternity consists of 30% freshmen and sophomores and 70% juniors and seniors. Bill is a member of the fraternity, he studies hard, he is well-liked by his fellow fraternity members, and he will probably be quite successful when he graduates. Is there any way to tell if Bill is more likely to be a lower classman (freshman or sophomore) or an upper classman (junior or senior)?

16.1   REVISITING PERSONAL PROBABILITY

In Chapter 15, we assumed that the probabilities of various outcomes were known or could be calculated using the relative-frequency interpretation of probability. But most decisions people make are in situations that require the use of personal probabilities. The situations are not repeatable, nor are there physical assumptions that can be used to calculate potential relative frequencies.

Personal probabilities, remember, are values assigned by individuals based on how likely they think events are to occur. By their very definition, personal probabilities do not have a single correct value. However, they should still follow the rules of probability, which we outlined in Chapter 15; otherwise, decisions based on them can be contradictory. For example, if you believe there is a very high probability that you will be killed in an automobile accident before you reach a certain age, but also believe there is a very high probability that you will live to be 100, then you will be conflicted over how well to protect your health. Your two personal probabilities are not consistent with each other and will lead to contradictory decisions.

In this chapter, we explore research that has shown how personal probabilities can be influenced by psychological factors in ways that lead to incoherent or inconsistent probability assignments. We also examine circumstances in which many people assign personal probabilities that can be shown to be incorrect based on the relative-frequency interpretation. Every day, you are required to make decisions that involve risks and rewards. Understanding the kinds of influences that can affect your decisions adversely should help you make more realistic judgments.

16.2   EQUIVALENT PROBABILITIES; DIFFERENT DECISIONS

People like a sure thing. It would be wonderful if we could be guaranteed that cancer would never strike us, for instance, or that we would never be in an automobile accident. For this reason, people are willing to pay a premium to reduce their risk of something to zero, but are not as willing to pay to reduce their risk by the same amount to a nonzero value. We consider two versions of this psychological reality.

The Certainty Effect

Suppose you are buying a new car. The salesperson explains that you can purchase an optional safety feature for $200 that will reduce your chances of death in a high-speed accident from 50% to 45%. Would you be willing to purchase the device?

Now suppose instead that the salesperson explains that you can purchase an optional safety feature for $200 that will reduce your chances of death in a high-speed accident from 5% to zero. Would you be willing to purchase the device?

In both cases, your chances of death are reduced by 5%, or 1/20. But research has shown that people are more willing to pay to reduce their risk from a fixed amount down to zero than they are to reduce their risk by the same amount when it is not reduced to zero. This is called the **certainty effect** (Plous, 1993, p. 99).

EXAMPLE 1	**PROBABILISTIC INSURANCE**

To test whether the certainty effect influences decisions, Kahneman and Tversky (1979) asked students if they would be likely to buy "probabilistic insurance." This insurance would cost half as much as regular insurance but would only cover losses with 50% probability. The majority of respondents (80%) indicated that they would not be interested in such insurance. Notice that the expected value for the return on this insurance is the same as on the regular policy. It is the lack of assurance of a payoff that makes it unattractive. ∎

The Pseudocertainty Effect

A related idea, used in marketing, is that of the **pseudocertainty effect** (Slovic, Fischhoff, and Lichtenstein, 1982, p. 480). Rather than being offered a reduced risk on a variety of problems, you are offered a complete reduction of risk on certain problems and no reduction on others.

As an example, consider the extended warranty plans offered on automobiles and appliances. You pay a price when you purchase the item and certain problems are covered completely for a number of years. Other problems are not covered. If you were offered a plan that covered all problems with 30% probability, you would probably not purchase it. But if you were offered a plan that completely covered 30% of the possible problems, you might consider it. Both plans have the same expected value over the long run, but most people prefer the plan that covers some problems with certainty.

EXAMPLE 2	**VACCINATION QUESTIONNAIRES**

To test the idea that the pseudocertainty effect influences decision making, Slovic and colleagues (1982) administered two different forms of a "vaccination questionnaire." The first form described what the authors called "probabilistic protection," in which a vaccine was available for a disease anticipated to afflict 20% of the population. However, the vaccine would protect people with only 50% probability. Respondents were asked if they would volunteer to receive the vaccine, and 40% indicated that they would.

The second form described a situation of "pseudocertainty," in which there were two equally likely strains of the disease, each anticipated to afflict 10% of the population. The available vaccine was completely effective against one strain but provided no protection at all against the other strain. This time, 57% of respondents indicated they would volunteer to receive the vaccine.

In both cases, receiving the vaccine would reduce the risk of disease from 20% to 10%. However, the scenario in which there was complete elimination of risk for a subset of problems was perceived much more favorably than the one for which there was the same reduction of risk overall. This is what the pseudocertainty effect predicts.

Plous (1993, p. 101) notes that a similar effect is found in marketing when items are given away free rather than having their price reduced. For example, rather than reduce all items by 50%, a merchandiser may instead advertise that you can "buy one, get one free." The overall reduction is the same, but the offer of free merchandise may be perceived as more desirable than the discount. ■

16.3   HOW PERSONAL PROBABILITIES CAN BE DISTORTED

Which do you think caused more deaths in the United States in 1993, homicide or diabetes? If you are like respondents to studies reported by Slovic and colleagues (1982, p. 467), you answered that it was homicide. The actual death rates were 9.9 per 100,000 for homicide compared with 21.4 per 100,000 for diabetes (*World Almanac and Book of Facts,* 1995, p. 959).

The distorted view that homicide is more probable results from the fact that homicide receives more attention in the media. Psychologists attribute this incorrect perception to the **availability heuristic.** It is just one example of how perceptions of risk can be influenced by reference points.

The Availability Heuristic

Tversky and Kahneman (1982a, p. 11) note that "there are situations in which people assess the . . . probability of an event by the ease with which instances or occurrences can be brought to mind. . . . This judgmental heuristic is called availability." In the study summarized by Slovic and colleagues (1982), media attention to homicides severely distorted judgments about their relative frequency. Slovic and colleagues noted that

> *Homicides were incorrectly judged more frequent than diabetes and stomach cancer deaths. Homicides were also judged to be about as frequent as death by stroke, although the latter actually claims about 11 times as many deaths. (1982, p. 467)*

Availability can cloud your judgment in numerous instances. For example, if you are buying a used car, you may be influenced more by the bad luck of a friend or relative who owned a particular model than by statistics provided by consumer groups based on the experience of thousands of owners. The memory of the one bad car in that class is readily available to you.

Similarly, most people know many smokers who don't have lung cancer. Fewer people know someone who has actually contracted lung cancer as a result of smoking. Therefore, it is easier to bring to mind the healthy smokers and, if you smoke, to believe that you too will continue to have good health.

Detailed Imagination

One way to encourage availability, in which the probabilities of risk can be distorted, is by having people vividly imagine an event. Salespeople use this trick when they try to sell you extended warranties or insurance. For example, they may convince you that $500 is a reasonable price to pay for an extended warranty on your new car by having you imagine that if your air conditioner fails it will cost you more than the price of the policy to get it fixed. They don't mention that it is extremely unlikely that your air conditioner will fail during the period of the extended warranty.

Anchoring

Psychologists have shown that people's risk perception can also be severely distorted when they are provided with a reference point, or an **anchor,** from which they then adjust up or down. Most people tend to stay relatively close to the anchor, or initial value, provided.

EXAMPLE 3 **NUCLEAR WAR**

Plous (1993, pp. 146–147) conducted a survey between January 1985 and May 1987 in which he asked respondents to assess the likelihood of a nuclear war between the United States and the Soviet Union. He gave three different versions of the survey, which he called the low-anchor, high-anchor, and no-anchor conditions. In the low-anchor case, he asked people if they thought the chances were higher or lower than 1% and then asked them to give their best estimate of the exact chances. In the high-anchor case, the 1% figure was replaced by 90% or 99%. In the no-anchor case, they were simply asked to give their own assessment.

According to Plous:

> In all variations of the survey, anchoring exerted a strong influence on likelihood estimates of a nuclear war. Respondents who were initially asked whether the probability of nuclear war was greater or less than 1 percent subsequently gave lower estimates than people who were not provided with an explicit anchor, whereas respondents who were first asked whether the probability of war was greater or less than 90 (or 99) percent later gave estimates that were higher than those given by respondents who were not given an anchor. (1993, p. 147) ■

Research has shown that anchoring influences real-world decisions as well. For example, jurors who are first told about possible harsh verdicts and then about more lenient ones are more likely to give a harsh verdict than jurors given the choices in reverse order.

EXAMPLE 4   **SALES PRICE OF A HOUSE**

Plous (1993) describes a study conducted by Northcraft and Neale (1987), in which real estate agents were asked to give a recommended selling price for a home. They were given a ten-page packet of information about the property and spent 20 minutes walking through it. Contained in the ten-page packet of information was a listing price. To test the effect of anchoring, four different listing prices, ranging from $119,900 to $149,900, were given to different groups of agents. The house had actually been appraised at $135,000. As the anchoring theory predicts, the agents were heavily influenced by the particular listing price they were given. The four listing prices and the corresponding mean recommended selling prices were

Apparent Listed Price	$119,900	$129,900	$139,900	$149,900
Mean Recommended Sales Price	$117,745	$127,836	$128,530	$130,981

As you can see, the recommended sales price differed by more than $10,000 just because the agents were anchored at different listing prices. Yet, when asked how they made their judgments, very few of the agents mentioned the listing price as one of their top factors.

Anchoring is most effective when the anchor is extreme in one direction or the other. It does not have to take the form of a numerical assessment either. Be wary when someone describes a worst- or best-case scenario and then asks you to make a decision. For example, an investment counselor may encourage you to invest in a certain commodity by describing how one year it had such incredible growth that you would now be rich if you had only been smart enough to invest in the commodity during that year. If you use that year as your anchor, you'll fail to see that, on average, the price of this commodity has risen no faster than inflation. ■

The Representativeness Heuristic and the Conjunction Fallacy

In some cases, the **representativeness heuristic** leads people to assign higher probabilities than are warranted to scenarios that are representative of how we imagine things would happen. For example, Tversky and Kahneman (1982a, p. 98) note that "the hypothesis 'the defendant left the scene of the crime' may appear less plausible than the hypothesis 'the defendant left the scene of the crime for fear of being accused of murder,' although the latter account is less probable than the former."

It is the representativeness heuristic that sometimes leads people to fall into the judgment trap called the **conjunction fallacy.** We learned in Chapter 15 (Rule 4) that the probability of two events occurring together, in *conjunction,* cannot be higher than the probability of either event occurring alone. The conjunction fallacy occurs when detailed scenarios involving the conjunction of events are given higher probability assessments than statements of one of the simple events alone.

EXAMPLE 5   **AN ACTIVE BANK TELLER**

A classic example, provided by Kahneman and Tversky (1982, p. 496), was a study in which they presented subjects with the following statement:

> Linda is 31 years old, single, outspoken, and very bright. She majored in philosophy. As a student, she was deeply concerned with issues of discrimination and social justice, and also participated in antinuclear demonstrations.

Respondents were then asked which of two statements was more probable:

1. Linda is a bank teller.
2. Linda is a bank teller who is active in the feminist movement.

Kahneman and Tversky report that "in a large sample of statistically naive undergraduates, 86% judged the second statement to be more probable" (1982, p. 496).

The problem with that judgment is that the group of people in the world who fit the second statement is a subset of the group who fit the first statement. If Linda falls into the second group (bank tellers who are active in the feminist movement), she must also fall into the first group (bank tellers). Therefore, the *first* statement must have a higher probability of being true.

The misjudgment is based on the fact that the second statement is much more representative of how Linda was described. This example illustrates that intuitive judgments can directly contradict the known laws of probability. In this example, it was easy for respondents to fall into the trap of the conjunction fallacy. ∎

The representativeness heuristic can be used to affect your judgment by giving you detailed scenarios about how an event is likely to happen. For example, Plous (1993, p. 4) asks readers of his book to:

Place a check mark beside the alternative that seems most likely to occur within the next 10 years:

- An all-out nuclear war between the United States and Russia

- An all-out nuclear war between the United States and Russia in which neither country intends to use nuclear weapons, but both sides are drawn into the conflict by the actions of a country such as Iraq, Libya, Israel, or Pakistan

Notice that the second alternative describes a subset of the first alternative, and thus the first one must be at least as likely as the second. Yet, according to the representativeness heuristic, most people would see the second alternative as more likely.

Be wary when someone describes a scenario to you in great detail in order to try to convince you of its likelihood. For example, lawyers know that jurors are much more likely to believe a person is guilty if they are provided with a detailed scenario of how the person's guilt could have occurred.

Forgotten Base Rates

The representativeness heuristic can lead people to ignore information they may have about the likelihood of various outcomes. For example, Kahneman and Tversky (1973) conducted a study in which they told subjects that a population consisted of 30 engineers and 70 lawyers. The subjects were first asked to assess the likelihood that a randomly selected individual would be an engineer. The average response was indeed close to the correct 30%.

Subjects were then given the following description, written to give no clues as to whether this individual was an engineer or a lawyer.

> *Dick is a 30-year-old man. He is married with no children. A man of high ability and high motivation, he promises to be quite successful in his field. He is well liked by his colleagues. (Kahneman and Tversky, 1973, p. 243)*

This time, the subjects ignored the base rate. When asked to assess the likelihood that Dick was an engineer, the median response was 50%. Because the individual in question did not appear to represent either group more heavily, the respondents concluded that there must be an equally likely chance that he was either. They ignored the information that only 30% of the population were engineers.

Neglecting base rates can cloud the probability assessments of experts as well. For example, physicians who are confronted with a patient's positive test results for a rare disease routinely overestimate the probability that the patient actually has the disease. They fail to take into account the extremely low base rate in the population.

16.4   OPTIMISM, RELUCTANCE TO CHANGE, AND OVERCONFIDENCE

Psychologists have also found that some people tend to have personal probabilities that are unrealistically optimistic. Further, people are often overconfident about how likely they are to be right and are reluctant to change their views even when presented with contradictory evidence.

Optimism

Slovic and colleagues (1982) cite evidence showing that most people view themselves as personally more immune to risks than other people. They note that "the great majority of individuals believe themselves to be better than average drivers, more likely to live past 80, less likely than average to be harmed by the products they use, and so on" (pp. 469–470).

<table>
<tr><td>**EXAMPLE 6**</td></tr>
</table>

OPTIMISTIC COLLEGE STUDENTS

Research on college students confirms that they see themselves as more likely than average to encounter positive life events and less likely to encounter negative ones. Weinstein (1980) asked students at Cook College (part of Rutgers, the State University in New Jersey) to rate how likely certain life events were to happen to them compared to other Cook students of the same sex. Plous (1993) summarizes Weinstein's findings:

> On the average, students rated themselves as 15 percent more likely than others to experience positive events and 20 percent less likely to experience negative events. To take some extreme examples, they rated themselves as 42 percent more likely to receive a good starting salary after graduation, 44 percent more likely to own their own home, 58 percent less likely to develop a drinking problem, and 38 percent less likely to have a heart attack before the age of 40. (p. 135)

Notice that if all the respondents were accurate, the median response for each question should have been 0 percent more or less likely because approximately half of the students should be more likely and half less likely than average to experience any event. ■

The tendency to underestimate one's probability of negative life events can lead to foolish risk taking. Examples are driving while intoxicated and having unprotected sex. Plous (1993, p. 134) calls this phenomenon, "It'll never happen to me," whereas Slovic and colleagues (1982, p. 468) title it, "It won't happen to me." The point is clear: If *everyone* underestimates his or her own personal risk of injury, someone has to be wrong . . . it *will* happen to someone.

Reluctance to Change

In addition to optimism, most people are also guilty of **conservatism.** As Plous (1993) notes, "Conservatism is the tendency to change previous probability estimates more slowly than warranted by new data" (p. 138). This explains the reluctance of the scientific community to accept new paradigms or to examine compelling evidence for phenomena such as extrasensory perception. As noted by Hayward (1984):

> There seems to be a strong need on the part of conventional science to exclude such phenomena from consideration as legitimate observation. Kuhn and Feyerabend showed that it is always the case with "normal" or conventional science that observations not confirming the current belief system are ignored or dismissed. The colleagues of Galileo who refused to look through his telescope because they "knew" what the moon looked like are an example. (pp. 78–79)

This reluctance to change one's personal-probability assessment or belief based on new evidence is not restricted to scientists. It is notable there only because science is supposed to be "objective."

Overconfidence

Consistent with the reluctance to change personal probabilities in the face of new data is the tendency for people to place too much confidence in their own assessments. In other words, when people venture a guess about something for which they are uncertain, they tend to overestimate the probability that they are correct.

....

EXAMPLE 7   **HOW ACCURATE ARE YOU?**

Fischhoff, Slovic, and Lichtenstein (1977) conducted a study to see how accurate assessments were when people were sure they were correct. They asked people to answer hundreds of questions on general knowledge, such as whether *Time* or *Playboy* had a larger circulation or whether absinthe is a liqueur or a precious stone. They also asked people to rate the odds that they were correct, from 1:1 (50% probability) to 1,000,000:1 (virtually certain).

The researchers found that the more confident the respondents were, the more the true proportion of correct answers deviated from the odds given by the respondents. For example, of those questions for which the respondents gave even odds of being correct (50% probability), 53% of the answers were correct. However, of those questions for which they gave odds of 100:1 (99% probability) of being correct, only 80% of the responses were actually correct. ■

Researchers have found a way to help eliminate overconfidence. As Plous (1993) notes, "The most effective way to improve calibration seems to be very simple: *Stop to consider reasons why your judgment might be wrong*" (p. 228). In a study by Koriat, Lichtenstein, and Fischhoff (1980), respondents were asked to list reasons to support and to oppose their initial judgments. The authors found that when subjects were asked to list reasons to oppose their initial judgment, their probabilities became extremely well-calibrated. In other words, respondents were much better able to judge how much confidence they should have in their answers when they considered reasons why they might be wrong.

....

16.5   CALIBRATING PERSONAL PROBABILITIES OF EXPERTS

Professionals who need to help others make decisions often use personal probabilities themselves, and their personal probabilities are sometimes subject to the same distortions discussed in this chapter. For example, your doctor may observe your symptoms and give you an assessment of the likelihood that you have a certain disease—but fail to take into account the baseline rate for the disease.

Weather forecasters routinely use personal probabilities to deliver their predictions of tomorrow's weather. They attach a number to the likelihood that it will rain in a certain area, for example. Those numbers are a composite of information about what has happened in similar circumstances in the past and the forecaster's knowledge of meteorology.

Using Relative Frequency to Check Personal Probabilities

As consumers, we would like to know how accurate the probabilities delivered by physicians, weather forecasters, and similar professionals are likely to be. To discuss what we mean by accuracy, we need to revert to the relative-frequency interpretation of probability. For example, if we routinely listen to the same professional weather forecaster, we could check his or her predictions using the relative-frequency measure. Each evening, we could record the forecaster's probability of rain for the next day, and then the next day we could record whether it actually rained.

For a *perfectly calibrated* forecaster, of the many times he or she gave a 30% chance of rain, it would actually rain 30% of the time. Of the many times the forecaster gave a 90% chance of rain, it would rain 90% of the time, and so on. We can assert that personal probabilities are *well-calibrated* if they come close to meeting this standard.

Notice that we can assess whether probabilities are well-calibrated only if we have enough repetitions of the event to apply the relative-frequency definition. For instance, we will never be able to ascertain whether the late Carl Sagan was well-calibrated when he made the assessment we saw in Section 15.3 that "the probability that the Earth will be hit by a civilization-threatening small world in the next century is a little less than one in a thousand." This event is obviously not one that will be repeated numerous times.

CASE STUDY 16.1 ## Calibrating Weather Forecasters and Physicians

Studies have been conducted of how well calibrated various professionals are. Figure 16.1 displays the results of two such studies, one for weather forecasters and one for physicians. The open circles indicate actual relative frequencies of rain, plotted against various forecast probabilities. The dark circles indicate the relative frequency with which a patient actually had pneumonia versus his or her physician's personal probability that the patient had it.

The plot indicates that the weather forecasters were generally quite accurate but that, at least for the data presented here, the physicians were not. The weather forecasters were slightly off at the very high end, when they predicted rain with almost certainty. For example, of the times they were sure it was going to rain, and gave a

FIGURE 16.1
Calibrating weather forecasters and physicians
Source: Plous, 1993, p. 223, using data from Murphy and Winkler (1984) for the weather forecasters and Christensen-Szalanski and Bushyhead (1981) for the physicians.

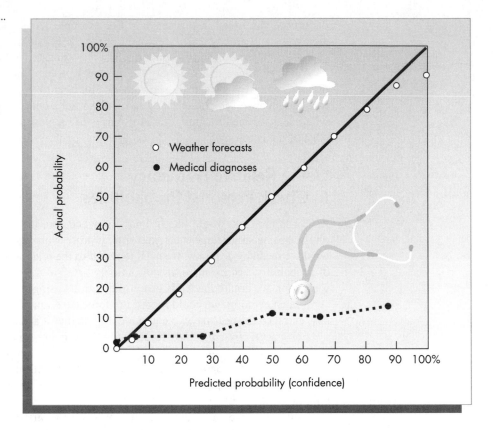

probability of 1, it rained only about 91% of the time. Still, the weather forecasters were well calibrated enough that you could use their assessments to make reliable decisions about how to plan tomorrow's events.

The physicians were not at all well calibrated. The actual probability of pneumonia rose only slightly and remained under 15% even when physicians placed it almost as high as 90%. As we will see in an example in Section 17.4, physicians tend to overestimate the probability of disease, especially when the baseline risk is low. When your physician quotes a probability to you, you should ask if it is a personal probability or one based on data from many individuals in your circumstances. ■

16.6   TIPS FOR IMPROVING YOUR PERSONAL PROBABILITIES AND JUDGMENTS

The research summarized in this chapter suggests methods for improving your own decision making when uncertainty and risks are involved. Here are some tips to consider when making judgments:

1. Think of the big picture, including risks and rewards that are not presented to you. For example, when comparing insurance policies, be sure to compare coverage as well as cost.

2. When considering how a decision changes your risk, try to find out what the baseline risk is to begin with. Try to determine risks on an equal scale, such as the drop in *number* of deaths per 100,000 people rather than the *percent* drop in death rate.

3. Don't be fooled by highly detailed scenarios. Remember that excess detail actually decreases the probability that something is true, yet the representativeness heuristic leads people to increase their personal probability that it is true.

4. Remember to list reasons why your judgment might be wrong, to provide a more realistic confidence assessment.

5. Do not fall into the trap of thinking that bad things only happen to other people. Try to be realistic in assessing your own individual risks, and make decisions accordingly.

6. Be aware that the techniques discussed in this chapter are often used in marketing. For example, watch out for the anchoring effect when someone tries to anchor your personal assessment to an unrealistically high or low value.

7. If possible, break events into pieces and try to assess probabilities using the information in Chapter 15 and in publicly available information. For example, Slovic and colleagues (1982, p. 480) note that, because the risk of a disabling injury on any particular auto trip is only about 1 in 100,000, the need to wear a seat belt on a specific trip would seem to be small. However, using the techniques described in Chapter 15, they calculated that over a lifetime of riding in an automobile the risk is about .33. It thus becomes much more reasonable to wear a seat belt at all times.

EXERCISES

1. Explain how the pseudocertainty effect differs from the certainty effect.

2. Suppose a television advertisement were to show viewers a product and then say, "You might expect to pay $25, $30, or even more than this. But we are offering it for only $16.99." Explain which of the ideas in this chapter is being used to try to exploit viewers.

3. There are many more termites in the world than there are mosquitoes, but most of the termites live in tropical forests. Using the ideas in this chapter, explain why most people would think there were more mosquitoes in the world than termites.

4. Suppose a defense attorney is trying to convince the jury that his client's wallet, found at the scene of the crime, was actually planted there by his client's gardener. Here are two possible ways he might present this to the jury:

Statement A: The gardener dropped the wallet when no one was looking.

Statement B: The gardener hid the wallet in his sock and when no one was looking he quickly reached down and pulled off his sock, allowing the wallet to drop to the ground.

    a. Explain why the second statement cannot have a higher probability of being true than the first statement.

    b. Based on the material in this chapter, to which statement are members of the jury likely to assign higher personal probabilities? Explain.

5. Explain why you should be cautious when someone tries to convince you of something by presenting a detailed scenario. Give an example.

6. A telephone solicitor recently contacted the author to ask for money for a charity in which typical contributions are in the range of $25 to $50. The solicitor said, "We are asking for as much as you can give, up to $300.00" Do you think the amount people give would be different if the solicitor said, "We typically get $25 to $50, but give as much as you can." Explain, using the relevant material from this chapter.

7. In this chapter, we learned that one way to lower personal-probability assessments that are too high is to list reasons why you might be wrong. Explain how the availability heuristic might account for this phenomenon.

8. Research by Slovic and colleagues (1982) found that people judged that accidents and diseases cause about the same number of deaths in the United States, whereas in truth diseases cause about 16 times as many deaths as accidents. Using the material from this chapter, explain why the researchers found this misperception.

9. Determine which statement (A or B) has a higher probability of being true and explain your answer. Using the material in this chapter, also explain which statement you think a statistically naive person would think had a higher probability.

    A. A car traveling 120 miles per hour on a two-lane highway will have an accident.

    B. A car traveling 120 miles per hour on a two-lane highway will have a major accident in which all occupants are instantly killed.

10. Explain how an insurance salesperson might try to use each of the following concepts to sell you insurance:

    a. Anchoring

    b. Pseudocertainty

    c. Availability

11. There are approximately 5 billion people in the world. Plous (1993, p. 5) asked readers to estimate how wide a cube-shaped tank would have to be to hold all of the human blood in the world. The correct answer is about 870 feet, but most people give much higher answers. Explain which of the concepts covered in this chapter leads people to give higher answers.

12. Barnett (1990) examined front page stories in the *New York Times* for 1 year, beginning with October 1, 1988, and found 4 stories related to automobile deaths but 51 related to deaths from flying on a commercial jet. These correspond to 0.08 story per thousand U.S. deaths by automobile and 138.2 stories per thousand U.S. deaths for commercial jets. He also reported a mid-August 1989 Gallup Poll finding that 63% of Americans had lost confidence in recent years in the safety of airlines. Discuss this finding in the context of the material in this chapter.

13. Explain how the concepts in this chapter account for each of the following scenarios:

    a. Most people rate death by shark attacks to be much more likely than death by falling airplane parts, yet the chances of dying from the latter are actually 30 times greater (Plous, 1993, p. 121).

    b. You are a juror on a case involving an accusation that a student cheated on an exam. The jury is asked to assess the likelihood of the statement, "Even though he knew it was wrong, the student copied from the person sitting next to him because he desperately wants to get into medical school." The other jurors give the statement a high probability assessment although they know nothing about the accused student.

    c. Research by Tversky and Kahneman (1982b) has shown that people think that words beginning with the letter k are more likely to appear in a typical segment of text in English than words with k as the third letter. In fact, there are about twice as many words with k as the third letter than words that begin with k.

    d. A 45-year-old man dies of a heart attack and does not leave a will.

14. Suppose you go to your doctor for a routine examination, without any complaints of problems. A blood test reveals that you have tested positive for a certain disease. Based on the ideas in this chapter, what should you ask your doctor in order to assess how worried you should be?

15. Give one example of how each of the following concepts has had or might have an unwanted effect on a decision or action in your daily life:

    a. Conservatism

    b. Optimism

    c. Forgotten base rates

    d. Availability

16. Explain which of the concepts in this chapter might contribute to the decision to buy a lottery ticket.

17. Suppose you have a friend who is willing to ask her friends a few questions and then, based on their answers, is willing to assess the probability that those friends will get an A in each of their classes. She always assesses the probability to be either .10 or .90. She has made hundreds of these assessments and has kept track of whether her friends actually received A's. How would you determine if she is well calibrated?

18. Guess at the probability that if you ask five people when their birthdays are, you will find someone born in the same month as you. For simplicity, assume that the probability that a randomly selected person will have the same birth month you have is 1/12. Now use the material from Chapter 15 to make a table listing the numbers from 1 to 5 and then fill in the probabilities that you will *first* encounter someone with your birth month by asking that many people. Determine the *accumulated* probability that you will have found someone with your birth month by the time you ask the fifth person. How well calibrated was your initial guess?

....

MINI-PROJECTS

1. Design and conduct an experiment to try to elicit misjudgments based on one of the phenomena described in this chapter. Explain exactly what you did and your results.

2. Find and explain an example of a marketing strategy that uses one of the techniques in this chapter to try to increase the chances that someone will purchase something. Do not use an exact example from the chapter, such as "buy one, get one free."

3. Find a journal article that describes an experiment designed to test the kinds of biases described in this chapter. Summarize the article and discuss what conclusions can be made from the research. You can find such articles by searching appropriate bibliographic databases and trying key words from this chapter.

4. Estimate the probability of some event in your life using a personal probability, such as the probability that a person who passes you on the street will be wearing a hat. Use an event for which you can keep a record of the relative frequency of occurrence over the next week. How well calibrated were you?

....

REFERENCES

Barnett, A. (1990). Air safety: End of the golden age? *Chance* 3, no. 2, pp. 8–12.

Christensen-Szalanski, J. J. J., and J. B. Bushyhead. (1981). Physicians' use of probabilistic information in a real clinical setting. *Journal of Experimental Psychology: Human Perception and Performance* 7, pp. 928–935.

Fischhoff, B., P. Slovic, and S. Lichtenstein. (1977). Knowing with certainty: The appropriateness of extreme confidence. *Journal of Experimental Psychology: Human Perception and Performance* 3, pp. 552–564.

Hayward, J. W. (1984). *Perceiving ordinary magic. Science and intuitive wisdom.* Boulder, CO: New Science Library.

Kahneman, D., and A. Tversky. (1973). On the psychology of prediction. *Psychological Review* 80, pp. 237–251.

Kahneman, D., and A. Tversky. (1979). Prospect theory: An analysis of decision under risk. *Econometrica* 47, pp. 263–291.

Kahneman, D., and A. Tversky. (1982). On the study of statistical intuitions. In D. Kahneman, P. Slovic, and A. Tversky (eds.), *Judgment under uncertainty: Heuristics and biases* (Chapter 34). Cambridge, England: Cambridge University Press.

Koriat, A., S. Lichtenstein, and B. Fischhoff. (1980). Reasons for confidence. *Journal of Experimental Psychology: Human Learning and Memory* 6, pp. 107–118.

Murphy, A. H., and R. L. Winkler. (1984). Probability forecasting in meteorology. *Journal of the American Statistical Association* 79, pp. 489–500.

Northcraft, G. B., and M. A. Neale. (1987). Experts, amateurs, and real estate: An anchoring and adjustment perspective on property pricing decisions. *Organizational Behavior and Human Decision Processes* 39, pp. 84–97.

Plous, S. (1993). *The psychology of judgment and decision making.* New York: McGraw-Hill.

Slovic, P., B. Fischhoff, and S. Lichtenstein. (1982). Facts versus fears: Understanding perceived risk. In D. Kahneman, P. Slovic, and A. Tversky (eds.), *Judgment under uncertainty: Heuristics and biases* (Chapter 33). Cambridge, England: Cambridge University Press.

Tversky, A. and D. Kahneman. (1982a). Judgment under uncertainty: Heuristics and biases. In D. Kahneman, P. Slovic, and A. Tversky (eds.), *Judgment under uncertainty: Heuristics and biases* (Chapter 1). Cambridge, England: Cambridge University Press.

Tversky, A., and D. Kahneman. (1982b). Availability: A heuristic for judging frequency and probability. In D. Kahneman, P. Slovic, and A. Tversky (eds.), *Judgment under uncertainty. Heuristics and biases* (Chapter 11). Cambridge, England: Cambridge University Press.

Weinstein, N. D. (1980). Unrealistic optimism about future life events. *Journal of Personality and Social Psychology* 39, pp. 806–820.

World almanac and book of facts. (1995). Edited by Robert Famighetti. Mahwah, NJ: Funk and Wagnalls.

CHAPTER SEVENTEEN

When Intuition Differs from Relative Frequency

THOUGHT QUESTIONS

1. Do you think it likely that anyone will ever win a state lottery twice in a lifetime?

2. How many people do you think would need to be in a group in order to be at least 50% certain that two of them will have the same birthday?

3. Suppose you test positive for a rare disease, and your original chances of having the disease are no higher than anyone else's—say, close to 1 in 1000. You are told that the test has a 10% false positive rate and a 10% false negative rate. In other words, whether you have the disease or not, the test is 90% likely to give a correct answer. Given that you tested positive, what do you think is the probability that you actually have the disease? Do you think the chances are higher or lower than 50%?

4. If you were to flip a fair coin six times, which sequence do you think would be most likely: HHHHHH or HHTHTH or HHHTTT?

5. If you were faced with the following sets of alternatives, which one would you choose in each set? (Choose either A or B and either C or D.) Explain your answer.

   A. A gift of $240, guaranteed

   B. A 25% chance to win $1000 and a 75% chance of getting nothing

   C. A sure loss of $740

   D. A 75% chance to lose $1000 and a 25% chance to lose nothing

17.1   REVISITING RELATIVE FREQUENCY

Recall that the relative-frequency interpretation of probability provides a precise answer to certain probability questions. As long as we agree on the physical assumptions underlying an uncertain process, we should also agree on the probabilities of various outcomes. For example, if we agree that lottery numbers are drawn fairly, then we should agree on the probability that a particular ticket will be a winner.

In many instances, the physical situation lends itself to computing a relative-frequency probability, but people ignore that information. In this chapter, we examine how probability assessments that should be objective are instead confused by incorrect thinking.

17.2   COINCIDENCES

When I was in college in upstate New York, I visited Disney World in Florida during summer break. While there, I ran into three people I knew from college, none of whom were there together. A few years later, I visited the top of the Empire State Building in New York City and ran into two friends (who were there together) and two additional unrelated acquaintances. Years later, I was traveling from London to Stockholm and ran into a friend at the airport in London while waiting for the flight. Not only did the friend turn out to be taking the same flight, but we had been assigned adjacent seats.

Are Coincidences Improbable?

These events are all examples of what would commonly be called *coincidences*. They are certainly surprising, but are they improbable? Most people think that coincidences have low probabilities of occurring, but we shall see that our intuition can be quite misleading regarding such phenomena.

We will adopt the definition of **coincidence** proposed by Diaconis and Mosteller:

> *A coincidence is a surprising concurrence of events, perceived as meaningfully related, with no apparent causal connection. (1989, p. 853)*

The mathematically sophisticated reader may wish to consult the article by Diaconis and Mosteller, in which they provide some instructions on how to compute probabilities for coincidences. For our purposes, we need nothing more sophisticated than the simple probability rules we encountered in Chapter 15.

Here are some examples of coincidences that at first glance seem highly improbable:

EXAMPLE 1   **TWO GEORGE D. BRYSONS**

"My next-door neighbor, Mr. George D. Bryson, was making a business trip some years ago from St. Louis to New York. Since this involved weekend travel and he was in no hurry, . . . and since his train went through Louisville, he asked the conductor, after he had boarded the train, whether he might have a stopover in Louisville.

"This was possible, and on arrival at Louisville he inquired at the station for the leading hotel. He accordingly went to the Brown Hotel and registered. And then, just as a lark, he stepped up to the mail desk and asked if there was any mail for him. The girl calmly handed him a letter addressed to Mr. George D. Bryson, Room 307, that being the number of the room to which he had just been assigned. It turned out that the preceding resident of Room 307 was another George D. Bryson" (Weaver, 1963, pp. 282–283). ■

EXAMPLE 2   **IDENTICAL CARS AND MATCHING KEYS**

Plous (1993, p. 154) reprinted an Associated Press news story describing a coincidence in which a man named Richard Baker and his wife were shopping on April Fool's Day at a Wisconsin shopping center. Mr. Baker went out to get their car, a 1978 maroon Concord, and drove it around to pick up his wife. After driving for a short while, they noticed items in the car that did not look familiar. They checked the license plate, and sure enough, they had someone else's car. When they drove back to the shopping center (to find the police waiting for them), they discovered that the owner of the car they were driving was a Mr. Thomas Baker, no relation to Richard Baker. Thus, both Mr. Bakers were at the same shopping center at the same time, with identical cars and with matching keys. The police estimated the odds as "a million to one." ■

EXAMPLE 3   **WINNING THE LOTTERY TWICE**

Moore (1991, p. 278) reported on a *New York Times* story of February 14, 1986, about Evelyn Marie Adams, who won the New Jersey lottery twice in a short time period. Her winnings were $3.9 million the first time and $1.5 million the second time. Then, in May 1988, Robert Humphries won a second Pennsylvania state lottery, bringing his total winnings to $6.8 million. When Ms. Adams won for the second time, the *New York Times* claimed that the odds of one person winning the top prize twice were about 1 in 17 trillion. ■

Someone, Somewhere, Someday

Most people think that the events just described are exceedingly improbable, and they are. *What is not improbable is that someone, somewhere, someday will experience those events or something similar.*

When we examine the probability of what appears to be a startling coincidence, we ask the wrong question. For example, the figure quoted by the *New York Times* of 1 in 17 trillion is the probability that a *specific* individual who plays the New Jersey state lottery exactly twice will win both times (Diaconis and Mosteller, 1989, p. 859). However, millions of people play the lottery every day, and it is not surprising that someone, somewhere, someday would win twice.

In fact, Purdue professors Stephen Samuels and George McCabe (cited in Diaconis and Mosteller, 1989, p. 859) calculated those odds to be practically a sure thing. They calculated that there was at least a 1 in 30 chance of a double winner in a 4-month period and better than even odds that there would be a double winner in a 7-year period somewhere in the United States. Further, they used conservative assumptions about how many tickets past winners purchase.

When you experience a coincidence, remember that there are over 5 billion people in the world and over 250 million in the United States. If something has a 1 in 1 million probability of occurring to each individual on a given day, it will occur to an average of over 250 people in the United States *each day* and over 5000 people in the world *each day*. Of course, probabilities of specific events depend on individual circumstances, but you can see that, quite often, it is not unlikely that something surprising will happen.

EXAMPLE 4   **SHARING THE SAME BIRTHDAY**

Here is a famous example that you can use to test your intuition about surprising events. How many people would need to be gathered together to be at least 50% sure that two of them share the same birthday? Most people provide answers that are much higher than the correct one, which is only 23 people.

There are several reasons why people have trouble with this problem. If your answer was somewhere close to 183, or half the number of birthdays, then you may have confused the question with another one, such as the probability that someone in the group has *your* birthday or that two people have a specific date as their birthday.

It is not difficult to see how to calculate the appropriate probability, using our rules from Chapter 15. Notice that the only way to avoid two people having the same birthday is if all 23 people have different birthdays. To find that probability, we simply use the rule that applies to the word *and* (Rule 3), thus multiplying probabilities.

Hence, the probability that the first three people have different birthdays is the probability that the second person does not share a birthday with the first, which is 364/365 (ignoring February 29), and the third person does not share a birthday with either of the first two, which is 363/365. (Two dates were already taken.)

Continuing this line of reasoning, the probability that none of the 23 people share a birthday is

$$(364)(363)(362) \bullet \cdots \bullet (343)/(365)^{22} = .493$$

Therefore, the probability that at least two people share a birthday is what's left of the probability, or $1 - .493 = .507$.

If you find it difficult to follow the arithmetic line of reasoning, simply consider this. Imagine each of the 23 people shaking hands with the remaining 22 people and asking them about their birthday. There would be 253 handshakes and birthday conversations. Surely there is a relatively high probability that at least one of those pairs would discover a common birthday.

By the way, the probability of a shared birthday in a group of 10 people is already better than 1 in 9, at .117. (There would be 45 handshakes.) With only 50 people, it is almost certain, with a probability of .97. (There would be 1225 handshakes.) ■

EXAMPLE 5   **UNUSUAL HANDS IN CARD GAMES**

As a final example, consider a card game, like bridge, in which a standard 52-card deck is dealt to four players, so they each receive 13 cards. Any specific set of 13 cards is equally likely, each with a probability of about 1 in 635 billion. You would probably not be surprised to get a mixed hand—say, the 4, 7, and 10 of hearts; 3, 8, 9, and jack of spades; 2 and queen of diamonds; and 6, 10, jack, and ace of clubs. Yet, that specific hand is just as unlikely as getting all 13 hearts. The point is that any very specific event, surprising or not, has extremely low probability; however, there are many such events, and their combined probability is quite high.

Magicians sometimes exploit the fact that many small probabilities add up to one large probability by doing a trick in which they don't tell you what to expect in advance. They set it up so that *something* surprising is almost sure to happen. When it does, you are likely to focus on the probability of *that particular outcome*, rather than realizing that a multitude of other outcomes would have also been surprising and that one of them was likely to happen. ■

Most Coincidences Only Seem Improbable

To summarize, most coincidences seem improbable only if we ask for the probability of that specific event occurring at that time to us. If, instead, we ask the probability of it occurring some time, to someone, the probability can become quite large.

Further, because of the multitude of experiences we each have every day, it is not surprising that some of them may appear to be improbable. That specific event is undoubtedly improbable. What is not improbable is that something "unusual" will happen to each of us once in a while.

17.3 THE GAMBLER'S FALLACY

Another common misperception about random events is that they should be self-correcting. Another way to state this is that people think the long-run frequency of an event should apply even in the short run. This misperception has classically been called the **gambler's fallacy.**

A related misconception is what Tversky and Kahneman (1982) call the **belief in the law of small numbers,** "according to which [people believe that] even small samples are highly representative of the populations from which they are drawn" (p. 7). They report that "in considering tosses of a coin for heads and tails, for example, people regard the sequence HTHTTM to be more likely than the sequence HHHTTT, which does not appear random, and also more likely than the sequence HHHHTH, which does not represent the fairness of the coin" (p. 7). Remember that any specific sequence of heads and tails is just as likely as any other if the coin is fair, so the idea that the first sequence is more likely is a misperception.

Independent Chance Events Have No Memory

The gambler's fallacy can lead to poor decision making, especially if applied to gambling. For example, people tend to believe that a string of good luck will follow a string of bad luck in a casino. Unfortunately, independent chance events have no such memory. Making ten bad gambles in a row doesn't change the probability that the next gamble will also be bad.

When the Gambler's Fallacy May Not Apply

Notice that the gambler's fallacy applies to independent events. Recall from Chapter 15 that independent events are those for which occurrence on one occasion gives no information about occurrence on the next occasion, as with successive flips of a coin. The gambler's fallacy may not apply to situations where knowledge of one outcome affects probabilities of the next. For instance, in card games using a single deck, knowledge of what cards have already been played provides information about what cards are likely to be played next. If you normally receive lots of mail but have received none for two days, you would probably (correctly) assess that you are likely to receive more than usual the next day.

17.4   CONFUSION OF THE INVERSE

Consider the following scenario, discussed by Eddy (1982). You are a physician. One of your patients has a lump in her breast. You are almost certain that it is benign; in fact, you would say there is only a 1% chance that it is malignant. But just to be sure, you have the patient undergo a mammogram, a breast X ray designed to detect cancer.

You know from the medical literature that mammograms are 80% accurate for malignant lumps and 90% accurate for benign lumps. In other words, if the lump is truly malignant, the test results will say that it is malignant 80% of the time and will falsely say it is benign 20% of the time. If the lump is truly benign, the test results will say so 90% of the time and will falsely declare that it is malignant only 10% of the time.

Sadly, the mammogram for your patient is returned with the news that the lump is malignant. What are the chances that it is truly malignant?

Eddy posed this question to 100 physicians. Most of them thought the probability that the lump was truly malignant was about 75% or .75. In truth, given the probabilities as described, *the probability is only .075*. The physicians' estimates were ten times too high!

When he asked them how they arrived at their answers, Eddy realized that the physicians were confusing the actual question with a different question: "When asked about this, the erring physicians usually report that they assumed that the probability of cancer given that the patient has a positive X ray was approximately equal to the probability of a positive X ray in a patient with cancer" (1982, p. 254).

Robyn Dawes has called this phenomenon **confusion of the inverse** (Plous, 1993, p. 132). The physicians were confusing the probability of cancer *given a positive X ray* with its inverse, the probability of a positive X ray, *given that the patient has cancer.*

Determining the Actual Probability

It is not difficult to see that the correct answer to the question posed to the physicians by Eddy (in the previous section) is indeed .075. Let's construct a hypothetical table of 100,000 women who fit this scenario. In other words, these are women who would present themselves to the physician with a lump for which the probability that it was malignant seemed to be about 1%. Thus, of the 100,000 women, about 1%, or 1000 of them, would have a malignant lump. The remaining 99%, or 99,000, would have a benign lump.

Further, given that the test was 80% accurate for malignant lumps, it would show a malignancy for 800 of the 1000 women who actually had one. Given that it was 90% accurate for the 99,000 women with benign lumps, it would show benign for 90%, or 89,100 of them and malignant for the remaining 10%, or 9900 of them. Table 17.1 shows how the 100,000 women would fall into these possible categories.

Let's return to the question of interest. Our patient has just received a positive test for malignancy. Given that her test showed malignancy, what is the actual prob-

TABLE 17.1   Breakdown of Actual Status versus Test Status for a Rare Disease

	Test Shows Malignant	Test Shows Benign	Total
Actually Malignant	800	200	1,000
Actually Benign	9,900	89,100	99,000
Total	10,700	89,300	100,000

ability that her lump is malignant? Of the 100,000 women, 10,700 of them would have an X ray show malignancy. But of those 10,700 women, only 800 of them actually have a malignant lump. Thus, given that the test showed a malignancy, the probability of malignancy is just 800/10,700 = 8/107 = .075.

The Probability of False Positives

Many physicians are guilty of confusion of the inverse. Remember, in a situation where the *base rate* for a disease is very low and the test for the disease is less than perfect, there will be a relatively high probability that a positive test result is a false positive.

If you ever find yourself in a situation similar to the one just described, you may wish to construct a table like Table 17.1.

> *To determine the probability of a positive test result being accurate, you need only three pieces of information:*
>
> **1.** The base rate or probability that you are likely to have the disease, without any knowledge of your test results
>
> **2.** The **sensitivity** of the test, which is the proportion of people who correctly test positive when they actually have the disease
>
> **3.** The **specificity** of the test, which is the proportion of people who correctly test negative when they don't have the disease

Notice that items 2 and 3 are measures of the accuracy of the test. They do not measure the probability that someone has the disease when they test positive or the probability that they do not have the disease when they test negative. Those probabilities, which are obviously the ones of interest to the patient, can be computed by constructing a table similar to Table 17.1. They can also be computed by using a formula called **Bayes' Rule,** given in the For Those Who Like Formulas section at the end of this chapter.

Streak Shooting in Basketball: Reality or Illusion?

SOURCE: Tversky and Gilovich, Winter 1989.

We have learned in this chapter that people's intuition, when it comes to assessing probabilities, is not very good, particularly when their wishes for certain outcomes are motivated by outside factors. Tversky and Gilovich (Winter 1989) decided to compare basketball fans' impressions of "streak shooting" with the reality evidenced by the records.

First, they generated phony sequences of 21 alleged "hits and misses" in shooting baskets, and showed them to 100 knowledgeable basketball fans. Without telling them the sequences were faked, they asked the fans to classify each sequence as "chance shooting," in which the probability of a hit on each shot was unrelated to previous shots; "streak shooting," in which the runs of hits and misses were longer than would be expected by chance; or "alternating shooting," in which runs of hits and misses were shorter than would be expected by chance. They found that people tended to think that streaks had occurred when they had not. In fact, 65% of the respondents thought the sequence that had been generated by "chance shooting" was in fact "streak shooting."

To give you some idea of the task involved, decide which of the following two sequences of 10 successes (S) and 11 failures (F) you think is more likely to be the result of "chance shooting":

Sequence 1: FFSSSFSFFFSSSSFSFFFSF

Sequence 2: FSFFSFSFFFSFSSFSFSSSF

Notice that each sequence represents 21 shots. In "chance shooting," the proportion of throws on which the result is different from the previous throw should be about one-half. If you thought sequence 1 was more likely to be due to chance shooting, you're right. Of the 20 throws that have a preceding throw, exactly 10 are different. In sequence 2, 14 of 20, or 70%, of the shots differ from the previous shot. If you selected sequence 2, you are like the fans tested by Tversky and Gilovich. The sequences with 70% and 80% alternating shots were most likely to be selected (erroneously) as being the result of "chance shooting."

To further test the idea that basketball fans (and players) see patterns in shooting success and failure, Tversky and Gilovich asked questions about the probability of successful hitting after hitting versus after missing. For example, they asked the following question of 100 basketball fans:

When shooting free throws, does a player have a better chance of making his second shot after making his first shot than after missing his first shot? (1989, p. 20)

Sixty-eight percent of the respondents said yes; 32% said no. They asked members of the Philadelphia 76ers basketball team the same question, with similar results. A similar question about ordinary shots elicited even stronger belief in streaks, with 91% responding that the probability of making a shot was higher after having just made the last two or three shots than after having missed them.

What about the data on shooting? The researchers examined data from several NBA teams, including the Philadelphia 76ers, the New Jersey Nets, the New York Knicks, and the Boston Celtics. In this case study, we examine the data they reported for free throws. These are throws in which action stops and the player stands in a fixed position, usually for two successive attempts to put the ball in the basket. Examining free-throw shots removes the possible confounding effect that members of the other team would more heavily guard a player they perceive as being "hot."

Tversky and Gilovich reported free-throw data for nine members of the Boston Celtics basketball team. They examined the long-run frequency of a hit on the second free throw after a hit on the first one, and after a miss on the first one. Of the nine players, five had a higher probability of a hit after a miss, whereas four had a higher probability of a hit after a hit. In other words, the perception of 65% of the fans that the probability of a hit was higher after just receiving a hit was not supported by the actual data.

Tversky and Gilovich looked at other sequences of hits and misses from the NBA teams, in addition to generating their own data in a controlled experiment using players from Cornell University's varsity basketball teams. They analyzed the data in a variety of ways, but they could find no evidence of a "hot hand" or "streak shooting." They conclude:

> Our research does not tell us anything in general about sports, but it does suggest a generalization about people, namely that they tend to "detect" patterns even where none exist, and to overestimate the degree of clustering in sports events, as in other sequential data. We attribute the discrepancy between the observed basketball statistics and the intuitions of highly interested and informed observers to a general misconception of the laws of chance that induces the expectation that random sequences will be far more balanced than they generally are, and creates the illusion that there are patterns of streaks in independent sequences. (1989, p. 21)

The research by Tversky and Gilovich has not gone unchallenged. For additional reading, see the articles by Hooke (1989) and by Larkey, Smith, and Kadane (1989). They argue that just because Tversky and Gilovich did not find evidence of "streak shooting" in the data they examined doesn't mean that it doesn't exist, sometimes. ∎

17.5   USING EXPECTED VALUES TO MAKE WISE DECISIONS

In Chapter 15, we learned how to compute the expected value of numerical outcomes when we know the outcomes and their probabilities. Using this information, you would think that people would make decisions that allowed them to maximize their expected monetary return. But people don't behave this way. If they did, they would not buy lottery tickets or insurance.

Businesses like insurance companies and casinos rely on the theory of expected value to stay in business. Insurance companies know that young people are more likely than middle-aged people to have automobile accidents and that older people are more likely to die of nonaccidental causes. They determine the prices of automobile and life insurance policies accordingly.

If individuals were solely interested in maximizing their monetary gains, they would use expected value in a similar manner. For example, in Chapter 15, Example 15, we illustrated that for the California Decco lottery game, there was an average loss of 35 cents for each ticket purchased. Most lottery players know that there is an expected loss for every ticket purchased, yet they continue to play. Why? Probably because the excitement of playing and possibly winning has intrinsic, nonmonetary value that compensates for the expected monetary loss.

Social scientists have long been intrigued with how people make decisions, and much research has been conducted on the topic. The most popular theory among early researchers, in the 1930s and 1940s, was that people made decisions to maximize their expected *utility*. This may or may not correspond to maximizing their expected dollar amount. The idea was that people would assign a worth or utility to each outcome and choose whatever alternative yielded the highest expected value.

More recent research has shown that decision making is influenced by a number of factors and can be a complicated process. (Plous [1993] presents an excellent summary of much of the research on decision making.) The way in which the decision is presented can make a big difference. For example, Plous (1993, p. 97) discusses experiments in which respondents were presented with scenarios similar to the following:

If you were faced with the following alternatives, which would you choose? Note that you can choose either A or B and either C or D.

A. A gift of $240, guaranteed

B. A 25% chance to win $1000 and a 75% chance of getting nothing

C. A sure loss of $740

D. A 75% chance to lose $1000 and a 25% chance to lose nothing

When asked to choose between A and B, the majority of people chose the sure gain represented by choice A. Notice that the expected value under choice B is $250, which is higher than the sure gain of $240 from choice A, yet people prefer choice A.

When asked to choose between C and D, the majority of people chose the gamble rather than the sure loss. Notice that the expected value under choice D is $750, representing a larger expected loss than the $740 presented in choice C. For dollar amounts and probabilities of this magnitude, people tend to value a sure gain, but are willing to take a risk to prevent a loss.

The second set of choices (C and D) is similar to the decision people must make when deciding whether to buy insurance. The cost of the premium is the sure loss. The probabilistic choice represented by alternative D is similar to gambling on whether you will have a fire, burglary, accident, and so on. Why then do people choose to gamble in the scenario just presented, yet tend to buy insurance?

As Plous (1993) explains, one factor seems to be the magnitudes of the probabilities attached to the outcomes. People tend to give small probabilities more weight than they deserve for their face value. Losses connected with most insurance policies have a low probability of actually occurring, yet people worry about them. Plous (1993, p. 99) reports on a study in which people were presented with the following two scenarios:

Alternative A: A 1 in 1000 chance of winning $5000

Alternative B: A sure gain of $5

Alternative C: A 1 in 1000 chance of losing $5000

Alternative D: A sure loss of $5

About three-fourths of the respondents presented with scenario A and B chose the risk presented by alternative A. This is similar to the decision to buy a lottery ticket, where the sure gain corresponds to keeping the money rather than using it to buy a ticket.

For scenario C and D, nearly 80% of respondents chose the sure loss (D). This is the situation that results in the success of the insurance industry. Of course, the dollar amounts are also important. A sure loss of $5 may be easy to absorb, while the risk of losing $5000 may be equivalent to the risk of bankruptcy.

CASE STUDY 17.2 # How Bad Is a Bet on the British Open?

SOURCE: Larkey, 1990, pp. 24–26.

Betting on sports events is big business all over the world, yet if people made decisions solely on the basis of maximizing expected dollar amounts, they would not bet. Larkey (1990) decided to look at the odds given for the 1990 British Open golf tournament "by one of the largest betting shops in Britain" (p. 25) to see how much the betting shop stood to gain and the bettors stood to lose.

Here is how betting on sports events works. The bookmaker sets odds on each of the possible outcomes, which in this case were individual players winning the tournament. For example, for the 1990 British Open, the bookmaker we will examine set odds of 50 to 1 for Jack Nicklaus. You pay one dollar, pound, or whatever to play. If your outcome happens, you win the amount given in the odds, plus get your money back. For example, if you placed a $1 bet on Jack Nicklaus winning and he won, you would receive $50 (minus a handling fee, which we will ignore for this discussion), in addition to getting your $1 back. Thus, the two possible outcomes are that you gain $50 or you lose $1.

Table 17.2 shows the 13 players who were assigned the highest odds by the betting shop we are using, along with the odds the shop assigned to each of the players. The table also lists the probability of winning that would be required for each player, in order for someone who bet on that player to have a break-even expected value.

Let's look at how the probabilities in Table 17.2 are computed. Suppose you bet on Nick Faldo. The odds given for him were 6 to 1. Therefore, if you bet $1 on him and won, you would have a gain of $6. If you lost, you would have a net "gain" of

---

TABLE 17.2	Odds Given on the Top Ranked Players in the 1990 British Open

Player	Odds	Probability
1. Nick Faldo	6 to 1	.1429
2. Greg Norman	9 to 1	.1000
3. Jose-Maria Olazabal	14 to 1	.0667
4. Curtis Strange	14 to 1	.0667
5. Ian Woosnam	14 to 1	.0667
6. Seve Ballesteros	16 to 1	.0588
7. Mark Calcavecchia	16 to 1	.0588
8. Payne Stewart	16 to 1	.0588
9. Bernhard Langer	22 to 1	.0435
10. Paul Azinger	28 to 1	.0345
11. Ronan Rafferty	33 to 1	.0294
12. Fred Couples	33 to 1	.0294
13. Mark McNulty	33 to 1	.0294

SOURCE: Larkey, 1990.

−$1. Let's call the probability that Faldo wins p and the probability that he doesn't win $1 - p$. What value of p would allow you to break even—that is, have an expected value of zero?

$$EV = (\$6) \times p + (-\$1) \times (1 - p) = \$7 \times p - \$1$$

It should be obvious that if $p = 1/7$, the expected value would be zero. The value listed in Table 17.2 for Faldo is $1/7 = .1429$. Probabilities for other players are derived using the same method, and the general formula should be obvious. If the odds are n to 1 for a particular player, someone who bets on that player will have an expected gain (or loss) of zero if the probability of the player winning is $1/(n + 1)$. In other words, for the odds presented, this would be a fair bet if the player's actual probability of winning was the probability listed in the table.

If the bookmaker had set fair odds, so that both the house and those placing bets had expected values of zero, then the probabilities for all of the players should sum to 1.00. The probabilities listed in Table 17.2 sum to .7856. But those weren't the only players for whom bets could be placed. Larkey (1990, p. 25) lists a total of 40 players, which is still apparently only a subset of the 156 choices. The 40 players listed by Larkey have probabilities summing to 1.27. With the odds set by the bookmaker, the house has a definite advantage, even after taking off the "handling fee."

It is impossible to compute the true expected value for the house because we would need to know both the true probabilities of winning for each player and the number of bets placed on each player. Also, notice that just because the house has a positive expected value does not mean that it will come out ahead. The winner of the tournament was Nick Faldo. If everyone had bet $1 on Nick Faldo, the house would have to pay each bettor $6 in addition to the $1 bet, which would clearly be a losing proposition. Bookmakers rely on the fact that many thousands of bets are made, so the aggregate win (or loss) per bet for them should be very close to the expected value. ■

FOR THOSE WHO LIKE FORMULAS

Conditional Probability
The *conditional probability* of event A, given knowledge that event B happened, is denoted by P(A|B).

Bayes' Rule
Suppose A_1 and A_2 are complementary events with known probabilities. In other words, they are mutually exclusive and their probabilities sum to 1. For example, they might represent presence and absence of a disease in a randomly chosen individual. Suppose B is another event such that the conditional probabilities $P(B \mid A_1)$ and $P(B \mid A_2)$ are both known. For example, B might be the probability of testing positive for the disease. We do not need to know $P(B)$.

Then Bayes' Rule determines the conditional probability in the other direction:

$$P(A_1 \mid B) = \frac{P(A_1)P(B \mid A_1)}{P(A_1)P(B \mid A_1) + P(A_2)P(B \mid A_2)}$$

For example, Bayes' Rule can be used to determine the probability of having a disease, *given* that the test is positive. The base rate, sensitivity, and specificity would all need to be known. Bayes' Rule is easily extended to more than two mutually exclusive events, as long as the probability of each one is known, and the probability of B conditional on each one is known.

EXERCISES

1. Although it's not quite true, suppose the probability of having a male child (M) is equal to the probability of having a female child (F). A couple has four children.
   a. Are they more likely to have FFFF or to have MFFM? Explain your answer.
   b. Explain which sequence in part a of this exercise a *belief in the law of small numbers* would lead people to *say* had higher probability.
   c. Is a couple with four children more likely to have four girls or to have two children of each sex? Explain. (Assume the decision to have four children was independent of the sex of the children.)

2. Give an example of a sequence of events to which the gambler's fallacy would not apply because the events are not independent.

3. Explain why it is not at all unlikely that in a class of 50 students two of them will have the same last name.

4. Suppose two sisters are reunited after not seeing each other since they were 3 years old. They are amazed to find out that they are both married to men named James and that they each have a daughter named Jennifer. Explain why this is not so amazing.

5. Why is it not surprising that the night before a major airplane crash several people will have dreams about an airplane disaster? If you were one of those people, would you think that something amazing had occurred?

6. Find a dollar bill or other item with a serial number. Write down the number. I predict that there is something unusual about it or some pattern to it. Explain what is unusual about it and how I was able to make that prediction.

7. The U.C. Berkeley *Wellness Encyclopedia* (1991) contains the following statement in its discussion of HIV testing: "In a high-risk population, virtually all people who test positive will truly be infected, but among people at low risk the false positives will outnumber the true positives. Thus, for every infected person correctly identified in a low-risk population, an estimated ten noncarriers [of the HIV virus] will test positive" (p. 360).

   a. Suppose you have a friend who is part of this low-risk population but who has just tested positive. Using the numbers in the statement, calculate the probability that the person actually carries the virus.

   b. Your friend is understandably upset and doesn't believe that the probability of being infected with HIV isn't really near 1. After all, the test is accurate and it came out positive. Explain to your friend how the *Wellness Encyclopedia* statement can be true, even though the test is very accurate both for people with HIV and for people who don't carry it. If it's easier, you can make up numbers to put in a table to support your argument.

8. In financial situations, are businesses or individuals more likely to make use of expected value for making decisions? Explain.

9. Many people claim that they can often predict who is on the other end of the phone when it rings. Do you think that phenomenon has a normal explanation? Explain.

10. Suppose a rare disease occurs in about 1 out of 1000 people who are like you. A test for the disease has sensitivity of 95% and specificity of 90%. Using the technique described in this chapter, compute the probability that you actually have the disease, given that your test results are positive.

11. You are at a casino with a friend, playing a game in which dice are involved. Your friend has just lost six times in a row. She is convinced that she will win on the next bet because she claims that, by the law of averages, it's her turn to win. She explains to you that the probability of winning this game is 40%, and because she has lost six times, she has to win four times to make the odds work out. Is she right? Explain.

12. Using the data in Table 17.1 about a hypothetical population of 100,000 women tested for breast cancer, find the probability of each of the following events:

    a. A woman whose test shows a malignant lump actually has a benign lump.

    b. A woman who actually has a benign lump has a test that shows a malignant lump.

    c. A woman with unknown status has a test showing a malignant lump.

13. Using the data in Table 17.1, give numerical values and explain the meaning of the sensitivity and the specificity of the test.

14. Explain why the story about George D. Bryson, reported in Example 1 in this chapter, is not all that surprising.

15. A statistics professor once made a big blunder by announcing to his class of about 50 students that he was fairly certain that someone in the room would share his birthday. We have already learned that there is a 97% chance that there will be 2 people in a room of 50 with a common birthday. Given that information, why was the professor's announcement a blunder? Do you think he was successful in finding a match? Explain.

16. Suppose the sensitivity of a test is .90. Give either the false positive or the false negative rate for the test, and explain which you are providing. Could you provide the other one without additional information? Explain.

17. Suppose a friend reports that she has just had a string of "bad luck" with her car. She had three major problems in as many months and now has replaced many of the worn parts with new ones. She concludes that it is her turn to be lucky and that she shouldn't have any more problems for a while. Is she making the gambler's fallacy? Explain.

18. If you wanted to pretend that you could do psychic readings, you could perform "cold readings" by inviting people you do not know to allow you to tell them about themselves. You would then make a series of statements like

    > "I see that there is some distance between you and your mother that bothers you."

    > "It seems that you are sometimes less sure of yourself than you indicate."

    > "You are thinking of two men in your life [or two women, for a male client], one of whom is sort of light-complexioned and the other of whom is slightly darker. Do you know who I mean?"

    In the context of the material in this chapter, explain why this trick would often work to convince people that you are indeed psychic.

19. Explain why it would be much more surprising if someone were to flip a coin and get six heads in a row *after* telling you they were going to do so than it would be to simply watch them flip the coin six times and observe six heads in a row.

20. We learned in this chapter that one idea researchers have tested was that when forced to make a decision, people choose the alternative that yields the highest expected value.

a. If that were the case, explain which of the following two choices people would make:

> *Choice A:* Accept a gift of $10
>
> *Choice B:* Take a gamble with probability 1/1000 of winning $9000 and 999/1000 of winning nothing

b. Explain how the situation in part a resembles the choices people have when they decide whether to buy lottery tickets.

21. It is time for the end-of-summer sales. One store is offering bathing suits at 50% of their usual cost, and another store is offering to sell you two for the price of one. Assuming the suits originally all cost the same amount, which store is offering a better deal? Explain.

22. Refer to Case Study 17.2, in which the relationship between betting odds and probability of occurrence is explained.

a. Suppose you are offered a bet on an outcome for which the odds are 2 to 1 and there is no handling fee. For you to have a break-even expected value of zero, what would the probability of the outcome occurring have to be?

b. Suppose you believe that the probability that your team will win a game is 1/4. What odds should you be offered in order to place a bet in which you think you have a break-even expected value?

c. Explain what the two possible outcomes would be for the situation in part b, assuming you were offered the break-even odds and bet $1. Show that the expected value would indeed be zero.

23. Suppose you are trying to decide whether to park illegally while you attend class. If you get a ticket, the fine is $25. If you assess the probability of getting a ticket to be 1/100, what is the expected value for the fine you will have to pay? Under those circumstances, explain whether you would be willing to take the risk and why. (Note that there is no correct answer to the last part of the question; it is designed to test your reasoning.)

MINI-PROJECTS

1. Find out the sensitivity and specificity of a common medical test. Calculate the probability of a true positive for someone who tests positive with the test, assuming the rate in the population is 1 per 100; then calculate the probability assuming the rate in the population is 1 per 1000.

2. Ask four friends to tell you their most amazing coincidence story. Use the material in this chapter to assess how surprising each of the stories is to you. Pick one of the stories and try to approximate the probability of that specific event happening to your friend.

3. Conduct a survey in which you ask 20 people the two scenarios presented in Thought Question 5 at the beginning of this chapter and discussed in Section

17.5. Record the percentage who choose alternative A over B and the percentage who choose alternative C over D.

a. Report your results. Are they consistent with what other researchers have found? (Refer to p. 306.) Explain.

b. Explain how you conducted your survey. Discuss whether you overcame the potential difficulties with surveys that were discussed in Chapter 4.

REFERENCES

Diaconis, P., and F. Mosteller. (1989). Methods for studying coincidences. *Journal of the American Statistical Association* 84, pp. 853–861.

Eddy, D. M. (1982). Probabilistic reasoning in clinical medicine: Problems and opportunities. In D. Kahneman, P. Slovic, and A. Tversky (eds.), *Judgment under uncertainty: Heuristics and biases* (Chapter 18). Cambridge, England: Cambridge University Press.

Hooke, R. (1989). Basketball, baseball, and the null hypothesis, *Chance* 2, no. 4, pp. 35–37.

Larkey, Patrick D. (1990). Fair bets on winners in professional golf. *Chance* 3, no. 4, pp. 24–26.

Larkey, P. D., R. A. Smith, and J. B. Kadane. (1989). It's okay to believe in the "hot hand." *Chance* 2, no. 4, pp. 22–30.

Moore, D. S. (1991). *Statistics: Concepts and controversies.* 3d ed., New York: W. H. Freeman.

Plous, S. (1993). *The psychology of judgment and decision making.* New York: McGraw Hill.

Tversky, A., and T. Gilovich. (Winter 1989). The cold facts about the "hot hand" in basketball. *Chance* 2, no. 1, pp. 16–21.

Tversky, A., and D. Kahneman. (1982). Judgment under uncertainty: Heuristics and biases. In D. Kahneman, P. Slovic, and A. Tversky (eds.), *Judgment under uncertainty: Heuristics and biases* (Chapter 1). Cambridge, England: Cambridge University Press.

University of California, Berkeley. (1991). *The Wellness Encyclopedia.* Boston: Houghton Mifflin.

Weaver, W. (1963). *Lady luck: The theory of probability.* Garden City, NY: Doubleday.

Making Judgments from Surveys and Experiments

In Part 1, you learned how data should be collected in order to be meaningful. In Part 2, you learned some simple things you could do with data, and in Part 3, you learned that uncertainty can be quantified and can lead to worthwhile information about the aggregate.

In Part 4, you will learn about the final steps that allow us to turn data into useful information. You will learn how to use samples collected in surveys and experiments to say something intelligent about what is probably happening in an entire population.

Chapters 18 to 24 are somewhat more technical than previous chapters. Try not to get bogged down in the details. Remember that the purpose of this material is to enable you to say something about a whole population after examining just a small piece of it in the form of a sample. The book concludes with Chapter 25, which provides ten case studies that will reinforce your awareness that you have indeed become an educated consumer of statistical information.

CHAPTER EIGHTEEN

The Diversity of Samples from the Same Population

THOUGHT QUESTIONS

1. Suppose that 40% of a large population disagree with a proposed new law. In parts a and b, think about the role of the sample size when you answer the question.

   a. If you randomly sample ten people, will exactly four (40%) disagree with the law? Would you be surprised if only two of the people in the sample disagreed with the law? How about if none of the sample disagreed with it?

   b. Now suppose you randomly sample 1000 people. Will exactly 400 (40%) disagree with the law? Would you be surprised if only 200 of the people in the sample disagreed with the law? How about if none of the sample disagreed with it?

   c. Explain how the long-run relative-frequency interpretation of probability and the gambler's fallacy helped you answer parts a and b.

2. Suppose the mean weight of all women at a large university is 135 pounds, with a standard deviation of 10 pounds.

   a. Recalling the material from Chapter 8 about bell-shaped curves, in what range would you expect 95% of the women's weights to fall?

   b. If you were to randomly sample 10 women at the university, how close do you think their *average* weight would be to 135 pounds? If you sample 1000 women, would you expect the average weight to be closer to 135 pounds than it would be for the sample of only 10 women?

3. Recall from Chapter 4 that a survey of 1000 randomly selected individuals has a margin of error of about 3%, so that the results are accurate to within plus or minus 3% most of the time. Suppose 25% of adults believe in reincarnation. If ten polls are taken, each asking a different random sample of 1000 adults about belief in reincarnation, would you expect each poll to find exactly 25% of

respondents expressing belief in reincarnation? If not, into what range would you expect the ten sample proportions to reasonably fall?

18.1   SETTING THE STAGE

This chapter serves as an introduction to the reasoning that allows pollsters and researchers to make conclusions about entire populations on the basis of a relatively small sample of individuals. The reward for understanding the material presented in this chapter will come in the remaining chapters of this book, as you begin to realize the power of the statistical tools in use today.

Working Backward from Samples to Populations

The first step in this process is to work backward: from a sample to a population. We start with a question about a population, such as: How many teenagers are infected with HIV? At what average age do left-handed people die? What is the average income of all students at a large university? We collect a sample from the population about which we have the question, and we measure the variable of interest. We can then answer the question of interest for the sample. Finally, based on what statisticians have worked out, we will be able to determine how close the answer from our sample is to what we really want to know: the actual answer for the population.

Understanding Dissimilarity among Samples

The secret to understanding how things work is to understand what kind of dissimilarity we should expect to see in various samples from the same population. For example, suppose we knew that most samples were likely to provide an answer that is within 10% of the population answer. Then we would also know the reverse—the population answer should be within 10% of whatever our specific sample gave. Armed only with our sample value, we could make a good guess about the population value. You have already seen this idea at work in Chapter 4, when we used the margin of error for a sample survey to estimate results for the entire population. Statisticians have worked out similar techniques for a variety of sample measurements. In this and the next two chapters, we will cover some of these techniques in detail.

18.2   WHAT TO EXPECT OF SAMPLE PROPORTIONS

Suppose we want to know what proportion of a population carries the gene for a certain disease. We sample 25 people, and from that sample we make an estimate of the true answer.

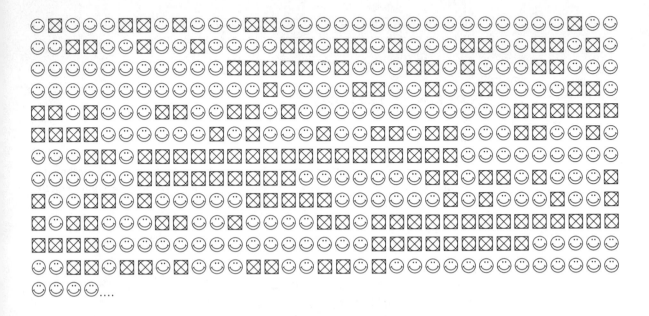

FIGURE 18.1
A slice of a population where 40% are ⊠

Suppose that 40% of the population actually carries the gene. We can think of the population as consisting of two types of people: those who do not carry the gene, represented as ☺, and those who do carry the gene, represented as ⊠. Figure 18.1 is a conceptual illustration of part of such a population.

Possible Samples

What would we find if we randomly sampled 25 people from this population? Would we always find 10 people (40%) with the gene and 15 people (60%) without? You should know from our discussion of the gambler's fallacy in Chapter 17 that we would not.

Each person we chose for our sample would have a 40% probability of carrying the gene. But remember that the relative-frequency interpretation of probability only ensures that we would see 40% of our sample with the gene *in the very long run.* A sample of only 25 people does not qualify as "the very long run."

What should we expect to see? Figure 18.2 shows four different random samples of 25 people taken from the population shown in Figure 18.1. Here is what we would have concluded about the proportion of people who carry the gene, given each of those samples:

Sample 1: Proportion with gene = 12/25 = .48 = 48%

Sample 2: Proportion with gene = 9/25 = .36 = 36%

Sample 1:

Sample 2:

Sample 3:

Sample 4:

FIGURE 18.2
Four possible random samples from Figure 18.1

> *Sample 3:* Proportion with gene = 10/25 = .40 = 40%
>
> *Sample 4:* Proportion with gene = 7/25 = .28 = 28%

Notice that each sample gives a different answer, and the sample answer may or may not actually match the truth about the population.

In practice, when a researcher conducts a study similar to this one or a pollster randomly samples a group of people to measure public opinion, only one sample is collected. There is no way to determine whether the sample is an accurate reflection of the population. However, statisticians have calculated what to expect for possible samples. We call the applicable rule the **Rule for Sample Proportions.**

Conditions for Which the Rule for Sample Proportions Applies

The following three conditions must all be met for the Rule for Sample Proportions to apply:

1. There exists an actual population with a fixed proportion who have a certain trait, opinion, disease, and so on.

   or

   There exists a repeatable situation for which a certain outcome is likely to occur with a fixed relative-frequency probability.

2. A random sample is selected from the population, thus ensuring that the probability of observing the characteristic is the same for each sample unit.

   or

   The situation is repeated numerous times, with the outcome each time independent of all other times.

> **3.** The size of the sample or the number of repetitions is relatively large. The necessary size depends on the proportion or probability under investigation. It must be large enough so that we are likely to see at least five of each of the two possible responses or outcomes.

Examples of Situations for Which the Rule for Sample Proportions Applies

Here are some examples of situations that meet these conditions.

EXAMPLE 1   **ELECTION POLLS**

A pollster wants to estimate the proportion of voters who favor a certain candidate. The voters are the population units, and favoring the candidate is the opinion of interest. ■

EXAMPLE 2   **TELEVISION RATINGS**

A television rating firm wants to estimate the proportion of households with television sets that are tuned to a certain television program. The collection of all households with television sets makes up the population, and being tuned to that particular program is the trait of interest. ■

EXAMPLE 3   **CONSUMER PREFERENCES**

A manufacturer of soft drinks wants to know what proportion of consumers prefers a new mixture of ingredients compared with the old recipe. The population consists of all consumers, and the response of interest is preference of the new formula over the old one. ■

EXAMPLE 4   **TESTING ESP**

A researcher studying extrasensory perception (ESP) wants to know the probability that people can successfully guess which of five symbols is on a hidden card. Each card is equally likely to contain each of the five symbols. There is no physical population. The repeatable situation of interest is a guess, and the response of interest is a successful guess. The researcher wants to see if the probability of a correct guess is higher than 20%, which is what it would be if there were no such thing as extrasensory perception. ■

Defining the Rule for Sample Proportions

The following is what statisticians have determined to be approximately true for the situations that have just been described in Examples 1–4 and for similar ones.

> If numerous samples or repetitions of the same size are taken, the frequency curve made from proportions from the various samples will be approximately bell-shaped. The mean of those sample proportions will be the true proportion from the population. The standard deviation will be:
>
> the square root of:
>
> (true proportion) × (1 − true proportion)/(sample size)

EXAMPLE 5   **USING THE RULE FOR SAMPLE PROPORTIONS**

Suppose of all voters in the United States, 40% are in favor of Candidate X for president. Pollsters take a sample of 2400 people. What proportion of the sample would be expected to favor Candidate X? The rule tells us that the proportion of the sample who favor Candidate X could be anything from a bell-shaped curve with mean of .40 (40%) and standard deviation of:

the square root of:

$(.40) \times (1 − .40)/2400 = (.4)(.6)/2400 = .24/2400 = .0001$

Thus, the mean is .40 and the standard deviation is .01 or 1/100 or 1%.

Figure 18.3 shows what we can expect of the sample proportion in this situation. Recalling the rule we learned in Chapter 8 about bell-shaped distributions, we can also specify that, for our sample of 2400 people:

There is a *68% chance that the sample proportion is between 39% and 41%.*

There is a *95% chance that the sample proportion is between 38% and 42%.*

It is almost certain that the sample proportion is between *37% and 43%.* ■

In practice, we have only one sample proportion and we don't know the true population proportion. However, we do know how far apart the sample proportion and the true proportion are likely to be. That information is contained in the standard deviation, which can be estimated using the sample proportion combined with the known sample size. Therefore, when all we have is a sample proportion, we can indeed say something about the true proportion.

FIGURE 18.3
*Possible sample
proportions when
n = 2400 and truth = .4*

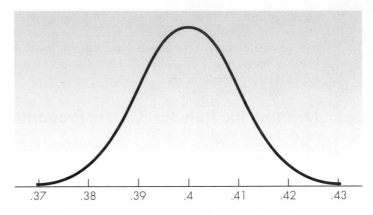

18.3   WHAT TO EXPECT OF SAMPLE MEANS

In the previous section, the question of interest was the proportion falling into one category of a categorical variable. We saw that we could determine an interval of values that was likely to cover the sample proportion if we knew the size of the sample and the magnitude of the true proportion.

We now turn to the case where the information of interest involves the mean or means of measurement variables. For example, researchers might want to compare the mean age at death for left- and right-handed people. A company that sells oat products might want to know the mean cholesterol level people would have if everyone had a certain amount of oat bran in their diet. To help determine financial aid levels, a large university might want to know the mean income of all students on campus who work.

Possible Samples

Suppose a population consists of thousands or millions of individuals, and we are interested in estimating the mean of a measurement variable. If we sample 25 people and compute the mean of the variable, how close will that sample mean be to the population mean we are trying to estimate? Each time we take a sample we will get a different sample mean. Can we say anything about what we expect those means to be?

For example, suppose we are interested in estimating the average weight loss for everyone who attends a national weight-loss clinic for 10 weeks. Suppose, unknown to us, the weight losses for everyone have a mean of 8 pounds, with a standard deviation of 5 pounds. If the weight losses are approximately bell-shaped, we know from Chapter 8 that 95% of the individuals will fall within 2 standard deviations, or 10 pounds, of the mean of 8 pounds. In other words, 95% of the individual weight losses will fall between –2 (a gain of 2 pounds) and 18 pounds lost.

Figure 18.4 lists some possible samples that could result from randomly sampling 25 people from this population; these were indeed the first four samples produced by a computer that is capable of simulating such things. The weight losses

FIGURE 18.4
Four potential samples from a population with mean = 8, standard deviation = 5

Sample 1: 1,1,2,3,4,4,4,5,6,7,7,7,8,8,9,9,11,11,13,13,14,14,15,16,16

Sample 2: –2,–2,0,0,3,4,4,4,5,5,6,6,8,8,9,9,9,9,9,10,11,12,13,13,16

Sample 3: –4,–4,2,3,4,5,7,8,8,9,9,9,9,9,10,10,11,11,11,12,12,13,14,16,18

Sample 4: –3,–3,–2,0,1,2,2,4,4,5,7,7,9,9,10,10,10,11,11,12,12,14,14,14,19

have been put into increasing order for ease of reading. A negative value indicates a weight gain.

Following are the sample means and standard deviations, computed for each of the four samples. You can see that the sample means, although all different, are relatively close to the population mean of 8. You can also see that the sample standard deviations are relatively close to the population standard deviation of 5.

Sample 1: Mean = 8.32 pounds, standard deviation = 4.74 pounds

Sample 2: Mean = 6.76 pounds, standard deviation = 4.73 pounds

Sample 3: Mean = 8.48 pounds, standard deviation = 5.27 pounds

Sample 4: Mean = 7.16 pounds, standard deviation = 5.93 pounds

Conditions to Which the Rule for Sample Means Applies

As with sample proportions, statisticians have developed a rule to tell us what to expect of sample means.

The Rule for Sample Means *applies in both of the following situations:*

1. The population of the measurements of interest is bell-shaped, and a random sample of any size is measured.

2. The population of measurements of interest is not bell-shaped, but a *large* random sample is measured. A sample of size 30 is usually considered "large," but if there are extreme outliers, it is better to have a larger sample.

There are only a limited number of situations for which the Rule for Sample Means does *not* apply. It does not apply at all if the sample is not random, and it does not apply for small random samples unless the original population is bell-shaped. In practice, it is often difficult to get a random sample.

Researchers are usually willing to use the Rule for Sample Means as long as they can get a representative sample with no obvious sources of confounding or bias.

Examples of Situations to Which the Rule for Sample Means Applies

Following are some examples of situations that meet the conditions for applying the Rule for Sample Means.

EXAMPLE 6

AVERAGE WEIGHT LOSS

A weight-loss clinic is interested in measuring the average weight loss for participants in its program. The clinic makes the assumption that the weight losses will be bell-shaped, so the Rule for Sample Means will apply for any sample size. The population of interest is all current and potential clients, and the measurement of interest is weight loss. ■

EXAMPLE 7

AVERAGE AGE AT DEATH

A researcher is interested in estimating the average age at which left-handed adults die, assuming they have lived to be at least 50. Because ages at death are not bell-shaped, the researcher should measure at least 30 such ages at death. The population of interest is all left-handed people who live to be at least 50 years old. The measurement of interest is age at death. ■

EXAMPLE 8

AVERAGE STUDENT INCOME

A large university wants to know the mean monthly income of students who work. The population consists of all students at the university who work. The measurement of interest is monthly income. Because incomes are not bell-shaped and there are likely to be outliers (a few people with high incomes), the university should use a large random sample of students. The researchers should take particular care to reach the people who are actually selected to be in the sample. A large bias could be created if, for example, they were willing to replace the desired respondent with a roommate who happened to be home when the researchers called. The students working the longest hours, and thus making the most money, would probably be hardest to reach by phone and the least likely to respond to a mail questionnaire. ■

Defining the Rule for Sample Means

The Rule for Sample Means is simple:
If numerous samples of the same size are taken, the frequency curve of means from the various samples will be approximately bell-shaped. The mean of this collection of sample means will be the same as the mean of the population. The standard deviation will be

population standard deviation / square root of sample size

EXAMPLE 9   **USING THE RULE FOR SAMPLE MEANS**

For our hypothetical weight-loss example, the population mean and standard deviation were 8 pounds and 5 pounds, respectively, and we were taking random samples of size 25. The rule tells us that potential sample means are represented by a bell-shaped curve with a mean of 8 pounds and standard deviation of 5/5 = 1.0. (We divide the population standard deviation of 5 by the square root of 25, which also happens to be 5.)

Therefore, we know the following facts about possible sample means in this situation, based on intervals extending 1, 2, and 3 standard deviations from the mean of 8:

There is a *68%* chance that the sample mean will be between *7 and 9*.

There is a *95%* chance that the sample mean will be between *6 and 10*.

It is *almost certain* that the sample mean will be between *5 and 11*.

If you look at the four hypothetical samples we chose (see Figure 18.4), you will see that the sample means range from 6.76 to 8.48, well within the range we expect to see using these criteria ■

Increasing the Size of the Sample

Suppose we had taken a sample of 100 people instead of 25. Notice that the mean of the possible sample means would not change; it would still be 8 pounds, but the standard deviation would decrease. It would now be 5/10 = .5, instead of 1.0. Therefore, for samples of size 100, here is what we would expect of sample means for the weight-loss situation:

There is a *68%* chance that the sample mean will be between *7.5 and 8.5*.

There is a *95%* chance that the sample mean will be between *7 and 9*.

It is *almost certain* that the sample mean will be between *6.5 and 9.5.*

It's obvious that the Rule for Sample Means tells us the same thing our common sense tells us: Larger samples tend to result in more accurate estimates of population values than do smaller samples.

This discussion presumed that we know the population mean and the standard deviation. Obviously, that's not much use to us in real situations when the population mean is what we are trying to determine. In Chapter 20, we will see how to use the Rule for Sample Means to accurately estimate the population mean when all we have available is a single sample for which we can compute the mean and the standard deviation.

18.4   WHAT TO EXPECT IN OTHER SITUATIONS

We have discussed two common situations that arise in assessing public opinion, conducting medical research, and so on. The first situation arises when we want to know what proportion of a population falls into one category of a categorical variable. The second situation occurs when we want to know the mean of a population for a measurement variable.

There are numerous other situations for which researchers would like to use results from a sample to say something about a population or to compare two or more populations. Statisticians have determined rules similar to those in this chapter for most of the situations researchers are likely to encounter. Those rules are too complicated for a book of this nature. However, once you understand the basic ideas for the two common scenarios covered here, you will be able to understand the results researchers present in more complicated situations. The basic ideas we explore apply equally to most other situations. You may not understand exactly how researchers determined their results, but you will understand the terminology and some of the potential misinterpretations.

In the next several chapters, we explore the two basic techniques researchers use to summarize their statistical results: confidence intervals and hypothesis testing.

Confidence Intervals

One basic technique researchers use is to create a **confidence interval,** which is an interval of values that the researcher is fairly sure covers the true value for the population.

We encountered confidence intervals in Chapter 4, when we learned about the margin of error. Adding and subtracting the margin of error to the reported sample proportion creates an interval that we are 95% "confident" covers the truth. That interval is the confidence interval. We will explore confidence intervals further in Chapters 19 and 20.

Hypothesis Testing

The second statistical technique researchers use is called **hypothesis testing** or **significance testing.** Hypothesis testing uses sample data to attempt to reject the hypothesis that nothing interesting is happening—that is, to reject the notion that chance alone can explain the sample results.

We encountered this idea in Chapter 12, when we learned how to determine whether the relationship between two categorical variables is "statistically significant." The hypothesis that researchers set about to reject in that setting was that two categorical variables are unrelated to each other. In most research settings, the *desired* conclusion is that the variables under scrutiny are related. Achieving statistical significance is equivalent to rejecting the idea that chance alone can explain the observed results. We will explore hypothesis testing further in Chapters 21, 22, and 23.

CASE STUDY 18.1 ## Do Americans Really Vote When They Say They Do?

On November 8, 1994, a historic election took place, in which the Republican Party won control of both houses of Congress for the first time since 1952. But how many people actually voted? On November 28, 1994, *Time* magazine (p. 20) reported that in a telephone poll of 800 adults taken during the two days following the election, 56% reported that they had voted. Considering that only about 68% of adults are registered to vote, that isn't a bad turnout.

But, along with these numbers, *Time* reported another disturbing fact. They reported that, in fact, only 39% of American adults had voted, based on information from the Committee for the Study of the American Electorate.

Could it be the case that the results of the poll simply reflected a sample that, by chance, voted with greater frequency than the general population? The Rule for Sample Proportions can answer that question. Let's suppose that the truth about the population is, as reported by *Time,* that only 39% of American adults voted. Then the Rule for Sample Proportions tells us what kind of sample proportions we can expect in samples of 800 adults, the size used by the *Time* magazine poll. The mean of the possibilities is .39, or 39%. The standard deviation is the square root of $(.39)(.61)/800 = .017$, or 1.7%.

Therefore, we are almost certain that the sample proportion based on a sample of 800 adults should fall within $3 \times 1.7\% = 5.1\%$ of the truth of 39%. In other words, if respondents were telling the truth, the sample proportion should be no higher than 44.1%, nowhere near the reported percentage of 56%.

In fact, if we combine the Rule for Sample Proportions with what we learned about bell-shaped curves in Chapter 8, we can say even more about how unlikely this sample result would be. If, in truth, only 39% of the population voted, the standardized score for the sample proportion of 56% is $(.56 - .39)/.017 = 10$. We know from Chapter 8 that it is virtually impossible to obtain a standardized score of 10.

Another example of the fact that *reported* voting tends to exceed *actual* voting occurred in the 1992 U.S. presidential election. According to the *World Almanac and Book of Facts* (1995, p. 631), 61.3% of American adults reported voting in the 1992 election. In a footnote, the Almanac explains:

> *Total reporting voting compares with 55.9 percent of population actually voting for president, as reported by Voter News Service. Differences between data may be the result of a variety of factors, including sample size, differences in the respondents' interpretation of the questions, and the respondents' inability or unwillingness to provide correct information or recall correct information.*

Unfortunately, because figures are not provided for the size of the sample, we cannot assess whether the difference between the actual percentage of 55.9 and the reported percentage of 61.3 can be explained by the natural variability among possible sample proportions. ■

FOR THOSE WHO LIKE FORMULAS

Notation for Population and Sample Proportions

Sample size = n

Population proportion = p

Sample proportion = $\hat{p}$, which is read "p-hat" because the p appears to have a little hat on it.

The Rule for Sample Proportions

If numerous samples or repetitions of size n are taken, the frequency curve of the $\hat{p}$'s from the various samples will be approximately bell-shaped. The mean of those $\hat{p}$'s will be p. The standard deviation will be

$$\sqrt{\frac{p(1-p)}{n}}$$

Notation for Population and Sample Means and Standard Deviations

Population mean = μ (read "mu"), population standard deviation = σ (read "sigma")

Sample mean = $\overline{X}$, sample standard deviation = s

The Rule for Sample Means

If numerous samples of size n are taken, the frequency curve of the $\overline{X}$'s from the various samples is approximately bell-shaped with mean μ and standard deviation $\sigma/\sqrt{n}$.

Another way to write these rules is using the notation for normal distributions from Chapter 8:

$$\hat{p} \sim N(p, \frac{p(1-p)}{n}) \quad \text{and} \quad \overline{X} \sim N(\mu, \frac{\sigma^2}{n})$$

....

EXERCISES

1. Suppose you want to estimate the proportion of students at your college who are left-handed. You decide to collect a random sample of 200 students and ask them which hand is dominant. Go through the conditions for which the rule for sample proportions applies (pp. 319–320) and explain why the rule would apply to this situation.

2. Refer to Exercise 1. Suppose the truth is that .12 or 12% of the students are left-handed, and you take a random sample of 200 students. Use the Rule for Sample Proportions to draw a picture similar to Figure 18.3, showing the possible sample proportions for this situation.

3. According to the *Sacramento Bee* (2 April 1998, p. F5), "A 1997-98 survey of 1027 Americans conducted by the National Sleep Foundation found that 23% of adults say they have fallen asleep at the wheel in the last year."

   a. Conditions 2 and 3 needed to apply the Rule for Sample Proportions are met because this result is based on a large random sample of adults. Explain how condition 1 is also met.

   b. The article also said that (based on the same survey) "37 percent of adults report being so sleepy during the day that it interferes with their daytime activities." If, in truth, 40% of all adults have this problem, find the interval in which about 95% of all sample proportions should fall, based on samples of size 1027. Does the result of this survey fall into that interval?

   c. Suppose a survey based on a random sample of 1027 college students was conducted and 25% reported being so sleepy during the day that it interferes with their daytime activities. Would it be reasonable to conclude that the population proportion of college students who have this problem differs from the proportion of all adults who have the problem? Explain.

4. A recent Gallup poll found that of 800 randomly selected drivers surveyed, 70% thought they were better than average drivers. In truth, in the population, only 50% of all drivers can be "better than average."

   a. Draw a picture of the possible sample proportions that would result from samples of 800 people from a population with a true proportion of .50.

   b. Would we be unlikely to see a sample proportion of .70, based on a sample of 800 people, from a population with a proportion of .50? Explain, using your picture from part a.

   c. Explain the results of this survey using the material from Chapter 16.

5. Suppose you are interested in estimating the average number of miles per gallon of gasoline your car can get. You calculate the miles per gallon for each of the next nine times you fill the tank. Suppose, in truth, the values for your car are bell-shaped, with a mean of 25 miles per gallon and a standard deviation of 1. Draw a picture of the possible sample means you are likely to get based on your sample of nine observations. Include the intervals into which 68%, 95%, and almost all of the potential sample means will fall.

6. Refer to Exercise 5. Redraw the picture under the assumption that you will collect 100 measurements instead of only 9. Discuss how the picture differs from the one in Exercise 5.

7. Give an example of a situation of interest to you for which the Rule for Sample Proportions would apply. Explain why the conditions allowing the rule to be applied are satisfied for your example.

8. Suppose the population of IQ scores in the town or city where you live is bell-shaped, with a mean of 105 and a standard deviation of 15. Describe the frequency curve for possible sample means that would result from random samples of 100 IQ scores.

9. Suppose that 35% of the students at a university favor the semester system, 60% favor the quarter system, and 5% have no preference. Is a random sample of 100 students large enough to provide convincing evidence that the quarter system is favored? Explain.

10. According to *USA Today* (20 April 1998, Snapshot), a poll of 8709 adults taken in 1976 found that 9% believed in reincarnation, whereas a poll of 1000 adults taken in 1997 found that 25% held that belief.

   a. Assuming a proper random sample was used, verify that the sample proportion for the poll taken in 1976 almost certainly represents the population proportion to within about 1%.

   b. Based on these results, would you conclude that the proportion of all adults who believe in reincarnation was higher in 1997 than it was in 1976? Explain.

11. Suppose 20% of all television viewers in the country watch a particular program.

   a. For a random sample of 2500 households measured by a rating agency, describe the frequency curve for the possible sample proportions who watch the program.

   b. The program will be canceled if the ratings show less than 17% watching in a random sample of households. Given that 2500 households are used for the ratings, is the program in danger of getting canceled? Explain.

12. Use the Rule for Sample Means to explain why it is desirable to take as large a sample as possible when trying to estimate a population value.

13. According to the *Sacramento Bee* (2 April 1998, p. F5), Americans get an average of 6 hours and 57 minutes of sleep per night. A survey of a class of 190 sta-

tistics students at a large university found that they averaged 7.1 hours of sleep the previous night, with a standard deviation of 1.95 hours.

    a. Assume that the population average for adults is 6 hours and 57 minutes, or 6.95 hours of sleep per night, with a standard deviation of 2 hours. Draw a picture illustrating how the Rule for Sample Means would apply to sample means for random samples of 190 adults.

    b. Would the mean of 7.1 hours of sleep obtained from the statistics students be a reasonable value to expect for the sample mean of a random sample of 190 adults? Explain.

    c. Can the sample taken in the statistics class be considered to be a representative sample of all adults? Explain.

14. Explain whether each of the following situations meets the conditions for which the Rule for Sample Proportions applies. If not, explain which condition is violated.

    a. Unknown to the government, 10% of all cars in a certain city do not meet appropriate emissions standards. The government wants to estimate that percentage, so they take a random sample of 30 cars and compute the sample proportion that do not meet the standards.

    b. The Census Bureau would like to estimate what proportion of households have someone at home between 7 P.M. and 7:30 P.M. on weeknights, to determine whether that would be an efficient time to collect census data. The Bureau surveys a random sample of 2000 households and visits them during that time to see whether someone is at home.

    c. You are interested in knowing what proportion of days in typical years have rain or snow in the area where you live. For the months of January and February, you record whether there is rain or snow each day, and then you calculate the proportion.

    d. A large company wants to determine what proportion of its employees are interested in on-site day care. The company asks a random sample of 100 employees and calculates the sample proportion who are interested.

15. Explain whether you think the Rule for Sample Means applies to each of the following situations. If it does apply, specify the population of interest and the measurement of interest. If it does not apply, explain why not.

    a. A researcher is interested in what the average cholesterol level would be if people restricted their fat intake to 30% of calories. He gets a group of patients who have had heart attacks to volunteer to participate, puts them on a restricted diet for a few months, and then measures their cholesterol.

    b. A large corporation would like to know the average income of the spouses of its workers. Rather than go to the trouble to collect a random sample, they post someone at the exit of the building at 5 P.M. Everyone who leaves between 5 P.M. and 5:30 P.M. is asked to complete a short questionnaire on the issue; there are 70 responses.

     c. A university wants to know the average income of its alumni. Staff members select a random sample of 200 alumni and mail them a questionnaire. They follow up with a phone call to those who do not respond within 30 days.

     d. An automobile manufacturer wants to know the average price for which used cars of a particular model and year are selling in a certain state. They are able to obtain a list of buyers from the state motor vehicle division, from which they select a random sample of 20 buyers. They make every effort to find out what those people paid for the cars and are successful in doing so.

16. In Case Study 18.1, we learned that about 56% of American adults actually voted in the presidential election of 1992, whereas about 61% of a random sample claimed that they had voted. The size of the sample was not specified, but suppose it were based on 1600 American adults, a common size for such studies.

     a. Into what interval of values should the sample proportion fall 68%, 95%, and almost all of the time?

     b. Is the observed value of 61% reasonable, based on your answer to part a?

     c. Now suppose the sample had been of only 400 people. Compute a standard-ized score to correspond to the reported percentage of 61%. Comment on whether you believe people in the sample could all have been telling the truth, based on your result.

17. Suppose the population of grade-point averages (GPAs) for students at the end of their first year at a large university has a mean of 3.1 and a standard deviation of .5. Draw a picture of the frequency curve for the mean GPA of a random sample of 100 students.

18. The administration of a large university wants to use a random sample of students to measure student opinion of a new food service on campus. Administrators plan to use a continuous scale from 1 to 100, where 1 is complete dissatisfaction and 100 is complete satisfaction. They know from past experience with such questions that the standard deviation for the responses is going to be about 5, but they do not know what to expect for the mean. They want to be almost sure that the sample mean is within plus or minus 1 point of the true population mean value. How large will their random sample have to be?

MINI-PROJECTS

1. The goal of this mini-project is to help you verify the Rule for Sample Proportions firsthand. You will use the population represented in Figure 18.1 to do so. It contains 400 individuals, of whom 160 (40%) are ⊠—that is, carry the gene for a disease—and the remaining 240 (60%) are ☺—that is, do not carry the gene. You are going to draw 20 samples from this population. Here are the steps you should follow:

Step 1: Develop a method for drawing simple random samples from this population. One way to do this is to cut up the symbols and put them all into a paper bag, shake well, and draw from the bag. There are less tedious methods, but make sure you actually get random samples. *Explain your method.*

Step 2: Draw a random sample of size 15 and record the number and percentage who carry the gene.

Step 3: Repeat step 2 a total of 20 times, thus accumulating 20 samples, each of size 15. Make sure to start over each time; for example, if you used the method of drawing symbols from a paper bag, then put the symbols back into the bag after each sample of size 15 is drawn so they are available for the next sample as well.

Step 4: Create a stemplot or histogram of your 20 sample proportions. Compute the mean.

Step 5: Explain what the Rule for Sample Proportions tells you to expect for this situation.

Step 6: Compare your results with what the Rule for Sample Proportions tells you to expect. Be sure to mention mean, standard deviation, shape, and the intervals into which you expect 68%, 95%, and almost all of the sample proportions to fall.

2. The purpose of this mini-project is to help you verify the Rule for Sample Means. Suppose you are interested in measuring the average amount of blood contained in the bodies of adult women, in ounces. Suppose, in truth, the population consists of the values listed below. (Each value would be repeated millions of times, but in the same proportions as they exist in this list.) The actual mean and standard deviation for these numbers are 110 ounces and 5 ounces, respectively. The values are bell-shaped.

Population Values for Ounces of Blood in Adult Women

97	100	101	102	103	103	104	104	104	105	106
106	106	107	107	108	108	109	109	109	110	110
110	110	110	111	112	112	112	112	113	113	113
113	113	113	114	114	114	114	114	114	115	115
116	116	116	117	118	118					

Step 1: Develop a method for drawing simple random samples from this population. One way to do this is to write each number on a slip of paper, put all the slips into a paper bag, shake well, and draw from the bag. If a number occurs multiple times, make sure you include it that many times. Make sure you actually get random samples. *Explain your method.*

Step 2: Draw a random sample of size 9. Calculate and record the mean for your sample.

Step 3: Repeat step 2 a total of 20 times, thus accumulating 20 samples, each of size 9. Make sure to start over each time; for example, if you drew numbers from a paper bag, put the numbers back after each sample of size 9 so they are available for the next sample as well.

Step 4: Create a stemplot or histogram of your 20 sample means. Compute the mean of those sample means.

Step 5: Explain what the Rule for Sample Means tells you to expect for this situation.

Step 6: Compare your results with what the Rule for Sample Means tells you to expect. Be sure to mention mean, standard deviation, shape, and the intervals into which you expect 68%, 95%, and almost all of the sample means to fall.

REFERENCE

World almanac and book of facts. (1995). Edited by Robert Famighetti. Mahwah, NJ: Funk and Wagnalls.

CHAPTER NINETEEN

Estimating Proportions with Confidence

THOUGHT QUESTIONS

1. One example we see in this chapter is a 95% confidence interval for the proportion of British couples in which the wife is taller than the husband. The interval extends from .02 to .08, or 2% to 8%. What do you think it means to say that the interval from .02 to .08 represents a 95% confidence interval for the proportion of couples in which the wife is taller than the husband?

2. Do you think a 99% confidence interval for the proportion described in Question 1 would be wider or narrower than the 95% interval given? Explain.

3. In a Yankelovich Partners poll of 1000 adults (*USA Today*, 20 April 1998), 45% reported that they believed in "faith healing." Based on this survey, a "95% confidence interval" for the proportion in the population who believe is about 42% to 48%. If this poll had been based on 5000 adults instead, do you think the "95% confidence interval" would be wider or narrower than the interval given? Explain.

4. How do you think the concept of *margin of error*, explained in Chapter 4, relates to confidence intervals for proportions? As a concrete example, can you determine the margin of error for the situation in Question 1 from the information given? In Question 3?

19.1   CONFIDENCE INTERVALS

In the previous chapter, we saw that we get a different answer each time we take a sample from a population. We also learned that statisticians have been able to quantify the amount by which those sample values are likely to differ from each other and from the population.

In practice, statistical methods are used in situations where only one sample is taken, and that sample is used to make a conclusion or an inference about the population from which it was taken. One of the most common types of inferences is to construct what is called a **confidence interval,** which is defined as

> *an interval of values computed from sample data that is almost sure to cover the true population number.*

The most common level of confidence used is 95%. In other words, researchers define "almost sure" to mean that they are 95% certain. They are willing to take a 5% risk that the interval does not actually cover the true value.

It would be impossible to construct an interval in which we could be 100% confident unless we actually measured the entire population. Sometimes, as we shall see in one of the examples in the next section, researchers employ only 90% confidence. In other words, they are willing to take a 10% chance that their interval will not cover the truth.

Methods for actually constructing confidence intervals differ, depending on the type of question asked and the type of sample used. In this chapter, we learn to construct confidence intervals for proportions, and in the next chapter, we learn to construct confidence intervals for means. If you understand the kinds of confidence intervals we study in this chapter and the next, you will understand any other type of confidence interval as well.

In most applications, we never know whether the confidence interval covers the truth; we can only apply the long-run frequency interpretation of probability. All we can know is that, in the long run, 95% of all confidence intervals tagged with 95% confidence will be correct and 5% of them will be wrong. There is no way to know for sure which kind we have in any given situation. A common and humorous phrase among statisticians is: "Being a statistician means never having to say you're certain."

19.2   THREE EXAMPLES OF CONFIDENCE INTERVALS FROM THE MEDIA

When the media report the results of a statistical study, they often supply the information necessary to construct a confidence interval. Sometimes they even provide a confidence interval directly. The most commonly reported information that can be used to construct a confidence interval is the *margin of error.* Most public opinion polls report a margin of error along with the proportion of the sample that had each opinion. To use that information, you need to know this fact:

To construct a 95% confidence interval for a population proportion, simply add and subtract the margin of error to the sample proportion.

The margin of error is often reported using the symbol "±," which is read "plus or minus." The formula for a 95% confidence interval can thus be expressed as

sample proportion ± margin of error

Let's examine three examples from the media, in which confidence intervals are either reported directly or can easily be derived.

EXAMPLE 1 A PUBLIC OPINION POLL

In a poll reported in *Newsweek* (16 May 1994, p. 23), one of the questions asked was: "Is the media paying too much attention to [President] Clinton's private life, too little, or about the right amount of attention?" Results showed that 59% of the sample answered "too much," 5% answered "too little," and 31% answered "right amount." The article also gave this information about the poll:

For this Newsweek *poll, Princeton Survey Research Associates interviewed 518 adults by telephone May 6, 1994. The margin of error is ±5 percentage points. Some responses not shown. (Newsweek, 16 May 1994, p. 22)*

What proportion of the entire adult population believed that the media was paying too much attention to Clinton's private life? A 95% confidence interval for that proportion can be found by taking

sample proportion ± margin of error
59% ± 5%
54% to 64%

Notice that this interval does not cover 50%; it resides completely above 50%. Therefore, it would be fair to conclude, with high confidence, that a majority of Americans believed the media was paying too much attention to President Clinton's private life at that time. ∎

EXAMPLE 2 NUMBER OF AIDS CASES IN THE UNITED STATES

An Associated Press story reported in the *Davis* (CA) *Enterprise* (14 December 1993, p. A-5) was headlined, "Rate of AIDS infection in U.S. may be declining." The story noted that previous estimates of the number of cases of HIV infection in the United States were based on mathematical projections working backward from the known cases, rather than on a survey of the general population. The article reported that

> For the first time, survey data is now available on a randomly chosen cross-section of Americans. Conducted by the National Center for Health Statistics, it concludes that 550,000 Americans are actually infected.

One of the main purposes of the research reported in the story was to estimate how many Americans were currently infected with HIV, the virus thought to cause AIDS. Some estimates had been as high as 10 million people, but the article noted that the Centers for Disease Control (CDC) had estimated the number at about 1 million. Could the results of this survey rule out the fact that the number of people infected may be as high as 10 million? The way to answer the question is with a confidence interval for the true number who are infected, and the article proceeded to report exactly that:

> Dr. Geraldine McQuillan, who presented the analysis, said the statistical margin of error in the survey suggests that the true figure is probably between 300,000 and just over 1 million people. "The real number may be a little under or a bit over the CDC estimate" [of 1 million], she said, "but it is not 10 million."

Notice that the article reports a confidence interval, but does not call it by that name. The article also does not report the level of confidence, but it is probably 95%, which is the default value used by most statisticians. You may also have noted that the interval is not the simple form of "sample value ± margin of error." Although the methods used to form the confidence interval in this example were more complicated, the interpretation is just as simple. With high confidence, we can conclude that the true number of HIV-infected individuals at the time of the study was between 300,000 and 1 million. ■

EXAMPLE 3   **THE DEBATE OVER PASSIVE SMOKING**

On July 28, 1993, the *Wall Street Journal* featured an article by Jerry E. Bishop with the headline, "Statisticians occupy front lines in battle over passive smoking" (pp. B-1, B-4). The interesting feature of this article was that it not only reported a confidence interval, but it highlighted a debate between the U.S. Environmental Protection Agency (EPA) and the tobacco industry over what level of confidence should prevail in a confidence interval.

Here is the first side of the story, as reported in the article:

> The U.S. Environmental Protection Agency says there is a 90% probability that the risk of lung cancer for passive smokers is somewhere between 4% and 35% higher than for those who aren't exposed to environmental smoke. To statisticians, this calculation is called the "90% confidence interval."

Now, for the other side of the story:

And that, say tobacco-company statisticians, is the rub. "Ninety-nine percent of all epidemiological studies use a 95% confidence interval," says Gio B. Gori, director of the Health Policy Center in Bethesda, Md., who has frequently served as a consultant and an expert witness for the tobacco industry.

The problem underlying this controversy is that the amount of data available at the time did not allow an extremely accurate estimate of the true change in risk of lung cancer for passive smokers. The EPA statisticians were afraid that the public would not understand the interpretation of a confidence interval. If a 95% confidence interval actually went below zero percent, which it might do, the tobacco industry could argue that passive smoke might *reduce* the risk of lung cancer. As noted by one of the EPA's statisticians:

Dr. Gori is correct in saying that using a 95% confidence interval would hint that passive smoking might reduce the risk of cancer. But, he says, this is exactly why it wasn't used. The EPA believes it is inconceivable that breathing in smoke containing cancer-causing substances could be healthy and any hint in the report that it might be would be meaningless and confusing. (p. B-4)

In Chapter 23, we study in more detail the issue of making erroneous conclusions based on too little data. It is often the case that the amount of data available does not allow us to conclusively detect a problem, but that is not the same as concluding that no problem exists. ■

19.3   CONSTRUCTING A CONFIDENCE INTERVAL FOR A PROPORTION

You can easily learn to construct your own confidence intervals for some simple situations. One of those situations is the one we encountered in the previous chapter, in which a simple random sample is taken for a categorical variable. It is easy to construct a confidence interval for the proportion of the population who fall into one of the categories. Following are some examples of situations where this would apply. After presenting the examples, we develop the method, and then return to the examples to compute confidence intervals.

EXAMPLE 4   **HOW OFTEN IS THE WIFE TALLER THAN THE HUSBAND?**

In Chapter 10, we displayed data representing the heights of husbands and wives for a random sample of 200 British couples. From that set of data, we can count the number of couples for whom the wife is taller than the husband. We can then construct a confidence interval for the true proportion of British couples for whom that would be the case. ■

EXAMPLE 5	**AN EXPERIMENT IN EXTRASENSORY PERCEPTION**

In Chapter 21, we will describe in detail an experiment that was conducted to test for extrasensory perception (ESP). For one part of the experiment, subjects were asked to describe a video being watched by a "sender" in another room. The subjects were then shown four videos and asked to pick the one they thought the "sender" had been watching. Without ESP, the probability of a correct guess should be .25, or one-fourth, because there were four equally likely choices. We will use the data from the experiment to construct a confidence interval for the true probability of a correct guess and see if the interval includes the .25 value that would be expected if there were no ESP. ∎

EXAMPLE 6	**THE PROPORTION WHO WOULD QUIT SMOKING WITH A NICOTINE PATCH**

In Case Study 5.1, we examined an experiment in which 120 volunteers were given nicotine patches. After 8 weeks, 55 of them had quit smoking. Although the volunteers were not a random sample from a population, we can estimate the proportion of people who would quit if they were recruited and treated exactly as these individuals were treated. ∎

Developing the Formula for a 95% Confidence Interval

We develop the formula for a 95% confidence interval only and discuss what we would do differently if we wanted higher or lower confidence.

> *The formula will follow directly from the Rule for Sample Proportions:*
> If numerous samples or repetitions of the same size are taken, the frequency curve made from proportions from the various samples will be approximately bell-shaped. The mean will be the true proportion from the population. The standard deviation will be
>
> the square root of:
>
> (true proportion) × (1 – true proportion)/(sample size)

Because the possible sample proportions are bell-shaped, we can make the following statement:

In 95% of all samples, the sample proportion *will fall within* 2 standard deviations *of the mean, which is the* true proportion *for the population.*

This statement allows us to easily construct a 95% confidence interval for the true population proportion. Notice that we can rewrite it slightly, as follows:

In 95% of all samples, the true proportion *will fall* within 2 standard deviations of the sample proportion.

In other words, if we simply add and subtract 2 standard deviations to the sample proportion, in 95% of all cases we will have captured the true population proportion.

There is just one hurdle left. If you examine the Rule for Sample Proportions to find the standard deviation, you will notice that it uses the "true proportion." But we don't know the true proportion; in fact, that's what we are trying to estimate. There is a simple solution to this dilemma. We can get a fairly accurate answer if we substitute the sample proportion for the true proportion in the formula for the standard deviation.

> *Putting all of this together, here is the formula for a 95% confidence interval for a population proportion:*
>
> sample proportion ± 2(S.D.)
>
> where
>
> S.D. = the square root of:
>
> (sample proportion) × (1 − sample proportion)/(sample size)

A technical note: To be exact, we would actually add and subtract 1.96(S.D.) instead of 2(S.D.) because 95% of the values for a bell-shaped curve fall within 1.96 standard deviations of the mean. However, in most practical applications, rounding 1.96 off to 2.0 will not make much difference and this is common practice.

Revisiting the Examples

Let us now apply this formula to the examples presented at the beginning of this section.

EXAMPLE 4 REVISITED

HOW OFTEN IS THE WIFE TALLER THAN THE HUSBAND?

The data presented in Chapter 10, on the heights of 200 British couples, showed that in only 10 couples the wife was taller than the husband. Therefore, we find the following numbers:

sample proportion = 10/200 = .05 or 5%

standard deviation = square root of (.05)(.95)/200 = .015

confidence interval = .05 ± 2(.015) = .05 ± .03 = .02 to .08

In other words, we are 95% confident that of all British couples, between .02 (2%) and .08 (8%) are such that the wife is taller than her husband. ■

**EXAMPLE 5
REVISITED**

AN EXPERIMENT IN EXTRASENSORY PERCEPTION

The data we will examine in detail in Chapter 21 include 165 cases of experiments in which a subject tried to guess which of four videos the "sender" was watching in another room. Of the 165 cases, 61 resulted in successful guesses. Therefore, we find the following numbers:

sample proportion = 61/165 = .37 or 37%

standard deviation = square root of (.37)(.63)/165 = .038

confidence interval = .37 ± 2(.038) = .37 ± .08 = .29 to .45

In other words, we are 95% confident that the probability of a successful guess in this situation is between .29 (29%) and .45 (45%). Notice that this interval lies entirely above the 25% value expected if ESP were not at work. ■

**EXAMPLE 6
REVISITED**

THE PROPORTION WHO WOULD QUIT SMOKING WITH A NICOTINE PATCH

In Case Study 5.1, we learned that of 120 volunteers randomly assigned to use a nicotine patch, 55 of them had quit smoking after 8 weeks. We use this information to estimate the probability that a smoker recruited and treated in an identical fashion would quit smoking after 8 weeks:

sample proportion = 55/120 = .46 or 46%

standard deviation = square root of (.46)(.54)/120 = .045

confidence interval = .46 ± 2(.045) = .46 ± .09 = .37 to .55

In other words, we are 95% confident that between 37% and 55% of smokers treated in this way would quit smoking after 8 weeks. Remember that a placebo group was included for this experiment, in which 24 people, or 20%, quit smoking after 8 weeks. A confidence interval surrounding that value runs from 13% to 27% and thus does not overlap with the confidence interval for those using the nicotine patch. ■

Other Levels of Confidence

If you wanted to present a narrower interval, you would have to settle for less confidence. Applying the reasoning we used to construct the formula for a 95% confidence interval and using the information about bell-shaped curves from Chapter 8, we could have constructed a 68% confidence interval, for example. We would simply add and subtract 1 standard deviation to the sample proportion instead of 2. Similarly, if we added and subtracted 3 standard deviations, we would have a 99.7% confidence interval.

Although 95% confidence intervals are by far the most common, you will sometimes see 90% or 99% intervals as well. To construct those, you simply replace the value 2 in the formula with 1.645 for a 90% confidence interval or with the value 2.576 for a 99% confidence interval.

How the Margin of Error Was Derived in Chapter Four

We have already noted that you can construct a 95% confidence interval for a proportion if you know the margin of error. You simply add and subtract the margin of error to the sample proportion.

Polls, such as the *Newsweek* poll quoted earlier in this chapter, generally use a multistage sample (see Chapter 4). Therefore, the simple formulas for confidence intervals given in this chapter, which are based on simple random samples, do not give exactly the same answers as those using the margin of error stated. For polls based on multistage samples, it is more appropriate to use the stated margin of error than to use the formula given in this chapter.

In Chapter 4, we presented an approximate method for computing the margin of error. Using the letter n to represent the sample size, we then said that a conservative way to compute the margin of error was to use $1/\sqrt{n}$. Thus, we now have two apparently different formulas for finding a 95% confidence interval:

sample proportion ± margin of error = sample proportion ± $1/\sqrt{n}$

or

sample proportion ± 2(S.D.)

How do we reconcile the two different formulas? In order to reconcile them, it should follow that:

margin of error = $1/\sqrt{n}$ = 2(S.D.)

It turns out that these two formulas are equivalent when the proportion used in the formula for S.D. is .50. In other words, when:

standard deviation = S.D. = square root of $(.5)(.5)/n$ = $(.5)/\sqrt{n}$

In that case, 2(S.D.) is simply $1/\sqrt{n}$, which is our conservative formula for margin of error. This is called a *conservative* formula because the true margin of error is actually likely to be smaller. If you use any value other than (.5) as the proportion in the formula for standard deviation, you will get a smaller answer than you get using .5. You will be asked to confirm this fact in Exercise 10 at the end of this chapter.

CASE STUDY 19.1   # A Winning Confidence Interval Loses in Court

Gastwirth (1988, p. 495) describes a court case in which Sears, Roebuck and Company, a large department store chain, tried to use a confidence interval to determine the amount by which they had overpaid city taxes at their stores in Inglewood, California. Unfortunately, the judge did not think the confidence interval was appropriate and required Sears to examine all the sales records for the period in question. This case study provides an example of a situation where the answer became known, so we can compare the results from the sample with the true answer.

The problem arose because Sears had erroneously collected and paid city sales taxes for sales made to individuals outside the city limits. The company discovered the mistake during a routine audit, and asked the city for a refund of $27,000, the amount by which they estimated they had overpaid.

Realizing that they needed data to substantiate this amount, Sears decided to take a random sample of sales slips for the period in question and then, on the basis of the sample proportion, try to estimate the proportion of all sales that had been made to people outside of city limits. They used a multistage sampling plan, in which they divided the 33-month period into eleven 3-month periods to ensure that seasonal effects were considered. They then took a random sample of 3 days in each period, for a total of 33 days, and examined all sales slips for those days.

Based on the data, they derived a 95% confidence interval for the true proportion of all sales that were made to out-of-city customers. The confidence interval was .367 ± .03, or .337 to .397. To determine the amount of tax they believed they were owed, they multiplied the percentage of out-of-city sales by the total tax they had paid, which was $76,975. The result was $28,250, with a 95% confidence interval extending from $25,940 to $30,559.

The judge did not accept the use of sampling despite testimony from accounting experts who noted that it was common practice in auditing. The judge required Sears to examine all of the sales records. In doing so, Sears discovered that about one month's worth of slips were missing; however, based on the available slips, they had overpaid $26,750.22. This figure is slightly under the true amount due to the missing month, but you can see that the sampling method Sears had used provided a fairly accurate estimate of the amount they were owed. If we assume that the dollar amount from the missing month was similar to those for the months counted, we find that the total they were owed would have been about $27,586.

Sampling methods and confidence intervals are routinely used for financial audits. These techniques have two main advantages over studying all of the records. First, they are much cheaper. It took Sears about 300 person-hours to conduct the sample and 3384 hours to do the full audit. Second, a sample can be done more carefully than a complete audit. In the case of Sears, they could have two well-trained people conduct the sample in less than a month. The full audit would require either having those same two people work for ten months or training ten times as many people. As Gastwirth (1988, p. 496) concludes in his discussion of the Sears case, "A well designed sampling audit may yield a more accurate estimate than a less carefully carried out complete audit or census." In fairness, the judge in this case was simply following the law; the sales tax return required a sale-by-sale computation. ■

FOR THOSE WHO LIKE FORMULAS

Notation for Population and Sample Proportions (from Chapter 18)

Sample size = n

Population proportion = p

Sample proportion = $\hat{p}$

Notation for the Multiplier for a Confidence Interval

For reasons that will become clear in later chapters, we specify the *level* of confidence for a confidence interval as $(1 - \alpha$ (read "alpha")) 100%. For example, for a 95% confidence interval, $\alpha = .05$. Let $z_{\alpha/2}$ = standardized normal score with area $\alpha/2$ above it. Then the area between $z_{\alpha/2}$ and $-z_{\alpha/2}$ is $1 - \alpha$. For example, when $\alpha = .05$, as for a 95% confidence interval, $z_{\alpha/2} = 1.96$, or about 2.

Formula for a $(1 - \alpha)$ 100% Confidence Interval for a Proportion

$$\hat{p} \pm z_{\alpha/2} \sqrt{\frac{\hat{p}(1 - \hat{p})}{n}}$$

Common Values of $z_{\alpha/2}$

1.0   for a 68% confidence interval

1.96 or 2.0 for a 95% confidence interval

1.645 for a 90% confidence interval

2.576 for a 99% confidence interval

3.0 for a 99.7% confidence interval

EXERCISES

1. An advertisement for Seldane-D, a drug prescribed for seasonal allergic rhinitis, reported results of a double-blind study in which 374 patients took Seldane-D and 193 took a placebo (*Time,* 27 March 1995, p. 18). Headaches were reported as a side effect by 65 of those taking Seldane-D.

   a. What is the sample proportion of Seldane-D takers who reported headaches?

   b. What is the standard deviation for the proportion computed in part a?

   c. Construct a 95% confidence interval for the population proportion based on the information from parts a and b.

   d. Interpret the confidence interval from part c by writing a few sentences explaining what it means.

2. Refer to Exercise 1. Of the 193 placebo takers, 43 reported headaches.

   a. Compute a 95% confidence interval for the true population proportion that would get headaches after taking a placebo.

b. Notice that a higher proportion of placebo takers than Seldane-D takers reported headaches. Use that information to explain why it is important to have a group taking placebos when studying the potential side effects of medications.

3. On September 10, 1998, the "Starr Report," alleging impeachable offenses by President Bill Clinton, was released to Congress. That evening, the Gallup Organization conducted a poll of 645 adults nationwide to assess initial reaction (reported at www.gallup.com). One of the questions asked was: "Based on what you know at this point, do you think that Bill Clinton should or should not be impeached and removed from office?" The response "Yes, should" was selected by 31% of the respondents.

a. The Gallup web page said, "For results based on the total sample of adults nationwide, one can say with 95% confidence that the margin of sampling error is no greater than ± 4 percentage points." Explain what this means and verify that the statement is accurate.

b. Give a 95% confidence interval for the proportion of all adults who would have said President Clinton should be impeached had they been asked that evening.

c. A similar Gallup poll taken in early June 1998 found that 19% responded that President Clinton should be impeached. Do you think the difference between the results of the two polls can be attributed to chance variation in the samples taken, or does it represent a real difference of opinion in the population in June versus mid-September? Explain.

4. A telephone poll reported in *Time* magazine (6 February 1995, p. 24) asked 359 adult Americans the question, "Do you think Congress should maintain or repeal last year's ban on several types of assault weapons?" Seventy-five percent responded "maintain."

a. Compute the standard deviation for the sample proportion of .75.

b. *Time* reported that the "sampling error is ± 4.5%." Verify that 4.5% is approximately what would be added and subtracted to the sample proportion to create a 95% confidence interval.

c. Use the information reported by *Time* to create a 95% confidence interval for the population proportion. Interpret the interval in words that would be understood by someone with no training in statistics. Be sure to specify the population to which it applies.

5. What level of confidence would accompany each of the following intervals?

a. Sample proportion ± 1.0 (S.D.)

b. Sample proportion ± 1.645 (S.D.)

c. Sample proportion ± 1.96 (S.D.)

d. Sample proportion ± 2.576 (S.D.)

6. Explain whether the width of a confidence interval would increase, decrease, or remain the same as a result of each of the following changes:

    a. The sample size is doubled, from 400 to 800

    b. The population size is doubled, from 25 million to 50 million

    c. The level of confidence is lowered from 95% to 90%

7. *Parade Magazine* reported that "nearly 3200 readers dialed a 900 number to respond to a survey in our Jan. 8 cover story on America's young people and violence" (19 February 1995, p. 20). Of those responding, "63.3% say they have been victims or personally know a victim of violent crime." Can the methods in this chapter legitimately be used to compute a 95% confidence interval for the proportion of Americans who fit that description? If so, compute the interval. If not, explain why not.

8. Refer to Example 6 in this chapter. It is claimed that a 95% confidence interval for the percentage of placebo-patch users who quit smoking by the eighth week covers 13% to 27%. There were 120 placebo-patch users, and 24 quit smoking by the eighth week. Verify that the confidence interval given is correct.

9. Find the results of a poll reported in a weekly newsmagazine such as *Newsweek* or *Time,* in a newspaper such as the *New York Times,* or on the Internet in which a margin of error is also reported. Explain what question was asked and what margin of error was reported; then present a 95% confidence interval for the results. Explain in words what the interval means for your example.

10. Confirm that the standard deviation is largest when the proportion used to calculate it is .50. Do this by using other values above and below .50 and comparing the answers to what you would get using .50. Try three values above and three values below .50.

11. A university is contemplating switching from the quarter system to the semester system. The administration conducts a survey of a random sample of 400 students and finds that 240 of them prefer to remain on the quarter system.

    a. Construct a 95% confidence interval for the true proportion of all students who would prefer to remain on the quarter system.

    b. Does the interval you computed in part a provide convincing evidence that the majority of students prefer to remain on the quarter system? Explain.

    c. Now suppose that only 50 students had been surveyed and that 30 said they preferred the quarter system. Compute a 95% confidence interval for the true proportion who prefer to remain on the quarter system. Does the interval provide convincing evidence that the majority of students prefer to remain on the quarter system?

    d. Compare the sample proportions and the confidence intervals found in parts a and c. Use these results to discuss the role sample size plays in helping make decisions from sample data.

12. In a special double issue of *Time* magazine, the cover story featured Pope John Paul II as "man of the year" (26 December 1994–2 January 1995, pp. 74–76). As part of the story, *Time* reported on the results of a survey of 507 adult American Catholics, taken by telephone on December 7–8. It was also reported that "sampling error is ± 4.4%."

a. One question asked was, "Do you favor allowing women to be priests?" to which 59% of the respondents answered yes. Using the reported margin of error of 4.4%, calculate a 95% confidence interval for the response to this question. Write a sentence interpreting the interval that could be understood by someone who knows nothing about statistics. Be careful about specifying the correct population.

b. Calculate a 95% confidence interval for the question in part a, using the formula in this chapter rather than the reported margin of error. Compare your answer to the answer in part a.

c. Another question in the survey was, "Is it possible to disagree with the Pope and still be a good Catholic?" to which 89% of respondents said yes. Using the formula in this chapter, compute a 95% confidence interval for the true proportion who would answer yes to the question. Now compute a 95% confidence interval using the reported margin of error of 4.4%. Compare your two intervals.

d. If you computed your intervals correctly, you would have found that the two intervals in parts a and b were quite similar to each other, whereas the two intervals in part c were not. In part c, the interval computed using the reported margin of error was wider than the one computed using the formula. Explain why the two methods for computing the intervals agreed more closely for the survey question in parts a and b than for the survey question in part c.

13. *U.S. News and World Report* (19 December 1994, pp. 62–71) reported on a survey of 1000 American adults, conducted by telephone on December 2–4, 1994, designed to measure beliefs about apocalyptic predictions. They reported that the margin of error was "±3 percentage points."

a. Verify that the margin of error for a sample of size 1000 is as reported.

b. One of the results reported was that 59% of Americans believe the world will come to an end. Construct a 95% confidence interval for the true percentage of Americans with that belief, using the margin of error given in the article. Interpret the interval in a way that could be understood by a statistically naive reader.

14. Refer to the article discussed in Exercise 13. The article continued by reporting that of those who do believe the world will come to an end, 33% believe it will happen within either a few years or a few decades. Respondents were only asked that question if they answered yes to the question about the world coming to an end, so about 590 respondents would have been asked the question.

a. Consider only those adult Americans who believe the world will come to an end. For that population, compute a 95% confidence interval for the proportion who believe it will come to an end within the next few years or few decades.

b. Explain why you could not use the margin of error of ±3% reported in the article to compute the confidence interval in part a.

15. A study first reported in the *Journal of the American Medical Association* (7 December 1994) received widespread attention as the first wide-scale study of the use of alcohol on American college campuses and was the subject of an article in *Time* magazine (19 December 1994, p. 16). The researchers surveyed 17,592 students at 140 4-year colleges in 40 states. One of the results they found was that about 8.8%, or about 1550 respondents, were frequent binge drinkers. They defined frequent binge drinking as having had at least four (for women) or five (for men) drinks at a single sitting at least three times during the previous 2 weeks.

  a. *Time* magazine (19 December 1994, p. 66) reported that of the approximately 1550 frequent binge drinkers in this study, 22% reported having had unprotected sex. Find a 95% confidence interval for the true proportion of all frequent binge drinkers who had unprotected sex, and interpret the interval for someone who has no knowledge of statistics.

  b. Notice that the results quoted in part a indicate that about 341 students out of the 17,592 interviewed said they were frequent binge drinkers and had unprotected sex. Compute a 95% confidence interval for the proportion of college students who are frequent binge drinkers and who also had unprotected sex.

  c. Using the results from parts a and b, write two short news articles on the problem of binge drinking and unprotected sex. In one, make the situation sound as disastrous as you can. In the other, try to minimize the problem.

16. In Example 5 in this chapter, we found a 95% confidence interval for the proportion of successes likely in a certain kind of ESP test. Construct a 99.7% confidence interval for that example. Explain why a skeptic of ESP would prefer to report the 99.7% confidence interval.

17. Refer to the formula for a confidence interval in the For Those Who Like Formulas section.

  a. Write the formula for a 90% confidence interval for a proportion.

  b. Refer to Example 6. Construct a 90% confidence interval for the proportion of smokers who would quit after 8 weeks using a nicotine patch.

  c. Compare the 90% confidence interval you found in part b to the 95% confidence interval used in the example. Explain which one you would present if your company were advertising the effectiveness of nicotine patches.

MINI-PROJECTS

1. You are going to use the methods discussed in this chapter to estimate the proportion of all cars in your area that are red. Stand on a busy street and count cars as they pass by. Count 100 cars and keep track of how many are red.

  a. Using your data, compute a 95% confidence interval for the proportion of cars in your area that are red.

    b. Based on how you conducted the survey, are any biases likely to influence your results? Explain.

2. Collect data and construct a confidence interval for a proportion for which you already know the answer. Use a sample of at least 100. You can select the situation for which you would like to do this. For example, you could flip a coin 100 times and construct a confidence interval for the proportion of heads, knowing that the true proportion is .5. Report how you collected the data and the results you found. Explain the meaning of your confidence interval and compare it to what you know to be the truth about the proportion of interest.

3. Choose a categorical variable for which you would like to estimate the true proportion that fall into a certain category. Conduct an experiment or a survey that allows you to find a 95% confidence interval for the proportion of interest. Explain exactly what you did, how you computed your results, and how you would interpret the results.

REFERENCE

Gastwirth, Joseph L. (1988). *Statistical reasoning in law and public policy*. Vol. 2. *Tort law, evidence and health*. Boston: Academic Press.

CHAPTER TWENTY

The Role of Confidence Intervals in Research

THOUGHT QUESTIONS

1. In this chapter, Example 1 compares weight loss (over 1 year) in men who diet but do not exercise and vice versa. The results show that a 95% confidence interval for the mean weight loss for men who diet but do not exercise extends from 13.4 to 18.0 pounds. A 95% confidence interval for the mean weight loss for men who exercise but do not diet extends from 6.4 to 11.2 pounds.

   a. Do you think this means that 95% of all men who diet will lose between 13.4 and 18.0 pounds? Explain.

   b. On the basis of these results, do you think you can conclude that men who diet without exercising lose more weight, on average, than men who exercise but do not diet?

2. The first confidence interval in Question 1 was based on results from 42 men. The confidence interval spans a range of almost 5 pounds. If the results had been based on a much larger sample, do you think the confidence interval for the mean weight loss would have been wider, narrower, or about the same? Explain your reasoning.

3. In Question 1, we compared average weight loss for dieting and for exercising by computing separate confidence intervals for the two means and comparing the intervals. What would be a more direct value to examine to make the comparison between the mean weight loss for the two methods?

4. In Case Study 5.3, we examined the relationship between baldness and heart attacks. Many of the results reported in the original journal article were expressed in terms of relative risk of a heart attack for men with severe vertex baldness compared to men with no hair loss. One result reported was that a 95%

confidence interval for the relative risk for men under 45 years of age extended from 1.1 to 8.2.

a. Recalling the material from Chapter 12, explain what it means to have a relative risk of 1.1 in this example.

b. Interpret the result given by the confidence interval.

20.1   CONFIDENCE INTERVALS FOR POPULATION MEANS

In Chapter 18, we learned what to expect of sample means, assuming we knew the mean and the standard deviation of the population from which the sample was drawn. In this section, we try to determine a population mean when all we have available is a sample. All we need from the sample are its mean, standard deviation, and number of observations.

EXAMPLE 1   **DO MEN LOSE MORE WEIGHT BY DIET OR BY EXERCISE?**

Wood and colleagues (1988), also reported by Iman (1994, p. 258), studied a group of 89 sedentary men for a year. Forty-two men were placed on a diet; the remaining 47 were put on an exercise routine.

The group on a diet lost an average of 7.2 kg, with a standard deviation of 3.7 kg. The men who exercised lost an average of 4.0 kg, with a standard deviation of 3.9 kg (Wood et al., 1988, Table 2). Before we discuss how to compare the groups, let's determine how to extend the sample results to what would happen if the entire population of men of this type were to diet or exercise exclusively. We will return to this example after we learn the general method. ■

The Rule for Sample Means, Revisited

In Chapter 18, we learned how sample means behave.

The Rule for Sample Means is
If numerous samples of the same size are taken, the frequency curve of means from the various samples will be approximately bell-shaped. The mean of this collection of sample means will be the same as the mean of the population. The standard deviation will be

population standard deviation/square root of sample size

Standard Error of the Mean

Before proceeding, we need to distinguish between the *population* standard deviation and the standard deviation for the sample means, which is the population standard deviation/$\sqrt{n}$. (Recall that n is the number of observations in the sample.) Consistent with the distinction made by most researchers, we use terminology as follows for these two different kinds of standard deviations.

> The standard deviation for the possible sample means is called the **standard error of the mean.** It is sometimes abbreviated by SEM or just "standard error." In other words:
>
> SEM = standard error = population standard deviation/$\sqrt{n}$

In practice, the population standard deviation is usually unknown and is replaced by the sample standard deviation, computed from the data. The term *standard error of the mean* or *standard error* is still used.

Population versus Sample Standard Deviation and Error

An example will help clarify the distinctions among these terms. In Chapter 18, we considered a hypothetical population of people who visited a weight-loss clinic. We said that the weight losses for the thousands of people in the *population* were bell-shaped, with a mean of 8 pounds and a standard deviation of 5 pounds. Further, we considered *samples* of $n = 25$ people. For one sample, we found the mean and standard deviation for the 25 people to be mean = 8.32 pounds, standard deviation = 4.74 pounds.

Thus, we have the following numbers:

population standard deviation = 5 pounds

sample standard deviation = 4.74 pounds

standard error of the mean (using population S.D.) = $5 / \sqrt{25} = 1$

standard error of the mean (using sample S.D.) = $4.74 / \sqrt{25} = 0.95$

Let's now return to our discussion of the Rule for Sample Means. It is important to remember the conditions under which this rule applies:

1. The population of measurements of interest is bell-shaped, and a random sample of any size is measured.

or

2. The population of measurements of interest is not bell-shaped, but a *large* random sample is measured. A sample of size 30 is usually considered "large," but if there are extreme outliers, it is better to have a larger sample.

Constructing a Confidence Interval for a Mean

We can use the same reasoning we used in Chapter 20, where we constructed a 95% confidence interval for a proportion, to construct a 95% confidence interval for a mean. The Rule for Sample Means and the Empirical Rule from Chapter 8 allow us to make the following statement:

> *In 95% of all samples, the* sample mean *will fall* within 2 standard errors *of the* true population mean.

Now let's rewrite the statement in a more useful form:

> *In 95% of all samples, the* true population mean *will be* within 2 standard errors *of the* sample mean.

In other words, if we simply add and subtract 2 standard errors to the sample mean, in 95% of all cases we will have captured the true population mean.

> Putting this all together, here is the formula for a **95% confidence interval for a population mean:**
>
> sample mean ± 2 standard errors
>
> where
>
> standard error = standard deviation/$\sqrt{n}$

Important note: This formula should be used only if there are at least 30 observations in the sample. To compute a 95% confidence interval for the population mean based on smaller samples, a multiplier larger than 2 is used, which is found from a "*t*-distribution." The technical details involved are beyond the scope of this book. However, if someone else has constructed the confidence interval for you based on a small sample, the interpretation discussed here is still valid.

EXAMPLE 1 REVISITED

COMPARING DIET AND EXERCISE FOR WEIGHT LOSS

Let's construct a 95% confidence interval for the mean weight losses for all men who might diet or who might exercise, based on the sample information given in Example 1. We will be constructing two separate confidence intervals, one for each condition. Notice the switch from kilograms to pounds at the end of the computation; 2.2 kg = 1 lb. The results could be expressed in either unit, but pounds are more familiar to many readers.

Diet Only
sample mean = 7.2 kg
sample standard deviation = 3.7 kg

Exercise Only
sample mean = 4.0 kg
sample standard deviation = 3.9 kg

number of participants = n = 42   number of participants = n = 47
standard error = $3.7/\sqrt{42}$ = 0.571   standard error = $3.9/\sqrt{47}$ = 0.569
2 × standard error = 2(0.571) = 1.1   2 × standard error = 2(0.569)=1.1

95% confidence interval for the mean:
sample mean ± 2 × standard error

Diet Only	**Exercise Only**
7.2 ± 1.1	4.0 ± 1.1
6.1 kg to 8.3 kg	2.9 kg to 5.1 kg
13.4 lb to 18.3 lb	6.4 lb to 11.2 lb

These results indicate that men similar to those in this study would lose an average of somewhere between 13.4 and 18.3 pounds on a diet but would lose only an average of 6.4 to 11.2 pounds with exercise alone. Notice that these intervals are trying to capture the true mean or average value for the population. They do not encompass the full range of weight loss that would be experienced by most individuals. (Also, remember that these intervals could be wrong. Ninety-five percent of intervals constructed this way will contain the correct population mean value, but 5% will not. We will never know which are which.)

Based on these results, it appears that dieting probably results in a larger weight loss than exercise because there is no overlap in the two intervals. Comparing the endpoints of these intervals, we are fairly certain that the average weight loss from dieting is no lower than 13.4 pounds and the average weight loss from exercising no higher than 11.2 pounds. In the next section, we learn a more efficient method for making the comparison, one that will enable us to estimate with 95% confidence the actual difference in the two averages for the population. ∎

20.2   CONFIDENCE INTERVALS FOR THE DIFFERENCE BETWEEN TWO MEANS

In many instances, such as in the preceding example, we are interested in comparing the population means under two conditions or for two groups. One way to do that is to construct separate confidence intervals for the two conditions and then compare them. That's what we did in Section 20.1 for the weight-loss example. A more direct and efficient approach is to construct a single confidence interval for the *difference* in the population means for the two groups or conditions. In this section, we learn how to do that.

You may have noticed that the formats are similar for the two types of confidence intervals we have discussed so far. That is, they were both used to estimate a population value, either a proportion or a mean. They were both built around the corresponding sample value, the sample proportion or the sample mean. They both had the form:

sample value ± 2 × measure of variability

This format was based on the fact that the "sample value" over repeated samples was predicted to follow a bell-shaped curve centered on the "population value." All we needed to know in addition was the "standard deviation" for that specific bell-shaped curve.

The same is true for calculating the difference in two means. Here is the recipe you follow:

Constructing a confidence interval for the difference in means

1. Collect a large sample of observations, independently, under each condition or from each group. Compute the mean and the standard deviation for each sample.

2. Compute the standard error of the mean (SEM) for each sample by dividing the sample standard deviation by the square root of the sample size.

3. Square the two SEMs and add them together. Then take the square root. This will give you the necessary, "measure of variability," which is called the **standard error of the difference in two means.** In other words:

measure of variability = square root of $[(SEM_1)^2 + (SEM_2)^2]$

4. A 95% confidence interval for the difference in the two population means is

difference in sample means $\pm$ 2 $\times$ measure of variability

or

difference in sample means $\pm$ 2 $\times$ square root of $[(SEM_1)^2+(SEM_2)^2]$

EXAMPLE 2 **A DIRECT COMPARISON OF DIET AND EXERCISE**

We are now in a position to compute a 95% confidence interval for the *difference* in population means for weight loss from dieting only and weight loss from exercising only. Let's follow the steps outlined, using the data from the previous section:

Steps 1 and 2. Compute sample means, standard deviations, and SEMs:

Diet Only
sample mean = 7.2 kg
sample standard deviation = 3.7 kg
number of participants = n = 42
standard error = SEM_1 =
   $3.7/\sqrt{42} = 0.571$

Exercise Only
sample mean = 4.0 kg
sample standard deviation = 3.9 kg
number of participants = n = 47
standard error = SEM_2 =
   $3.9/\sqrt{47} = 0.569$

Step 3. Square the two standard errors and add them together. Take the square root:

measure of uncertainty = square root of $[(0.571)^2 + (0.569)^2] = 0.81$

Step 4. Compute the interval. A 95% confidence interval for the difference in the two population means is

difference in sample means ± 2 × measure of variability

$$(7.2 - 4.0) \pm 2(0.81)$$
$$3.2 \pm 1.6$$
$$1.6 \text{ kg to } 4.8 \text{ kg}$$
$$3.5 \text{ lb to } 10.6 \text{ lb}$$

Notice that this interval is entirely above zero. Therefore, we can be highly confident that there really is a difference in average weight loss, with higher weight loss for dieting alone than for exercise alone. In other words, we are 95% confident that the interval captures the true difference in means and that the true difference is at least 3.5 pounds. Remember that this interval estimates the difference in *averages*, not in weight losses for individuals. ■

A Caution about Using This Method

The method described in this section is valid only when *independent* measurements are taken from the two groups. For instance, if matched pairs are used and one treatment is randomly assigned to each half of the pair, the measurements would not be independent. In that case, differences should be taken for each pair of measurements, and then a confidence interval computed for the mean of those differences. Using the method in this section would result in a measure of variability that was too large.

20.3

REVISITING CASE STUDIES: HOW JOURNALS PRESENT CONFIDENCE INTERVALS

Many of the case studies we examined in the early part of this book involved making conclusions about differences in means. The original journal articles from which those case studies were drawn each presented results in a slightly different way. Some provided confidence intervals directly; others gave the information necessary for you to construct your own interval. In this section, we revisit some of those case studies as examples of the kinds of information researchers provide for readers.

Direct Reporting of Confidence Intervals: Case Study 6.4

Case Study 6.4 examined the relationship between smoking during pregnancy and subsequent IQ of the child. The journal article in which that study was reported (Olds, Henderson, and Tatelbaum, 1994) provided 95% confidence intervals to accompany all the results it reported. Most of the confidence intervals were based on a comparison of the means for mothers who didn't smoke and mothers who smoked ten or more cigarettes per day, hereafter called "smokers." Some of the results were

| TABLE 20.1 | Some 95% Confidence Intervals from Case Study 6.4 |

| | Sample Means | | |
	0 Cigarettes	10+ Cigarettes	Difference (95% CI)
Maternal education, grades	11.57	10.89	0.67 (0.15, 1.19)
Stanford-Binet (IQ), 48 mo	113.28	103.12	10.16 (5.04, 15.30)
Birthweight, g	3416	3035	381.0 (167.1, 594.9)

SOURCE: Olds, Henderson, and Tatelbaum, 1994, Tables 1, 2, and 4.

presented in tables and others were presented in the text. Table 20.1 gives some of the results from the tables contained in the paper.

Let's interpret these results. The first confidence interval compares the mean educational levels for the smokers and nonsmokers. The result tells us that, in this sample, the average educational level for nonsmokers was 0.67 year higher than for smokers. The confidence interval extends this value to what the difference might be for the populations from which these samples were drawn. The interval tells us that the difference in the population is probably between 0.15 year and 1.19 years of education. In other words, mothers who did not smoke were also likely to have had more education. Maternal education was a *confounding variable* in this study and was part of what the researchers used to try to explain the differences observed in the children's IQs for the two groups.

The second row of Table 20.1 compares the mean IQs for the children of the nonsmokers and smokers at 48 months of age. The difference in means for the sample was 10.16 points. From this, the researchers inferred that there is probably a difference of somewhere between 5.04 and 15.30 points for the entire population. In other words, the children of nonsmokers in the population probably have IQs that are between 5.04 and 15.30 points higher than the children of mothers who smoke ten or more cigarettes per day.

The third row of Table 20.1 represents an example of the kind of explanatory confounding variables that may have been present. In other words, smoking may have caused lower birthweights, which in turn may have caused lower IQs. The result shown here is that the average difference in birthweight for babies of non-smokers and smokers in the sample was 381 grams. Further, with 95% confidence, we can state that there could be a difference as low as 167.1 grams or as high as 594.9 grams for the population from which this sample was drawn. Thus, we are fairly certain that mothers in the population who smoke are likely to have babies with lower birthweight, on average, than those who don't.

Olds and colleagues (1994) also included numerous confidence intervals in the text. For example, they realized that they needed to control for confounding variables to make a more realistic comparison between the IQs of children of smokers and nonsmokers. They wrote: "After control for confounding background variables

FIGURE 20.1
*Part of a table from the
journal article in Case
Study 6.2*

| TABLE II | Serum DHEA-S Concentrations (± SEM) in 606 Women for Comparison and TM Groups |

	Comparison Group		TM Group		
Age Group	N	DHEA-S Level (µg/dl)	N	DHEA-S Level (µg/dl)	% Elevation in TM Group
45 – 49	51	88 ± 12	30	117 ± 11	34

SOURCE: Glaser et al., 1992, p. 333.

(Table 3), the average difference observed at 12 and 24 months was 2.59 points (95% CI: –3.03, 8.20); the difference observed at 36 and 48 months was reduced to 4.35 points (95% CI: 0.02, 8.68)" (pp. 223–224).

The result, as reported in the news story quoted in Chapter 6, was that the gap in average IQ at 3 and 4 years of age narrowed to 4 points when "a wide range of inter-related factors were controlled." You are now in an excellent position to understand a much broader picture than that reported to you in the newspaper. For example, as you can see from the reported confidence intervals, we can't rule out the possibility that the differences in IQ at 1 and 2 years of age were in the other direction because the interval covers some negative values. Further, even at 3 and 4 years of age, the confidence interval tells us that the gap *could* have been just slightly above zero in the population.

Reporting Standard Errors of the Mean: Case Study 6.2

Case Study 6.2 involved a comparison in serum DHEA-S levels for practitioners and nonpractitioners of transcendental meditation. The results were presented as tables showing the mean DHEA-S level for each 5-year age group; there were separate tables for men and women. Confidence intervals were not presented, but values were given for standard errors of the means (SEMs). Therefore, confidence intervals could be computed from the information given.

To illustrate exactly how the results were presented, let's examine one of the tables. Figure 20.1 displays the heading and a typical row from Table II of Glaser and colleagues (1992, p. 333). In the heading for the table, we are told that the values following the ± sign are the standard errors of the means. Using that information, we can construct 95% confidence intervals for the mean DHEA-S levels in each group, or we can construct a single 95% confidence interval for the *difference* in the means for the meditators and nonmeditators. Constructing the latter interval, we find that a 95% confidence interval for the difference in means is

FIGURE 20.2
*Part of a table from
Case Study 5.1*

TABLE 1	Baseline Characteristics	
Mean ± SD (Range)	Active	Placebo
Age, y	42.8 ± 11.1(20–65)	43.6 ± 10.6(21–65)
Cigarettes/d(n = 119/119)*	28.8 ± 9.4(20–60)	30.6 ± 9.4(20–60)

*The notation (n = 119/119) means that there were 119 people in each group for these calculations.
SOURCE: Hurt et al., 23 February 1994, p. 596.

$$\text{difference in sample means} \pm 2 \times \text{square root of } [(SEM_1)^2 + (SEM_2)^2]$$
$$(117 - 88) \pm 2 \times \text{square root of } [(12)^2 + (11)^2]$$
$$29 \pm 2(16.3)$$
$$29 \pm 32.6$$
$$-3.6 \text{ to } 61.6$$

This interval is probably not easy to interpret if you are not familiar with DHEA-S values. Because the interval includes zero, we cannot say, even with 95% confidence that the observed difference in sample means represents a real difference in the population. However, because the interval extends so much further in the positive direction than the negative, the evidence certainly suggests that population DHEA-S levels are higher for the meditators than for others.

Reporting Standard Deviations: Case Study 5.1

In Case Study 5.1, we looked at the comparison of smoking cessation rates for patients using nicotine patches versus those using placebo patches. The main variables of interest were categorical and thus should be studied using methods for proportions. However, the authors also presented numerical summaries of a variety of other characteristics of the subjects. That information is useful to make sure that the randomization procedure distributed those variables fairly across the two treatment conditions.

The authors reported means, standard deviations (SD), and ranges (low to high) for a variety of characteristics. As an example, Figure 20.2 shows part of a table given by Hurt and colleagues (23 February 1994). In the table, the ages for the group wearing placebo patches had a mean of 43.6 years, a standard deviation of 10.6 years, and ranged from 21 to 65 years.

Notice that the intervals given in the table are *not* 95% confidence intervals, despite the fact that they are presented in the standard format used for such intervals. Be sure to read carefully when you examine results presented in journals so that you are not misled into thinking results presented in this format represent 95% confidence intervals.

From the information presented, we notice a slight difference in the mean ages for each group and in the mean number of cigarettes each group smoked per day at the start of the study. Let's use the information given in the table to compute a 95% confidence interval for the difference in number of cigarettes smoked per day, to find out if it represents a substantial difference in the populations represented by the two groups. (It should be obvious to you that if the placebo group started out smoking significantly more, the results of the study would be questionable.) To compute the confidence interval, we need the means, standard deviations, and the sample sizes (n) presented in the table. Here is how we would proceed with the computation:

Steps 1 and 2.   Compute sample means, standard deviations, and SEMs:

Active Group

sample mean = 28.8 cigarettes per day

sample standard deviation
   = 9.4 cigarettes

number of participants = n = 119

standard error = SEM_1
   = $9.4/\sqrt{119}$ = 0.86

Placebo Group

sample mean = 30.6 cigarettes per day

sample standard deviation
   = 9.4 cigarettes

number of participants = n = 119

standard error = SEM_2
   = $9.4/\sqrt{119}$ = 0.86

Step 3. Square the two standard errors and add them together. Take the square root:

measure of uncertainty = square root of $[(0.86)^2 + (0.86)^2] = 1.2$

Step 4. Compute the interval. A 95% confidence interval for the difference in the two population means is

difference in sample means $\pm 2 \times$ measure of variability

$$(28.8 - 30.6) \pm 2(1.2)$$
$$-1.8 \pm 2.4$$
$$-4.2 \text{ to } + 0.60$$

It appears that there could have been slightly fewer cigarettes smoked per day by the group that received the nicotine patches, but the interval covers zero, allowing for the possibility that the difference observed in the sample means was opposite in direction from the difference in the population means. In other words, we simply can't tell if the difference observed in the sample means represents a real difference in the populations.

Summary of the Variety of Information Given in Journals

There is no standard for how journal articles present results. However, you can determine confidence intervals for individual means or for the difference in two means (for large, independent samples) as long as you are given one of the following sets of information:

1. Direct confidence intervals
2. Means and standard errors of the means
3. Means, standard deviations, and sample sizes

20.4   UNDERSTANDING ANY CONFIDENCE INTERVAL

Not all results are reported as proportions, means, or differences in means. Numerous other statistics can be computed to make comparisons, almost all of which have corresponding formulas for computing confidence intervals. Some of those formulas are quite complex, however.

In cases where a complicated procedure is needed to compute a confidence interval, authors of journal articles usually provide the completed intervals. Your job is to be able to interpret those intervals. The principles you have learned for understanding confidence intervals for means and proportions are directly applicable to understanding any confidence interval. As an example, let's consider the confidence intervals reported in another of our earlier case studies.

Confidence Interval for Relative Risk: Case Study 5.3

In Case Study 5.3, we investigated a study relating baldness and heart disease. The measure of interest in that study was the relative risk of heart disease, based on degree of baldness. The investigators focused on the relative risk (RR) of myocardial infarction (MI)—that is, a heart attack—for men with baldness compared to men without any baldness. Here is how they reported some of the results:

> For mild or moderate vertex baldness, the age-adjusted RR estimates were approximately 1.3, while for extreme baldness the estimate was 3.4 (95% CI, 1.7 to 7.0). . . . For any vertex baldness (i.e., mild, moderate, and severe combined), the age-adjusted RR was 1.4 (95% CI, 1.2 to 1.9). (Lesko et al., 1993, p. 1000)

The confidence intervals for age-adjusted relative risk are not simple to compute. You may notice that they are not of the form "sample value ± 2 × (measure of uncertainty)," evidenced by the fact that they are not symmetric about the sample values given. However, these intervals can be interpreted in the same way as any other confidence interval. For instance, with 95% certainty we can say that men with extreme baldness are at higher risk of heart attack than men with no baldness and that the ratio of risks is probably between 1.7 and 7.0. In other words, men with extreme baldness are probably anywhere from 1.7 to 7 times more likely to experience a heart attack than men of the same age without any baldness. Of course, these results assume that the men in the study are representative of the larger population.

| TABLE 20.2 | Results for Case Study 20.1 |

	Symptom Complex Score: Mean ± SD	
	Placebo Group	Calcium-Treated Group
Baseline	0.92 ± 0.55 (n = 235)	0.90 ± 0.52 (n = 231)
Third Cycle	0.60 ± 0.52 (n = 228)	0.43 ± 0.40 (n = 212)

CASE STUDY 20.1   Premenstrual Syndrome? Try Calcium

It was front page news in the *Sacramento Bee*. The headline read "Study says calcium can help ease PMS," and the article continued: "Daily doses of calcium can reduce both the physical and psychological symptoms of premenstrual syndrome by at least half, according to new research that points toward a low-cost, simple remedy for a condition that affects millions of women" (Maugh, 26 August 1998). The article described a randomized, double-blind experiment in which women who suffered from premenstrual syndrome (PMS) were randomly assigned to take either a placebo or 1200 mg of calcium per day in the form of four Tums E-X tablets (Thys-Jacobs et al., 1998). Participants included 466 women with a history of PMS: 231 in the calcium treatment group and 235 in the placebo group.

The primary measure of interest was a composite score based on 17 PMS symptoms, including 6 that were mood-related, 5 involving water retention, 2 involving food cravings, 3 related to pain, and insomnia. Participants were asked to rate each of the 17 symptoms daily on a scale from 0(absent) to 3(severe). The actual "symptom complex score" was the mean rating for the 17 symptoms. Thus, a score of 0 would imply that all symptoms were absent, and a score of 3 would indicate that all symptoms were severe. The original article (Thys-Jacobs et al., 1998) presents results individually for each of the 17 symptoms plus the composite score.

One interesting outcome of this study was that the severity of symptoms was substantially reduced for both the placebo and the calcium-treated groups. Therefore, comparisons should be made between those two groups rather than examining the reduction in scores before and after taking calcium for the treatment group alone. In other words, part of the total reduction in symptoms for the calcium-treated group could be the result of a "placebo effect." We are interested in knowing the additional influence of taking calcium.

Let's compare the severity of symptoms as measured by the composite score for the placebo and calcium-treated groups. The treatments were continued for three menstrual cycles; we report the symptom scores for the premenstrual period (7 days) before treatments began (baseline) and before the third cycle.

Table 20.2 presents results as given in the journal article, including sample sizes and the mean symptom complex scores ± 1 standard deviation. Notice that sample sizes were slightly reduced by the third cycle due to patients dropping out of the study. Let's use the results in the table to compute a confidence interval for what the mean differences would be for the entire population of PMS sufferers.

The purpose of the experiment is to see if taking calcium diminishes symptom severity. Because we know that placebos alone can be responsible for reducing symptoms, the appropriate comparison is between the placebo and calcium-treated groups rather than between the baseline and third cycle symptoms for the calcium-treated group alone.

The difference in means (placebo – calcium) for the third cycle is (0.60 – 0.43) = 0.17. The "measure of uncertainty" is about 0.039, so a 95% confidence interval for the difference is 0.17 ± 2(0.039), or about 0.09 to 0.25.

To put this in perspective, remember that the scores are averages over the 17 symptoms. Therefore, a reduction from a mean of 0.60 to a mean of 0.43 would, for instance, correspond to a reduction from (0.6)(17) = 10 mild symptoms (rating of 1) to (0.43)(17) = 7.31, or just over 7 mild symptoms. In fact, examination of the full results shows that all 17 symptoms had reduced severity in the calcium-treated group compared with the placebo group. And, because this is a randomized experiment and not an observational study, we can conclude that the calcium actually *caused* the reduction in symptoms.

As a final note, Table 20.2 also indicates a striking drop in the mean symptom score from baseline to the third cycle for both groups. For the placebo group, the symptom scores dropped by about a third; for the calcium-treated group, they were more than cut in half. ∎

FOR THOSE WHO LIKE FORMULAS

Review of Notation from Previous Chapters

Population mean = μ, sample mean = $\overline{X}$, sample standard deviation = s

$z_{\alpha/2}$ = standardized normal score with area $\alpha/2$ above it

Standard Error of the Mean

Standard error of the mean = SEM = $s/\sqrt{n}$

Confidence Interval for a Single Population Mean μ

$$\overline{X} \pm z_{\alpha/2} \frac{s}{\sqrt{n}}$$

Notation for Two Populations and Samples

Population mean = μ_i, $i = 1$ or 2

Sample mean = $\overline{X}_i$, $i = 1$ or 2

Sample standard deviation = s_i, $i = 1$ or 2

Sample size = n_i, $i = 1$ or 2

Standard error of the mean = $\text{SEM}_i = s_i / \sqrt{n_i}$, $i = 1$ or 2

**Confidence Interval for the Difference in
Two Population Means, Independent Samples**

$$(\overline{X}_1 - \overline{X}_2) \pm z_{\alpha/2}\sqrt{\frac{s_1^2}{n_1} + \frac{s_2^2}{n_2}}$$

EXERCISES

1. In Chapter 19, we saw that to construct a confidence interval for a population proportion it was enough to know the sample proportion and the sample size. Is the same true for constructing a confidence interval for a population mean? That is, is it enough to know the sample mean and sample size? Explain.

2. Explain the difference between a population mean and a sample mean using one of the studies discussed in the chapter as an example.

3. The *Baltimore Sun* (Haney, 21 February 1995) reported on a study by Dr. Sara Harkness, in which she compared the sleep patterns of 6-month-old infants in the United States and the Netherlands. She found that the 36 U.S. infants slept an average of just under 13 hours out of every 24, whereas the 66 Dutch infants slept an average of almost 15 hours.

   a. The article did not report a standard deviation, but suppose it was 0.5 hour for each group. Compute the standard error of the mean (SEM) for the U.S. babies.

   b. Continuing to assume that the standard deviation is 0.5 hour, compute a 95% confidence interval for the mean sleep time for 6-month-old babies in the United States.

   c. Continuing to assume that the standard deviation for each group is 0.5 hour, compute a 95% confidence interval for the *difference* in average sleep time for 6-month-old Dutch and U.S. infants.

4. What is the probability that a 95% confidence interval will not cover the true population value?

5. Suppose a university wants to know the average income of its students who work, and all students supply that information when they register. Would the university need to use the methods in this chapter to compute a confidence interval for the population mean income? Explain. (*Hint:* What is the sample mean and what is the population mean?)

6. Suppose you were given a 95% confidence interval for the difference in two population means. What could you conclude about the population means if:

   a. The confidence interval *did not* cover zero

   b. The confidence interval *did* cover zero

7. Suppose you were given a 95% confidence interval for the relative risk of disease under two different conditions. What could you conclude about the risk of disease under the two conditions if:

a. The confidence interval *did not* cover 1.0

b. The confidence interval *did* cover 1.0

8. In Chapter 19, we learned that to compute a 90% confidence interval, we add and subtract 1.645 rather than 2 times the measure of uncertainty. In this chapter, we revisited Case Study 6.2 and found that a 95% confidence interval for the difference in mean DHEA-S levels for 45- to 49-year-old women who meditated compared with those who did not extended from −3.6 to 61.6.

   a. Compute a 90% confidence interval for the difference in mean DHEA-S levels for this group.

   b. Based on your result in part a, could you conclude with 90% confidence that the difference observed in the samples represents a real difference in the populations? Explain.

9. In revisiting Case Study 6.2, we computed a confidence interval for the difference in mean DHEA-S levels for 45- to 49-year-old women meditators and nonmeditators and concluded that there probably was a real difference in the population means because most of the interval was above zero. Now we will compute results for the two groups separately and compare them.

   a. Compute a 95% confidence interval for the mean DHEA-S level for 45- to 49-year-old women in the comparison group.

   b. Compute a 95% confidence interval for the mean DHEA-S level for 45- to 49-year-old women in the TM group.

   c. Based on the intervals computed in parts a and b, would you conclude that there is a difference in population means between the comparison and TM groups? Explain.

10. In Case Study 6.4, which examined maternal smoking and child's IQ, one of the results reported in the journal article was the average number of days the infant spent in the neonatal intensive care unit. The results showed an average of 0.35 day for infants of nonsmokers and an average of 0.58 day for the infants of women who smoked ten or more cigarettes per day. In other words, the infants of smokers spent an average of 0.23 day more in neonatal intensive care. A 95% confidence interval for the difference in the two means extended from −3.02 days to +2.57 days. Explain why it would have been misleading to report, "These results show that the infants of smokers spend more time in neonatal intensive care than do the infants of nonsmokers."

11. In a study comparing age of death for left- and right-handed baseball players, Coren and Halpern (1991, p. 93) provided the following information: "Mean age of death for strong right-handers was 64.64 years (SD = 15.5, n = 1472); mean age of death for strong left-handers [was] 63.97 years (SD = 15.4, n = 236)." The term "strong handers" applies to baseball players who both threw and batted with the same hand. The data were actually taken from entries in *The Baseball Encyclopedia* (6th ed., New York: Macmillan, 1985), but, for the purposes of this exercise, pretend that the data were from a sample drawn from a larger population.

a. Compute a 95% confidence interval for the mean age of death for the population of strong right-handers from which this sample was drawn.

b. Repeat part a for the strong left-handers.

c. Compare the results from parts a and b in two ways. First, explain why one confidence interval is substantially wider than the other. Second, explain whether you would conclude that there is a difference in the mean ages of death for left- and right-handers on the basis of these results.

d. Compute a 95% confidence interval for the difference in mean ages of death for the strong right- and left-handers. Interpret the result.

12. In revisiting Case Study 5.3, we quoted the original journal article as reporting that "for any vertex baldness (i.e., mild, moderate, and severe combined), the age-adjusted RR was 1.4 (95% CI, 1.2 to 1.9)" (Lesko et al., 1993, p. 1000). Interpret this result.

13. In a report titled, "Secondhand Smoke: Is It a Hazard?" (*Consumer Reports,* January 1995, pp. 27–33), 26 studies linking secondhand smoke and lung cancer were summarized by noting, "Those studies estimated that people breathing secondhand smoke were 8 to 150 percent more likely to get lung cancer sometime later" (p. 28). Although it is not explicit, assume that the statement refers to a 95% confidence interval and interpret what this means.

14. Refer to Case Study 20.1, illustrating the role of calcium in reducing the symptoms of PMS. Using the caution given at the end of the section, explain why we cannot use the method presented in Section 20.2 to compare baseline symptom scores with third cycle symptom scores for the calcium-treated group alone.

15. Parts a–d below provide additional results for Case Study 20.1. For each of the parts, compute a 95% confidence interval for the difference in mean symptom scores between the placebo and calcium-treated conditions for the symptom listed. In each case, the results given are mean ± standard deviation. There were 228 participants in the placebo group and 212 in the calcium-treated group.

a. Mood swings: placebo = 0.70 ± 0.75; calcium = 0.50 ± 0.58

b. Crying spells: placebo = 0.37 ± 0.57; calcium = 0.23 ± 0.40

c. Aches and pains: placebo = 0.49 ± 0.60, calcium = 0.31 ± 0.49

d. Craving sweets or salts: placebo = 0.60 ± 0.78; calcium = 0.43 ± 0.64

16. A story in *Newsweek* (14 November 1994, pp. 52–54) reported the results of a poll asking 756 American adults the question, "Do you think Clarence Thomas sexually harassed Anita Hill, as she charged three years ago?" The results of a similar poll 3 years earlier were also reported. The original poll was conducted in October 1991, during the time of congressional hearings in which Hill made the allegations against Supreme Court nominee Thomas. The proportion answering yes was 23% in 1991 and 40% in 1994. A 95% confidence interval for the *difference* in proportions answering yes in 1994 versus 1991 is 17% ± 5%. Compute and interpret the interval. Does it indicate that opinion on the issue among American adults had definitively changed during the 3 years between polls?

17. Using the data presented by Hand and colleagues (1994) and discussed in previous chapters, we would like to estimate the average age difference between husbands and wives in Britain. Recall that the data consisted of a random sample of 200 couples. Following are two methods that were used to construct a confidence interval for the difference in ages. Your job is to figure out which method is correct:

    Method 1: Take the difference between the husband's age and the wife's age for each couple, and use the differences to construct a 95% confidence interval for a single mean. The result was an interval from 1.6 to 2.9 years.

    Method 2: Use the method presented in this chapter for constructing a confidence interval for the difference in two means for two independent samples. The result was an interval from –0.4 to 4.3 years.

    Explain which method is correct, and why. Then interpret the confidence interval that resulted from the correct method.

18. Refer to Exercise 17. Suppose that from that same data set, we want to compute the average difference between the heights of adult British men and adult British women—not the average difference within married couples.

    a. Which of the two methods in Exercise 17 would be appropriate for this situation?

    b. The 200 men in the sample had a mean height of 68.2 inches, with a standard deviation of 2.7 inches. The 200 women had a mean height of 63.1 inches, with a standard deviation of 2.5 inches. Assuming these were independent samples, compute a 95% confidence interval for the mean difference in heights between British males and females. Interpret the resulting interval in words that a statistically naive reader would understand.

19. Refer to Case Study 20.1 and the material in Part 1 of this book.

    a. In their original report, Thys-Jacobs and colleagues (1998) noted that the study was "double-blind." Explain what that means in the context of this example.

    b. Explain why it is possible to conclude that, based on this study, calcium actually *causes* a reduction in premenstrual symptoms.

MINI-PROJECTS

1. Find a journal article that reports at least one 95% confidence interval. Explain what the study was trying to accomplish. Give the results as reported in the article in terms of 95% confidence intervals. Interpret the results. Discuss whether you think the article accomplished its intended purpose. In your discussion, include potential problems with the study, as discussed in Chapters 4 to 6.

2. Collect data on a measurement variable for which the mean is of interest to you. Collect at least 30 observations. Using the data, compute a 95% confidence

interval for the mean of the population from which you drew your observations. Explain how you collected your sample and note whether your method would be likely to result in any biases if you tried to extend your sample results to the population. Interpret the 95% confidence interval to make a conclusion about the mean of the population.

3. Collect data on a measurement variable for which the difference in the means for two conditions or groups is of interest to you. Collect at least 30 observations for each condition or group. Using the data, compute a 95% confidence interval for the difference in the means of the populations from which you drew your observations. Explain how you collected your samples and note whether your method would be likely to result in any biases if you tried to extend your sample results to the populations. Interpret the 95% confidence interval to make a conclusion about the difference in the means of the populations or conditions.

REFERENCES

Coren, S., and D. Halpern. (1991). Left-handedness: A marker for decreased survival fitness. *Psychological Bulletin* 109, no. 1, pp. 90–106.

Glaser, J. L., J. L. Brind, J. H. Vogelman, M. J. Eisner, M. C. Dillbeck, R. K. Wallace, D. Chopra, and N. Orentreich. (1992). Elevated serum dehydroepiandrosterone sulfate levels in practitioners of the transcendental meditation (TM) and TM-Sidhi programs. *Journal of Behavioral Medicine* 15, no. 4, pp. 327–341.

Hand, D. J., F. Daly, A. D. Lunn, K. J. McConway, and E. Ostrowski. (1994). *A handbook of small data sets.* London: Chapman and Hall.

Haney, Daniel Q. (21 February 1995). Highly stimulated babies may be sleeping less as a result. *Baltimore Sun,* pp. 1D, 5D.

Hurt, R., L. Dale, P. Fredrickson, C. Caldwell, G. Lee, K. Offord, G. Lauger, Z. Marusic, L. Neese, and T. Lundberg. (23 February 1994). Nicotine patch therapy for smoking cessation combined with physician advice and nurse follow-up. *Journal of the American Medical Association* 271, no. 8, pp. 595–600.

Iman, R. L. (1994). *A data-based approach to statistics.* Belmont, CA: Wadsworth.

Lesko, S. M., L. Rosenberg, and S. Shapiro. (1993). A case-control study of baldness in relation to myocardial infarction in men. *Journal of the American Medical Association* 269, no. 8, pp. 998–1003.

Maugh, Thomas H., II. (26 August 1998). Study says calcium can help ease PMS. *Sacramento Bee,* pp. A1, A9.

Olds, D. L., C. R. Henderson, Jr., and R. Tatelbaum. (1994). Intellectual impairment in children of women who smoke cigarettes during pregnancy. *Pediatrics* 93, no. 2, pp. 221–227.

Thys-Jacobs, S., P. Starkey, D. Bernstein, J. Tian, and the Premenstrual Syndrome Study Group. (1998). Calcium carbonate and the premenstrual syndrome: Effects on premenstrual and menstrual symptoms. *American Journal of Obstetrics and Gynecology* 179, no. 2, pp. 444–452.

Wood, R. D., M. L. Stefanick, D. M. Dreon, B. Frey-Hewitt, S. C. Garay, P. T. Williams, H. R. Superko, S. P. Fortmann, J. J. Albers, K. M. Vranizan, N. M. Ellsworth, R. B. Terry, and W. L. Haskell. (1988). Changes in plasma lipids and lipoproteins in overweight men during weight loss through dieting as compared with exercise. *New England Journal of Medicine* 319, no. 18, pp. 1173–1179.

CHAPTER TWENTY-ONE

Rejecting Chance— Testing Hypotheses in Research

THOUGHT QUESTIONS

1. In the courtroom, juries must make a decision about the guilt or innocence of a defendant. Suppose you are on the jury in a murder trial. It is obviously a mistake if the jury claims the suspect is guilty when in fact he or she is innocent. What is the other type of mistake the jury could make? Which is more serious?

2. Suppose exactly half, or 0.50, of a certain population would answer yes when asked if they support the death penalty. A random sample of 400 people results in 220, or 0.55, who answer yes. The Rule for Sample Proportions tells us that the potential sample proportions in this situation are approximately bell-shaped, with standard deviation of 0.025. Using the formula on page 136, find the standardized score for the observed value of 0.55. Then determine how often you would expect to see a standardized score at least that large or larger.

3. Suppose you are interested in testing a claim you have heard about the proportion of a population who have a certain trait. You collect data and discover that *if* the claim is true, the sample proportion you have observed is so large that it falls at the 99th percentile of possible sample proportions for your sample size. Would you believe the claim and conclude that you just happened to get a weird sample, or would you reject the claim? What if the result was at the 70th percentile? At the 99.99th percentile?

4. Which is generally more serious when getting results of a medical diagnostic test: a false positive, which tells you you have the disease when you don't, or a false negative, which tells you you do not have the disease when you do?

21.1   USING DATA TO MAKE DECISIONS

In Chapters 19 and 20, we computed confidence intervals based on sample data to learn something about the population from which the sample had been taken. We sometimes used those confidence intervals to make a decision about whether there was a difference between two conditions.

Examining Confidence Intervals

When we examined the confidence interval for the relative risk of heart attacks for men with vertex baldness compared with no baldness, we noticed that the interval (1.2 to 1.9) was entirely above a relative risk of 1.0. Remember that a relative risk of 1.0 is equivalent to equal risk for both groups, whereas a relative risk above 1.0 means the risk is higher for the first group. In this example, if the confidence interval had included 1.0, then we could not say whether the risk of heart attack is higher for men with vertex baldness or for men with no hair loss, even with 95% confidence.

Using another example from Chapter 20, we noticed that the confidence interval for the difference in mean weight loss resulting from dieting alone versus exercising alone was entirely above zero. From that, we concluded, with 95% confidence, that the mean weight loss using dieting alone would be higher than it would be with exercise alone. If the interval had covered (included) zero, we would not have been able to say, with high confidence, which method resulted in greater average weight loss in the population.

Hypothesis Tests

Researchers interested in answering direct questions often conduct *hypothesis tests*. We have already learned the basic question researchers ask when they conduct such a test:

Is the relationship observed in the sample large enough to be called statistically significant, or could it have been due to chance?

In this chapter, we learn about the basic thinking that underlies hypothesis testing. In the next chapter, we learn how to carry out some simple hypothesis tests and examine some in-depth case studies.

EXAMPLE 1   **DECIDING IF STUDENTS PREFER QUARTERS OR SEMESTERS**

To illustrate the idea behind hypothesis testing, let's look at a simple, hypothetical example. There are two basic academic calendar systems in the United States: the quarter system and the semester system. Universities using the quarter system generally have three 10-week terms of classes, whereas those using the semester system have classes for two 15-week terms.

Suppose a university is currently on the quarter system and is trying to decide whether to switch to the semester system. Administrators are leaning toward switching to the semester system, but they have heard that the majority of students may oppose the switch. They decide to conduct a survey to see if there is convincing evidence that a majority of students oppose the plan, in which case they will reconsider their proposed change.

Administrators must choose from two hypotheses:

1. There is no clear preference (or the switch is preferred), so there is no problem.
2. As rumored, a majority of students oppose the switch, so the administrators should stop their plan.

The administrators pick a random sample of 400 students and ask their opinions. Of the 400, 220 of them say they oppose the switch. Thus, a clear majority of the sample, 0.55 (55%), is opposed to the plan. Here is the question that would be answered by a hypothesis test:

If there is really no clear preference, how likely would we be to observe sample results of this magnitude or larger, just by chance?

We already have the tools to answer this question. From the Rule for Sample Proportions, we know what to expect if there is no clear preference—that is, if, in truth, 50% of the students prefer each system:

> If numerous samples of 400 students are taken, the frequency curve for the proportions from the various samples will be approximately bell-shaped. The mean will be the true proportion from the population—in this case, 0.50. The standard deviation will be:
>
> the square root of
>
> true proportion $\times$ (1 − true proportion)/sample size
>
> In this case, the square root of $[(0.5)(0.5)/400] = 0.025$.

In other words, *if* there is truly no preference, then the observed value of 0.55 must have come from a bell-shaped curve with a mean of 0.50 and a standard deviation of 0.025.

How likely would a value as large as 0.55 be from that particular bell-shaped curve? To answer that, we need to compute the standardized score corresponding to 0.55:

standardized score = z-score = $(0.55 − 0.50)/0.025 = 2.00$

From Table 8.1 in Chapter 8, we find that a standardized score of 2.00 falls between 1.96 and 2.05, the values for the 97.5th and 98th percentiles. That is, if there is truly no preference, then we would observe a sample proportion as high as this (or higher) between 2% and 2.5% of the time.

The administration must now make a decision. One of two things has happened:

1. There really is no clear preference, but by the "luck of the draw" this particular sample resulted in an unusually high proportion opposed to the switch. In fact, it is so high that chance would lead to such a high value only slightly more than 2% of the time.

2. There really is a preference against switching to the semester system. The proportion (of all students) against the switch is actually higher than 0.50.

Most researchers agree that, by convention, we can rule out chance if the "luck of the draw" would have produced such extreme results less than 5% of the time. Therefore, in this case, the administrators should probably decide to rule out chance. The proper conclusion is that, indeed, a majority is opposed to switching to the semester system.

When a relationship or value from a sample is so strong that we can effectively rule out chance in this way, we say that the result is *statistically significant*. In this case, we would say that *the proportion of students who are opposed to switching to the semester system is statistically significantly higher than 50%*. ■

21.2   THE BASIC STEPS FOR TESTING HYPOTHESES

Although the specific computational steps required to test hypotheses depend on the type of variables measured, the basic steps are the same for all hypothesis tests. In this section, we outline the basic steps.

The basic steps for testing hypotheses are

1. Determine the null hypothesis and the alternative hypothesis.

2. Collect and summarize the data and generate the test statistic.

3. Determine how unlikely the test statistic is if the null hypothesis is true.

4. Make a decision.

Step 1. *Determine the null hypothesis and the alternative hypothesis.* There are always two hypotheses. The first one is called the **null hypothesis**—the hypothesis that says nothing is happening. The exact interpretation varies, but it can generally be thought of as the status quo, no relationship, chance only, or some variation on that theme.

The second hypothesis is called the **alternative hypothesis** or the **research hypothesis.** This hypothesis is usually the reason the data are being collected in the first place. The researcher suspects that the status quo belief is incorrect or that there is indeed a relationship between two variables that has not been established before. The only way to conclude that the alternative hypothesis is the likely one is to have enough evidence to effectively rule out chance, as presented in the null hypothesis.

| EXAMPLE 2 | **A JURY TRIAL** |

If you are on a jury in the American Judicial system, you must presume that the defendant is innocent unless there is enough evidence to conclude that he or she is guilty. Therefore, the two hypotheses are

Null hypothesis: The defendant is innocent.

Alternative hypothesis: The defendant is guilty.

The trial is being held because the prosecution believes the status quo assumption of innocence is incorrect. The prosecution collects evidence, much like researchers collect data, in the hope that the jurors will be convinced that such evidence would be extremely unlikely if the assumption of innocence were true. ■

For Example 1, the two hypotheses were

Null hypothesis: There is no clear preference for quarters over semesters.

Alternative hypothesis: The majority opposes the switch to semesters.

The administrators are collecting data because they are concerned that the null hypothesis is incorrect. If, in fact, the results are extreme enough (many more than 50% of the sample oppose the switch), then the administrators must conclude that they will meet with opposition if they try to impose the new calendar.

Step 2. *Collect and summarize the data and generate the test statistic.* Recall how we summarized the data in Example 1, when we tried to determine whether a clear majority of students opposed the semester system. In the final analysis, we based our decision on only one number, the *standardized score* for our sample proportion.

In general, the decision in a hypothesis test is based on a single summary of the data. This summary is called the **test statistic.** We encountered this idea in Chapter 12, where we used the chi-squared statistic to determine whether the relationship between two categorical variables was statistically significant. In that kind of problem, the chi-squared statistic is the only summary of the data needed to make the decision. The chi-squared statistic is the *test statistic* in that situation. The standardized score was the *test statistic* for Example 1 in this chapter.

Step 3. *Determine how unlikely the test statistic is if the null hypothesis is true.* In order to decide whether the results could be just due to chance, we ask the following question:

If the null hypothesis is really true, how likely would we be to observe sample results of this magnitude or larger (in a direction supporting the alternative hypothesis) just by chance?

Answering that question usually requires special tables, such as Table 8.1 for standardized scores, a computer with Excel or other software, or a calculator with statistical functions. Fortunately, most researchers answer the question for you in their reports and you must simply learn how to interpret the answer. The numerical value giving the answer to the question is called the *p-value.*

> The *p-value* is computed by *assuming* the null hypothesis is true, and then asking how likely we would be to observe such extreme results (or even more extreme results) under that assumption.

Many statistical novices misinterpret the meaning of a *p*-value. *The p-value does not give the probability that the null hypothesis is true.* There is no way to do that. For example, in Case Study 1.2, we noticed that for the men in that sample, those who took aspirin had fewer heart attacks than those who took a placebo. The *p*-value for that example tells us the probability of observing a relationship that extreme or more so in a sample of that size *if* there really is no difference in heart attack rates for the two conditions in the population. There is *no way* to determine the probability that aspirin actually has no effect on heart attack rates. In other words, there is no way to determine the probability that the null hypothesis is true. Don't fall into the trap of believing that common misinterpretation of the *p*-value.

Step 4. *Make a decision.* Once we know how unlikely the results would have been if the null hypothesis were true, we must make one of two choices:

Choice 1: The *p*-value is not small enough to convincingly rule out chance. Therefore, we *cannot reject* the null hypothesis as an explanation for the results.

Choice 2: The *p*-value was small enough to convincingly rule out chance. We *reject* the null hypothesis and *accept* the alternative hypothesis.

Notice that it is not valid to actually *accept* that the null hypothesis is true. To do so would be to say that we are essentially *convinced* that chance alone produced the observed results. This is another common mistake, which we will explore in Chapter 23.

Making choice 2 is equivalent to declaring that the result is *statistically significant.* We can rephrase the two choices in terms of statistical significance as follows:

Choice 1: There is no statistically significant difference or relationship evidenced by the data.

Choice 2: There is a statistically significant difference or relationship evidenced by the data.

You may be wondering how small the *p*-value must be in order to be small enough to rule out the null hypothesis. The standard used by most researchers is 5%. However, that is simply a convention that has become accepted over the years, and there are situations for which that value may not be wise. (We explore this issue further in the next section.)

Let's return to the analogy of the jury in a courtroom. In that situation, the information provided is generally not summarized into a single number. However, the two choices are equivalent to those in hypothesis testing:

Choice 1: The evidence is not strong enough to convincingly rule out that the defendant is innocent. Therefore, we *cannot reject* the null hypothesis, or innocence of the defendant, based on the evidence presented.

Choice 2: The evidence was strong enough that we are willing to rule out the possibility that an innocent person (as stated in the null hypothesis) produced the observed data. We *reject* the null hypothesis, that the defendant is innocent, and assert the alternative hypothesis, that he or she is guilty.

Consistent with our thinking in hypothesis testing, in many cases we would not *accept* the hypothesis that the defendant is innocent. We would simply conclude that the evidence was not strong enough to rule out the possibility of innocence.

21.3 WHAT CAN GO WRONG: THE TWO TYPES OF ERRORS

Any time decisions are made in the face of uncertainty, mistakes can be made. In testing hypotheses, there are two potential decisions or choices and each one brings with it the possibility that a mistake, or error, has been made.

The Courtroom Analogy

It is important to consider the seriousness of the two possible errors before making a choice. Let's use the courtroom analogy as an illustration. Here are the potential choices and the error that could accompany each:

Choice 1: We cannot rule out that the defendant is innocent, so he or she is set free without penalty.

Potential error: A criminal has been erroneously freed.

Choice 2: We believe there is enough evidence to conclude that the defendant is guilty.

Potential error: An innocent person is falsely convicted and penalized and the guilty party remains free.

Although the seriousness of the two potential errors depends on the seriousness of the crime and the punishment, choice 2 is usually seen as more serious. Not only is an innocent person punished, but a guilty one remains completely free and the case is closed.

A Medical Analogy: False Positive versus False Negative

As another example, consider a medical scenario in which you are tested for a disease. Most tests for diseases are not 100% accurate. In reading your results, the lab technician or physician must make a choice between two hypotheses:

Null hypothesis: You do not have the disease.

Alternative hypothesis: You have the disease.

Notice the possible errors engendered by each of these decisions:

FIGURE 21.1
*Potential errors
in the courtroom,
in medical testing, and
in hypothesis testing*

DECISION MADE	TRUE STATE OF NATURE	
	Innocent, Healthy Null Hypothesis	Guilty, Diseased Alternative Hypothesis
Innocent Healthy Don't reject null hypothesis	Correct ☺	⎧ Undeserved freedom ⎨ False negative ⎩ Type 2 error
Guilty Diseased Accept alternative hypothesis	⎧ Undeserved punishment ⎨ False positive ⎩ Type 1 error	Correct ☺

Choice 1: In the opinion of the medical practitioner, you are healthy. The test result was weak enough to be called "negative" for the disease.

Potential error: You are actually diseased but have been told you are not. In other words, your test was a **false negative.**

Choice 2: In the opinion of the medical practitioner, you are diseased. The test results were strong enough to be called "positive" for the disease.

Potential error: You are actually healthy but have been told you are diseased. In other words, your test was a **false positive.**

Which error is more serious in medical testing? It depends on the disease and on the consequences of a negative or positive test result. For instance, a false negative in a screening test for cancer could lead to a fatal delay in treatment, whereas a false positive would probably only lead to a retest.

A more troublesome example occurs in testing for HIV, where it is very serious to report a false negative and tell someone they are not infected when in truth they are infected. However, it is quite frightening for the patient to be given a false positive test for HIV—that is, the patient is really healthy but is told otherwise. HIV testing tends to err on the side of erroneous false positives. However, before being given the results, people who test positive for HIV with an inexpensive screening test are generally retested using an extremely accurate but more expensive test. Unfortunately, those who test negative are generally not retested, so it is important for the initial test to have a very low false negative rate.

The Two Types of Error in Hypothesis Testing

You can see that there is a trade-off between the two types of errors just discussed. Being too lenient in making decisions in one direction or the other is not wise. Determining the best direction in which to err depends on the situation and the consequences of each type of potential error.

The courtroom and medical analogies are not much different from the scenario encountered in hypothesis testing in general. Two types of errors can be made. A **type 1 error** can *only* be made if the null hypothesis is actually true. A **type 2 error** can *only* be made if the alternative hypothesis is actually true. Figure 21.1 illustrates the errors that can happen in the courtroom, in medical testing, and in hypothesis testing.

In our example of switching from quarters to semesters, if the university administrators were to make a type 1 error, the decision would correspond to a false positive. The administrators would be creating a false alarm, in which they stopped their planned change for no good reason. A type 2 error would correspond to a false negative, in which the administrators would have decided there isn't a problem when there really is one. ■

Probabilities Associated with Type 1 and Type 2 Errors

It would be nice if we could specify the probability that we were indeed making an error with each potential decision. We could then weigh the consequence of the error against its probability. Unfortunately, in most cases, we can only specify the conditional probability of making a type 1 error, given that the null hypothesis is true. That probability is the *p*-value discussed earlier, and it accompanies almost all reports of hypothesis tests in scientific literature.

p-*Values and Type 1 Errors*

You should realize that a type 1 error is impossible if the alternative hypothesis is true, by definition. Because *p*-values are commonly misinterpreted, it is worthwhile to restate their meaning in the context of the error that can be made:

If the null hypothesis is true, the p-value *is the probability of making an error by choosing the alternative hypothesis instead. A type 1 error will have been made in that case.*

Type 2 Errors

Notice that making a type 2 error is possible only if the alternative hypothesis is true. A type 2 error is made *if* the alternative hypothesis is true, but you fail to choose it. The probability of doing that depends on exactly which part of the alternative hypothesis is true, so that computing the probability of making a type 2 error is not feasible.

For instance, university administrators in our hypothetical example did not specify what percentage of students would have to oppose switching to semesters in order for the alternative hypothesis to hold. They merely specified that it would be over half. If only 51% of the population of students oppose the switch, then a sample of 400 students may result in fewer than half of the sample in opposition, in which case the null hypothesis would not be rejected even though it is false. However, if 80% of the students in the population oppose the switch, the sample proportion would definitely be large enough to convince the administration that the alternative hypothesis holds. In the first case (51% in opposition), it would be very easy to make a type 2 error, whereas in the second case (80% in opposition), it would be almost impossible. Yet, both are legitimate cases where the alternative hypothesis is true. That's why we can't specify one single value for the probability of making a type 2 error.

The Power of a Test

Researchers should never conclude that the null hypothesis is true just because the data failed to provide enough evidence to reject it. There is no control over the probability that an erroneous decision has been made in that situation. The **power** of a test is the probability of making the correct decision when the alternative hypothesis is true. It should be clear to you that the *power* of the administrators to detect that a majority oppose switching to semesters will be much higher if that majority constitutes 80% of the population than if it constitutes just 51% of the population. In other words, if the population value falls close to the value specified as the null hypothesis, then it may be difficult to get enough evidence from the sample to conclusively choose the alternative hypothesis. There will be a relative high probability of making a type 2 error, and the test will have relatively low power in that case.

Sometimes news reports, especially in science magazines, will insightfully note that a study may have failed to find a relationship between two variables because the test had such low power. As we will see in Chapter 23, this is a common consequence of conducting research with samples that are too small, but it is one that is often overlooked in media reports.

When to Reject the Null Hypothesis

When you read the results of research in a journal, you often are presented with a *p*-value, and the conclusion is left to you. In deciding whether a null hypothesis should be rejected, you should consider the consequences of the two potential types of errors. If you think the consequences of a type 1 error are very serious, then you should only reject the null hypothesis and accept the alternative hypothesis if the *p*-value is very small. Conversely, if you think a type 2 error is more serious, you should be willing to reject the null hypothesis with a moderately large *p*-value, typically 0.05 to 0.10.

CASE STUDY 21.1   ## Testing for the Existence of Extrasensory Perception

For centuries, people have reported experiences of knowing or communicating information that cannot have been transmitted through normal sensory channels. There have also been numerous reports of people seeing specific events in visions and dreams that then came to pass. These phenomena are collectively termed *extrasensory perception.*

In Chapter 17, we learned that it is easy to be fooled by underestimating the probability that weird events could have happened just by chance. Therefore, many of these reported episodes of extrasensory perception are probably explainable in terms of chance phenomena and coincidences.

Scientists have been conducting experiments to test extrasensory perception in the laboratory for several decades. As with many experiments, those performed in the laboratory lack the "ecological validity" of the reported anecdotes, but they have the advantage that the results can be quantified and studied using statistics.

Description of the Experiments

In this case study, we focus on one branch of research that has been investigated for a number of years under very carefully controlled conditions. The experiments are reported in detail by Honorton and colleagues (1990) and have also been summarized by Utts (1991, 1996) and by Bem and Honorton (1994).

The experiments use an experimental setup called the *ganzfeld procedure*. This involves four individuals, two of whom are participants and two, researchers. One of the two participants is designated as the sender and the other as the receiver. One of the two researchers is designated as the experimenter and the other as the assistant.

Each session of this experiment produces a single yes or no data value and takes over an hour to complete. The session begins by sequestering both the receiver and the sender in separate sound-isolated, electrically shielded rooms. The receiver wears headphones over which "white noise" (which sounds like a continuous hissing sound) is played. He or she is also looking into a red light, with halved Ping-Pong balls taped over the eyes to produce a uniform visual field. The term *ganzfeld* means "total field" and is derived from this visual experience. The reasoning behind this setup is that the senses will be open and expecting meaningful input, but nothing in the room will be providing such input. The mind may therefore look elsewhere for input.

Meanwhile, in another room, the sender is looking at either a still picture (a "static target") or a short video (a "dynamic target") on a television screen and attempting to "send" the image (or images) to the receiver. Here is an example of a description of a static target, from Honorton and colleagues (1990, p. 123):

> *Flying Eagle. An eagle with outstretched wings is about to land on a perch; its claws are extended. The eagle's head is white and its wings and body are black.*

The receiver has a microphone into which he or she is supposed to provide a continuous monologue about what images or thoughts are present. The receiver has no idea what kind of picture the sender might be watching. Here is part of what the receiver said during the monologue for the session in which the "Flying Eagle" was the target (from Honorton et al., 1990, p. 123):

> *A black bird. I see a dark shape of a black bird with a very pointed beak with his wings down. . . . Almost needle-like beak. . . . Something that would fly or is flying . . . like a big parrot with long feathers on a perch. Lots of feathers, tail feathers, long, long, long. . . . Flying, a big huge, huge eagle. The wings of an eagle spread out. . . . The head of an eagle. White head and dark feathers. . . . The bottom of a bird.*

The experimenter monitors the whole procedure and listens to the receiver's monologue. The assistant has one task only, to randomly select the material the sender will view on the television screen. This is done by using a computer, and no one knows the identity of the "target" selected except the sender. For the particular set of experiments we will examine, there were 160 possibilities, of which half were static targets and half were dynamic targets.

Quantifying the Results

Although it may seem as though the receiver provided an excellent description of the Flying Eagle target, remember that the quote given was only a small part of what the receiver said. So far, there is no quantifiable result. Such results are secured only at the end of the session.

To provide results that can be analyzed statistically, the results must be expressed in terms that can be compared with chance. To provide a comparison to chance, a single categorical measure is taken. Three "decoy" targets are chosen from the set. They are of the same type as the real target, either static or dynamic. Note that due to the random selection, any of the four targets (the real one and the three decoys) could equally have been chosen to be the real target at the beginning of the session.

The receiver is shown the four targets. He or she is then asked to decide, on the basis of the monologue provided, which one the sender was watching. If the receiver picks the right one, the session is a success. The example provided, in which the target was a picture of an eagle, was indeed a success.

The Null and Alternative Hypotheses

By chance, one-fourth or 25%, of the sessions should result in a success. Therefore, the statistical question for this hypothesis test is: Did the sessions result in significantly more than 25% successes?

The hypotheses being tested are thus as follows:

Null hypothesis: There is no extrasensory perception, and the results are due to chance guessing. The probability of a successful session is 0.25.

Alternative hypothesis: The results are not due to chance guessing. The probability of a successful session is higher than 0.25.

The Results

Honorton and his colleagues, who reported their results in the *Journal of Parapsychology* in 1990, ran several experiments, using the setup described, between 1983 and 1989, when the lab closed. There were a total of 355 sessions, of which 122 were successful.

The sample proportion of successful results is 122/355 = 0.344, or 34.4%. If the null hypothesis were true, the true proportion would be 0.25 and the standard deviation would be:

standard deviation = square root of [(0.25)(0.75)/355] = 0.023

Therefore, the test statistic is:

standardized score = z-score = $(0.344 - 0.25)/0.023 = 4.09$

It is obvious that such a large standardized score would rarely occur by chance, and indeed the p-value is about 0.00005. In other words, if chance alone were operating, we would see results of this magnitude about 5 times in every 100,000 such experiments. Therefore, we would certainly declare this to be a statistically significant result.

Carl Sagan has said that "exceptional claims require exceptional proof," and for some nonbelievers, even a p-value this small may not increase their personal probability that extrasensory perception exists. However, as with any area of science, such a striking result should be taken as evidence that something out of the ordinary is definitely happening in these experiments. ■

EXERCISES

1. When we revisited Case Study 6.4 in Chapter 20, we learned that a 95% confidence interval for the difference in years of education for mothers who did not smoke compared with those who did extended from 0.15 to 1.19 years, with higher education for those who did not smoke. Suppose we had used the data to construct a test instead of a confidence interval to see if one group in the population was more educated than the other. What would the null and alternative hypotheses have been for the test?

2. Refer to Exercise 1. If we had conducted the hypothesis test, the resulting p-value would be 0.01. Explain what the p-value represents for this example.

3. Refer to Case Study 20.1, in which women were randomly assigned to receive either a placebo or calcium, and severity of premenstrual syndrome (PMS) symptoms was measured.

   a. What are the null and alternative hypotheses tested in this experiment?

   b. The researchers concluded that calcium helped reduce the severity of PMS symptoms. Which type of error could they have made?

   c. What would be the consequences of making a type 1 error in this experiment? What would be the consequences of making a type 2 error?

4. The journal article reporting the experiment described in Case Study 20.1 (see Thys-Jacobs et al., 1998, in Chapter 20) compared the placebo and calcium-treated groups for a number of PMS symptoms, both before the treatment began (baseline) and in the third cycle. A p-value was given for each comparison. For each of the following comparisons, state the null and alternative hypotheses and the appropriate conclusion:

   a. Baseline, mood swings, p-value = 0.484

   b. Third cycle, mood swings, p-value = 0.002

   c. Third cycle, insomnia, p-value = 0.213

5. An article in *Science News* reported on a study to compare treatments for reducing cocaine use. Part of the results are

short-term psychotherapy that offers cocaine abusers practical strategies for maintaining abstinence sparks a marked drop in their overall cocaine use. . . . In contrast, brief treatment with desipramine—a drug thought by some researchers to reduce cocaine cravings—generates much weaker drops in cocaine use. (Bower, 24, 31 December 1994, p. 421)

   a. The researchers were obviously interested in comparing the rates of cocaine use following treatment with the two methods. State the null and alternative hypotheses for this situation.

   b. Explain what the two types of error could be for this situation and what their consequences would be.

   c. Although no *p*-value is given, the researchers presumably concluded that the psychotherapy treatment was superior to the drug treatment. Which type of error could they have made?

6. State the null and alternative hypotheses for each of the following potential research questions:

   a. Does working 5 hours a day or more at a computer contribute to deteriorating eyesight?

   b. Does placing babies in an incubator during infancy lead to claustrophobia in adult life?

   c. Does placing plants in an office lead to fewer sick days?

7. For each of the situations in Exercise 6, explain the two errors that could be made and what the consequences would be.

8. Explain why we can specify the probability of making a type 1 error, given that the null hypothesis is true, but we cannot specify the probability of making a type 2 error, given that the alternative hypothesis is true.

9. Compute a 95% confidence interval for the probability of a successful session in the ganzfeld studies reported in Case Study 21.1.

10. Specify what a type 1 and a type 2 error would be for the ganzfeld studies reported in Case Study 21.1.

11. Given the convention of declaring that a result is statistically significant if the *p*-value is 0.05 or less, what decision would be made concerning the null and alternative hypotheses in each of the following cases? Be explicit about the wording of the decision.

   a. *p*-value = 0.35

   b. *p*-value = 0.04

12. In previous chapters, we learned that researchers have discovered a link between vertex baldness and heart attacks in men.

   a. State the null hypothesis and the alternative hypothesis used to investigate whether there is such a relationship.

   b. Discuss what would constitute a type 1 error in this study.

   c. Discuss what would constitute a type 2 error in this study.

13. A report in the *Davis* (CA) *Enterprise* (6 April 1994, p. A 11) was headlined, "Highly educated people are less likely to develop Alzheimer's disease, a new study suggests."

   a. State the null and alternative hypotheses the researchers would have used in this study.

   b. What do you think the headline is implying about statistical significance? Restate the headline in terms of statistical significance.

14. Suppose that a study is designed to choose between the hypotheses:

Null hypothesis: Population proportion is 0.25.

Alternative hypothesis: Population proportion is higher than 0.25.

On the basis of a sample of size 500, the sample proportion is 0.29. The standard deviation for the potential sample proportions in this case is about 0.02.

   a. Compute the standardized score corresponding to the sample proportion of 0.29, assuming the null hypothesis is true.

   b. What is the percentile for the standardized score computed in part a?

   c. Based on the results of parts a and b, make a conclusion. Be explicit about the wording of your conclusion and justify your answer.

   d. To compute the standardized score in part a, you assumed the null hypothesis was true. Explain why you could not compute a standardized score under the assumption that the *alternative* hypothesis was true.

15. An article in the *Los Angeles Times* (24 December 1994, p. A16) announced that a new test for detecting HIV had been approved by the Food and Drug Administration (FDA). The test requires the person to send a saliva sample to a lab. The article described the accuracy of the test as follows:

The FDA cautioned that the saliva test may miss one or two infected individuals per 100 infected people tested, and may also result in false positives at the same rate in the uninfected. For this reason, the agency recommended that those who test positive by saliva undergo confirmatory blood tests to establish true infection.

   a. Do you think it would be wise to use this saliva test to screen blood donated at a blood bank, as long as those who test positive were retested as suggested by the FDA? Explain your reasoning.

   b. Suppose that 10,000 students at a university were all tested with this saliva test and that, in truth, 100 of them were infected. Further, suppose the false positive and false negative rates were actually both 1 in 100 for this group. If someone tests positive, what is the probability that he or she is infected?

16. Consider medical tests in which the null hypothesis is that the patient does not have the disease and the alternative hypothesis is that he or she does.

a. Give an example of a medical situation in which a type 1 error would be more serious.

b. Give an example of a medical situation in which a type 2 error would be more serious.

17. Many researchers decide to reject the null hypothesis as long as the p-value is 0.05 or less. In a testing situation for which a type 2 error is much more serious than a type 1 error, should researchers require a higher or a lower p-value in order to reject the null hypothesis? Explain your reasoning.

18. In Case Study 1.2 and in Chapter 12, we examined a study showing that there appears to be a relationship between taking aspirin and incidence of heart attack. The null hypothesis in that study would be that there is no relationship between the two variables, and the alternative would be that there is a relationship. Explain what a type 1 error and a type 2 error would be for the study and what the consequences of each type would be for the public.

19. In Case Study 1.1, Lee Salk did an experiment to see if hearing the sound of a human heartbeat would help infants gain weight during the first few days of life. By comparing weight gains for two sample groups of infants, he concluded that it did. One group listened to a heartbeat and the other did not.

a. What are the null and alternative hypotheses for this study?

b. What would a type 1 and type 2 error be for this study?

c. Given the conclusion made by Dr. Salk, explain which error he could possibly have committed and which one he could not have committed.

d. Rather than simply knowing whether there was a difference in average weight gains for the two groups, what statistical technique would have provided additional information?

MINI-PROJECTS

1. Construct a situation for which you can test null and alternative hypotheses for a population proportion. For example, you could see whether you can flip a coin in a manner so as to bias it in favor of heads. Or you could conduct an ESP test in which you ask someone to guess the suits in a deck of cards. (To do the latter experiment properly, you must replace the card and shuffle each time so you don't change the probability of each suit by having only a partial deck, and you must separate the "sender" and "receiver" to rule out normal communication.) Collect data for a sample of at least size 100. Carry out the test. Make sure you follow the four steps given in Section 21.2 and be explicit about your hypotheses, your decision, and the reasoning behind your decision.

2. Find two newspaper articles reporting on the results of studies with the following characteristics. First, find one that reports on a study that failed to find a relationship. Next, find one that reports on a study that did find a relationship. For each study, state what hypotheses you think the researchers were trying to test. Then explain what you think the results really imply compared with what is implied in the newspaper reports for each study.

REFERENCES

Bem, D. J., and C. Honorton. (1994). Does psi exist? Replicable evidence for an anomalous process of information transfer. *Psychological Bulletin* 115, no. 1, pp. 4–18.

Bower, B. (24, 31 December 1994). Psychotherapy's road to cocaine control. *Science News* 146, nos. 26 and 27, p. 421.

Honorton, C., R. E. Berger, M. P. Varvoglis, M. Quant, P. Derr, E. I. Schechter, and D. C. Ferrari. (1990). Psi communication in the ganzfeld. *Journal of Parapsychology* 54, no. 2, pp. 99–139.

Utts, J. (1991). Replication and meta-analysis in parapsychology. *Statistical Science* 6, no. 4, pp. 363–403.

Utts, J. (1996). Exploring psychic functioning: Statistics and other issues. *Stats: The Magazine for Students of Statistics* 16, pp. 3–8.

CHAPTER TWENTY-TWO

Hypothesis Testing— Examples and Case Studies

THOUGHT QUESTIONS

1. In Chapter 20, we examined a study showing that the difference in sample means for weight loss based on dieting only versus exercising only was 3.2 kg. That same study showed that the difference in average amount of *fat weight* (as opposed to muscle weight) lost was 1.8 kg and that the corresponding standard error was 0.83 kg. Suppose the means are actually equal, so that the mean difference in fat that would be lost for the populations is actually zero. What is the standardized score corresponding to the observed difference of 1.8 kg? Would you expect to see a standardized score that large or larger very often?

2. In the journal article reported on in Case Study 6.4, comparing IQs for children of smokers and nonsmokers, one of the statements made was, "After control for confounding background variables, the average difference [in mean IQs] observed at 12 and 24 months was 2.59 points (95% CI: −3.03, 8.20; $P = 0.37$)" (Olds et al., 1994, p. 223). The reported value of 0.37 is the *p*-value. What do you think are the null and alternative hypotheses being tested?

3. In chi-squared tests for two categorical variables, introduced in Chapter 12, we were interested in whether a relationship observed in a sample reflected a real relationship in the population. What are the null and alternative hypotheses?

4. In Chapter 12, we found a statistically significant relationship between smoking (yes or no) and time to pregnancy (one cycle or more than one cycle). Explain what the type 1 and type 2 errors would be for this situation, and the consequences of making each type of error.

22.1

HOW HYPOTHESIS TESTS ARE REPORTED IN THE NEWS

When you read the results of hypothesis tests in a newspaper, you are given very little information about the details of the test. It is therefore important for you to remember the steps occurring behind the scenes, so you can translate what you read into the bigger picture. The basic steps to hypothesis testing are similar in any setting:

> *Step 1:* Determine the null and alternative hypotheses.
>
> *Step 2:* Collect and summarize the data into a test statistic.
>
> *Step 3:* Use the test statistic to determine the *p*-value.
>
> *Step 4:* Decide whether the result is statistically significant based on the *p*-value.

In the presentation of most research results in the media, you are simply told the results of step 4, which isn't usually even presented in statistical language. If the results are statistically significant, you are told that a relationship has been found between two variables or that a difference has been found between two groups. If the results are not statistically significant, you may simply be told that no relationship or difference was found. In Chapter 23, we will revisit the problems that can arise if you are only told whether a statistically significant difference emerged from the research, but not told how large the difference was or how many participants there were in the study.

EXAMPLE 1   A study, which will be examined in further detail in Chapter 25, found that cranberry juice lived up to its popular reputation of preventing bladder infections, at least in older women. Here is a newspaper article reporting on the study (*Davis* [CA] *Enterprise,* 9 March 1994, p. A9):

> CHICAGO *(AP) A scientific study has proven what many women have long suspected: Cranberry juice helps protect against bladder infections. Researchers found that elderly women who drank 10 ounces of a juice drink containing cranberry juice each day had less than half as many urinary tract infections as those who consumed a look-alike drink without cranberry juice. The study, which appeared today in the Journal of the American Medical Association, was funded by Ocean Spray Cranberries, Inc., but the company had no role in the study's design, analysis or interpretation, JAMA said. "This is the first demonstration that cranberry juice can reduce the presence of bacteria in the*

urine in humans," said lead researcher Dr. Jerry Avorn, a specialist in medication for the elderly at Harvard Medical School.

Reading the article, you should be able to determine the null and alternative hypotheses and the conclusion. But there is no way for you to determine the value of the test statistic or the *p*-value. The study is attempting to compare the odds of getting an infection for the population of elderly women if they were to follow one of two regimes: 10 ounces of cranberry juice per day or 10 ounces of a placebo drink. The null hypothesis is that the odds ratio is 1; that is, the odds are the same for both groups. The alternative hypothesis is that the odds of infection are higher for the group drinking the placebo. The article indicates that the odds ratio is under 50%. In fact, the original article (Avorn et al., 1994) sets it at 42% and reports that the associated *p*-value is 0.004. The newspaper article captured the most important aspect of the research, but did not make it clear that the *p*-value was extremely low. ∎

22.2   TESTING HYPOTHESES ABOUT PROPORTIONS AND MEANS

Performing the computations necessary to find the test statistic—and thus the *p*-value—requires a level of statistical expertise beyond the scope of this text. However, for some simple situations—such as those involving a proportion, a mean, or the difference between two means—you already have all the necessary tools to understand the steps that need to be taken in doing such computations.

The Standardized Score and the p-Value

If the null and alternative hypotheses can be expressed in terms of a population proportion, mean, or difference between two means, and if the sample sizes are large, then the test statistic is simply the standardized score associated with the sample proportion, mean, or difference between two means. The standardized score is computed assuming the null hypothesis is true. The *p*-value is found from a table of percentiles for standardized scores (such as Table 8.1) or with the help of a computer program like *Excel*. It gives us the percentile at which a particular sample would fall if the null hypothesis represented the truth. Rather than providing detailed formulas, we will reexamine two previous examples to illustrate how this works. These examples should help you understand the ideas behind hypothesis testing in general.

Illustrating a Two-Sided Hypothesis Test

EXAMPLE 2   **WEIGHT LOSS FOR DIET VERSUS EXERCISE**

In Chapter 20, we found a confidence interval for the mean difference in weight loss for men who participated in dieting only versus exercising only for the period of a year. However, weight loss occurs in two forms: lost fat and lost muscle. Did the dieters also lose more fat than the exercisers? These were the sample results for amount of fat lost after 1 year:

Diet Only
sample mean = 5.9 kg
sample standard deviation = 4.1 kg
number of participants = n = 42
standard error = SEM_1
    = $4.1/\sqrt{42}$ = 0.633

Exercise Only
sample mean = 4.1 kg
sample standard deviation = 3.7 kg
number of participants = n = 47
standard error = SEM_2
    = $3.7/\sqrt{47}$ = 0.540

measure of uncertainty = square root of $[(0.633)^2 + (0.540)^2]$ = 0.83

Here are the steps required to determine if there is a difference in average fat lost for the two methods:

Step 1. Determine the null and alternative hypotheses.

Null hypothesis: There is no difference in average fat lost in the population for the two methods. The population mean difference is zero.

Alternative hypothesis: There is a difference in average fat lost in the population for the two methods. The population mean difference is not zero.

Notice that we do not specify which method has higher fat loss in the population. In this study, the researchers were simply trying to ascertain whether there was a difference. They did not specify in advance which method they thought would lead to higher fat loss.

When the alternative hypothesis includes a possible difference in either direction, the test is called a **two-sided,** or **two-tailed, hypothesis test.** In this example, it made sense for the researchers to construct a two-sided test because they had no preconceived idea of which method would be more effective for fat loss. They were simply interested in knowing whether there was a difference. When the test is two-sided, the *p*-value must account for possible chance differences in both "tails" or both directions. In step 3 for this example, our computation of the *p*-value will take both possible directions into account.

Step 2. Collect and summarize the data into a test statistic. The test statistic is the standardized score for the sample value when the null hypothesis is true. If there is no difference in the two methods, then the mean *population* difference is zero. The *sample* value is the observed difference in the two sample means, namely 5.9 − 4.1 = 1.8 kg. The measure of uncertainty, or standard error for this difference, which we just computed, is 0.83. Thus, the test statistic is

standardized score = *z*-score = (1.8 − 0)/0.83 = 2.17

Step 3. Use the test statistic to determine the *p*-value. How extreme is this standardized score? First, we need to define extreme to mean both directions because a standardized score of −2.17 would have been equally informative against the null hypothesis. From Table 8.1, we see that 2.17 is between the 98th and 99th percentiles, at about the 98.5th percentile. Using a more detailed table, I found that this is exactly right, so the probability of a result of 2.17 or higher is 0.015. This is also the probability of an extreme result in the other direction—that is, below –2.17. Thus, the test statistic results in a *p*-value of 2(0.015), or 0.03.

Step 4. Decide whether the result is statistically significant based on the *p*-value.

If there were really no difference between dieting and exercise as fat loss methods, we would see such an extreme result only 3% of the time, or 3 times out of 100. We prefer to believe that the truth does not lie with the null hypothesis. We conclude that there is a statistically significant difference between average fat loss for the two methods. We reject the null hypothesis. We accept the alternative hypothesis. In the exercises, you will be given a chance to test whether the same thing holds true for lean body mass (muscle) weight loss. ■

Illustrating a One-Sided Hypothesis Test

EXAMPLE 3   **PUBLIC OPINION ABOUT THE PRESIDENT**

On May 16, 1994, *Newsweek* reported the results of a public opinion poll that asked: "From everything you know about Bill Clinton, does he have the honesty and integrity you expect in a president?" (p. 23). The poll surveyed 518 adults and 233, or 0.45 of them (clearly less than half), answered yes. Could Clinton's adversaries conclude from this that only a minority (less than half) of the *population* of Americans thought Clinton had the honesty and integrity to be president? Assume the poll took a simple random sample of American adults. (In truth, a multistage sample was used, but the results will be similar.) Here is how we would proceed with testing this question:

Step 1. Determine the null and alternative hypotheses.

Null hypothesis: There is no clear winning opinion on this issue; the proportions who would answer yes or no are each 0.50.

Alternative hypothesis: Fewer than 0.50, or 50%, of the population would answer yes to this question. The majority do not think Clinton has the honesty and integrity to be president.

Notice that, unlike the previous example, this alternative hypothesis includes values on only one side of the null hypothesis. It does not include the possibility that *more* than 50% would answer yes. If the data indicated an uneven split of opinion in that direction, we would not reject the null hypothesis because the data would not be supporting the alternative hypothesis, which is the direction of interest. Some researchers include the other direction as part of the null hypothesis, but others simply don't include that possibility in either hypothesis, as illustrated in this example.

When the alternative hypothesis includes values in one direction only, the test is called a **one-sided,** or **one-tailed, hypothesis test.** The *p*-value is computed using only the values in the direction specified in the alternative hypothesis, as we shall see for this example.

Step 2. Collect and summarize the data into a test statistic. The test statistic is the standardized score for the sample value when the null hypothesis is true. The sample value is 0.45. If the null hypothesis is true, the population proportion is

0.50. The corresponding standard deviation, assuming the null hypothesis is true, is

standard deviation = square root of $[(0.5)(0.5)/518] = 0.022$
standardized score = z-score = $(0.45 - 0.50)/0.022 = -2.27$

STEP 3. Use the test statistic to determine the *p*-value. The *p*-value is the probability of observing a standardized score of –2.27 or less, just by chance. From Table 8.1, we find that –2.27 is between the 1st and 2nd percentiles. In fact, using a more exact method, the statistical function NORMSDIST(–2.27) in *Excel*, the *p*-value is 0.0116. Notice that, unlike the two-sided hypothesis test, we *do not* double this value. If we were to find an extreme value in the other direction, a standardized score of 2.27 or more, we would *not* reject the null hypothesis. We are performing a one-sided hypothesis test and are only concerned with values in *one tail* of the normal curve when we find the *p*-value. In this example, it is the lower tail because those values would indicate that the true proportion is *less than* the null hypothesis value of 0.50, thus supporting the alternative hypothesis.

Step 4. Decide whether the result is statistically significant based on the *p*-value. Using the 0.05 criterion, we have found a statistically significant result. Therefore, we could conclude that the proportion of American adults who believe Bill Clinton has the honesty and integrity they expect in a president is significantly less than a majority. ■

22.3   REVISITING CHI-SQUARED TESTS FOR CATEGORICAL VARIABLES

In Chapter 12, we learned how to conduct hypothesis tests to determine whether there was a relationship between two categorical variables. Let's see how the methods we learned there fit into the formal structure of hypothesis testing. Here is how the basic steps for hypothesis testing apply to testing the relationship between two categorical variables.

Step 1. Determine the null and alternative hypotheses. The specific wording of the two hypotheses depends on the situation, but the general format is always the same when we are testing for a relationship between two categorical variables. The null hypothesis indicates that any relationship observed in the sample was due only to chance. Another way of stating this is that, for the population, knowing which category someone is in for one variable doesn't give you any information about which one they are in for the other variable. A simple way of stating these ideas is

Null hypothesis: There is no relationship between the two variables in the population.

Alternative hypothesis: There is a relationship between the two variables in the population.

Notice that we do not specify a direction for the relationship. For instance, in medical situations we do not specify whether a treatment or exposure increases or decreases the chance of disease, we ask only whether there is a difference in the chance of disease based on whether someone was treated or exposed. Also, no causal connection is implied when the alternative hypothesis is accepted. Existence of a relationship does not mean a cause-and-effect relationship.

Step 2. Collect and summarize the data into a test statistic. We learned in Chapter 12 that the appropriate test statistic for determining whether there is a relationship between two categorical variables is called the *chi-squared statistic,* and we saw how to compute it.

Step 3. Use the test statistic to determine the *p*-value. In Chapter 12, we answered only the simplest question related to this step. We learned that in the case of a table where each variable had two categories, the test was statistically significant if the chi-squared statistic was 3.84 or larger. In terms you can now understand, we learned that the null hypothesis can be rejected using the 0.05 criterion if the chi-squared statistic is 3.84 or larger.

You are now in a position to understand how to find the *p*-value for the case of a table with two categories for each variable. If you take the square root of the chi-squared statistic for a 2×2 table, it is equivalent to a *z*-score, or standardized score. Because the alternative hypothesis includes differences in either direction, you find the *p*-value by finding the proportion above the *z*-score and doubling it. In Example 4, we will see how to do this. You can also use some computer and calculator functions directly. For instance, using Excel, if the chi-squared statistic is *x,* then the *p*-value is CHIDIST$(x,1)$.

Step 4. Decide whether the result is statistically significant based on the *p*-value. This step is equivalent for all hypothesis tests. Remember to consider the consequences of the two types of errors before making a decision.

EXAMPLE 4

YOUNG DRIVERS, GENDER, AND DRIVING UNDER THE INFLUENCE OF ALCOHOL

In Case Study 6.3 and in Chapter 12, we studied a court case from Oklahoma concerning the ages at which young men and women could buy 3.2% beer. The case eventually went to the U.S. Supreme Court, which examined evidence from a "random roadside survey" that measured information on age, gender, and drinking behavior. Some of the data examined whether there was a relationship between gender and drinking alcohol in the previous 2 hours. In Chapter 12, we presented the results for drivers under 20 years of age; these are presented again in Table 22.1.

In Chapter 12, we learned that the observed difference in proportions of male and female drinkers could have been due to chance. Let's go through the steps of hypothesis testing for this example and determine the appropriate *p*-value. Based on our results from Chapter 12, all we know at this stage is that it exceeds 0.05.

Step 1. Determine the null and alternative hypotheses.

Null hypothesis: In the population of young drivers, there is no relationship between gender and whether the driver drank alcohol in the last 2 hours.

TABLE 22.1	Results of Random Roadside Survey			
	Drank Alcohol in Last 2 Hours?			Percentage Who Drank
	Yes	No	Total	
Males	77	404	481	16.0%
Females	16	122	138	11.6%
Total	93	526	619	

SOURCE: Gastwirth, 1988, p. 526.

Alternative hypothesis: In the population of young drivers, one of the two sexes is more likely than the other to have consumed alcohol in the 2 hours prior to driving.

Step 2. Collect and summarize the data into a test statistic. In Chapter 12, we used Minitab to determine that the chi-squared statistic for the data in Table 22.1 is 1.637.

Step 3. Use the test statistic to determine the *p*-value. To find the *p*-value, first take the square root of the chi-squared statistic, and then use the *z*-score table. The square root of 1.637 is 1.28. From Table 8.1, we find that 1.28 is at the 90th percentile. Therefore, there is a 10% chance of observing a value this large or larger just by chance. Because the test is two-tailed, we must consider values below −1.28 as well, so the *p*-value is $2 \times 0.10 = 0.20$. We find the same result using Excel; CHIDIST(1.637,1)= 0.2007. This *p*-value indicates that sample results showing a relationship this large or larger would occur by chance 20% of the time, assuming there really is no relationship between gender and drinking alcohol before driving.

Step 4. Decide whether the result is statistically significant based on the *p*-value. The *p*-value is 0.20, so we would probably not be willing to rule out chance as an explanation for the observed difference in proportions. Notice that about 16% of the males in the sample had been drinking, whereas only 11.6% of the females had been drinking. The results of the test are telling us that if there is no difference in the population, then a sample with a difference that large or larger, in either direction, would occur in about 20% of similarly conducted random roadside surveys. ■

22.4 REVISITING CASE STUDIES: HOW JOURNALS PRESENT HYPOTHESIS TESTS

Whereas newspapers and magazines tend to simply report the decision from hypothesis testing, journals tend to report *p*-values as well. This practice allows you to make your own decision, based on the severity of a type 1 error and the magnitude of the *p*-value. Newspaper reports leave you little choice but to accept the convention that the probability of a type 1 error is 5%.

To demonstrate the way in which journals report the results of hypothesis tests, we return to some of our earlier case studies.

CASE STUDY 6.1 REVISITED

Mozart, Relaxation, and Performance on Spatial Tasks

This case study reported that listening to Mozart for 10 minutes appeared to increase results on the spatial reasoning part of an IQ test. There were three listening conditions—Mozart, a relaxation tape, and silence—and all subjects participated in all three conditions.

Because three means were being compared, the appropriate test statistic to summarize the data is more complicated than those we have discussed. However, developing the hypotheses and interpreting the *p*-value are basically the same. Here are the hypotheses:

Null hypothesis: There are no differences in population mean spatial reasoning IQ scores after each of the three listening conditions.

Alternative hypothesis: Population mean spatial reasoning IQ scores do differ for at least one of the conditions compared with the others.

Notice that the researchers did not specify in advance which condition might result in higher IQ scores.

Here is how the results were reported:

A one-factor (listening condition) repeated measures analysis of variance . . . revealed that subjects performed better on the abstract/spatial reasoning tests after listening to Mozart than after listening to either the relaxation tape or to nothing ($F[2,35] = 7.08$, $p = 0.002$). (Rauscher et al., 14 October 1993, p. 611)

Because the *p*-value for this test is only 0.002, we can clearly reject the null hypothesis, accept the alternative, and conclude that at least one of the means differs from the others. If there were no population differences, sample mean results would vary as much as the ones in this sample did, or more, only 2 times in 1000 (0.002).

The researchers then reported the results, indicating that it was indeed the Mozart condition that resulted in higher scores than the others:

The music condition differed significantly from both the relaxation and silence conditions (Scheffé's $t = 3.41$, $p = 0.002$; $t = 3.67$, $p = 0.0008$, two-tailed, respectively). The relaxation and silence conditions did not differ ($t = 0.795$, $p = 0.432$, two-tailed). (Rauscher et al., 14 October 1993, p. 611)

Notice that three separate tests are being reported in this paragraph. Each pair of conditions has been compared. Significant differences, as determined by the extremely small *p*-values, were found between the music and relaxation conditions (*p*-value = 0.002) and between the music and silence condition (*p*-value = 0.0008). The difference between the relaxation and silence condition, however, was not statistically significant (*p* value = 0.432). ■

CASE STUDY 5.1 REVISITED

Quitting Smoking with Nicotine Patches

This study compared the smoking cessation rates for smokers randomly assigned to use a nicotine patch versus a placebo patch. In the summary at the beginning of the journal article, the results were reported as follows:

> *Higher smoking cessation rates were observed in the active nicotine patch group at 8 weeks (46.7% vs 20%) (P < .001) and at 1 year (27.5% vs 14.2%) (P = .011). (Hurt et al., 1994, p. 595)*

Two sets of hypotheses are being tested, one for the results after 8 weeks and one for the results after 1 year. In both cases, the hypotheses are

> *Null hypothesis:* The proportion of smokers in the population who would quit smoking using a nicotine patch and a placebo patch are the same.

> *Alternative hypothesis:* The proportion of smokers in the population who would quit smoking using a nicotine patch is higher than the proportion who would quit using a placebo patch.

In both cases, the reported *p*-values are quite small: less than 0.001 for the difference after 8 weeks and equal to 0.011 for the difference after a year. Therefore, we would conclude that rates of quitting are significantly higher using a nicotine patch than using a placebo patch after 8 weeks and after 1 year. ■

CASE STUDY 6.4 REVISITED

Smoking During Pregnancy and Child's IQ

In this study, researchers investigated the impact of maternal smoking on subsequent IQ of the child at ages 1, 2, 3, and 4 years of age. Earlier, we reported some of the confidence intervals provided in the journal article reporting the study. Those confidence intervals were actually accompanied by *p*-values. Here is the complete reporting of some of the results:

> *Children born to women who smoked 10+ cigarettes per day during pregnancy had developmental quotients at 12 and 24 months of age that were 6.97 points lower (averaged across these two time points) than children born to women who did not smoke during pregnancy (95% CI: 1.62,12.31, P = .01); at 36 and 48 months they were 9.44 points lower (95% CI: 4.52, 14.35, P = .0002). (Olds et al., 1994, p. 223)*

Notice that we are given more information in this report than in most because we are given both confidence interval and hypothesis testing results. This is excellent reporting because, with this information, we can determine the magnitude of the observed effects instead of just whether they are statistically significant or not.

Again, two sets of null and alternative hypotheses are being tested in the report, one set at 12 and 24 months (1 and 2 years) and another at 36 and 48 months (3 and 4 years of age). In both cases, the hypotheses are

Null hypothesis: The mean IQ scores for children whose mothers smoke 10 or more cigarettes a day during pregnancy are the same as the mean for those whose mothers do not smoke, in populations similar to the one from which this sample was drawn.

Alternative hypothesis: The mean IQ scores for children whose mothers smoke 10 or more cigarettes a day during pregnancy are not the same as the mean for those whose mothers do not smoke, in populations similar to the one from which this sample was drawn.

This is a *two-tailed test* because the researchers included the possibility that the mean IQ score could actually be *higher* for those whose mothers smoke. The confidence interval provides the evidence of the direction in which the difference falls. The *p*-value simply tells us that there is a statistically significant difference. ■

CASE STUDY 22.1 An Interpretation of a p-Value <u>Not</u> Fit to Print

In an article entitled, "Probability experts may decide Pennsylvania vote," the *New York Times* (Passell, 11 April 1994, p. A15) reported on the use of statistics to try to decide whether there had been fraud in a special election held in Philadelphia. Unfortunately, the newspaper account made a common mistake, misinterpreting a *p*-value to be the probability that the results *could* be explained by chance. The consequence was that readers who did not know how to spot the error would have been led to think that the election probably *was* a fraud.

It all started with the death of a state senator from Pennsylvania's Second Senate District. A special election was held to fill the seat until the end of the unexpired term. The Republican candidate, Bruce Marks, beat the Democratic candidate, William Stinson, in the voting booth but lost the election because Stinson received so many more votes in absentee ballots. The results in the voting booth were very close, with 19,691 votes for Mr. Marks and 19,127 votes for Mr. Stinson. But the absentee ballots were not at all close, with only 366 votes for Mr. Marks and 1391 votes for Mr. Stinson.

The Republicans charged that the election was fraudulent and asked that the courts examine whether the absentee ballot votes could be discounted on the basis of suspicion of fraud. In February 1994, 3 months after the election, Philadelphia Federal District Court Judge Clarence Newcomer disqualified all absentee ballots and overturned the election. The ruling was appealed, and statisticians were hired to help sort out what might have happened.

One of the statistical experts, Orley Ashenfelter, decided to examine previous senatorial elections in Philadelphia to determine the relationship between votes cast in the voting booth and those cast by absentee ballot. He computed the difference between the Republican and Democratic votes for those who voted in the voting booth, and then for those who voted by absentee ballot. He found there was a positive correlation between the voting booth difference and the absentee ballot difference. Using data from 21 previous elections, he calculated a regression equation to predict one from the other. Using his equation, the difference in votes for the two parties by absentee ballot could be predicted from knowing the difference in votes in the voting booth.

Ashenfelter then used his equation to predict what should have happened in the special election in dispute. There was a difference of 19,691 − 19,127 = 564 votes (in favor of the Republicans) in the voting booth. From that, he predicted a difference of 133 votes in favor of the Republicans in absentee ballots. Instead, a difference of 1025 votes in favor of the *Democrats* was observed in the absentee ballots of the disputed election.

Of course, everyone knows that chance events play a role in determining who votes in any given election. So Ashenfelter decided to set up and test two hypotheses. The null hypothesis was that, given past elections as a guide and given the voting booth difference, the overall difference observed in this election could be explained by chance. The alternative hypothesis was that something other than chance influenced the voting results in this election.

Ashenfelter reported that *if* chance alone was responsible, there was a 6% chance of observing results as extreme as the ones observed in this election, given the voting booth difference. In other words, the *p*-value associated with his test was about 6%. That is not how the result was reported in the *New York Times*. When you read their report, see if you can detect the mistake in interpretation:

> *There is some chance that random variations alone could explain a 1,158-vote swing in the 1993 contest—the difference between the predicted 133-vote Republican advantage and the 1,025-Democratic edge that was reported. More to the point, there is some larger probability that chance alone would lead to a sufficiently large Democratic edge on the absentee ballots to overcome the Republican margin on the machine balloting. And the probability of such a swing of 697 votes from the expected results, Professor Ashenfelter calculates, was about 6 percent. Putting it another way, if past elections are a reliable guide to current voting behavior, there is a 94 percent chance that irregularities in the absentee ballots, not chance alone,* swung the election to the Democrat, *Professor Ashenfelter concludes. (Passell, 11 April 1994, p. A15; emphasis added)*

The author of this article has mistakenly interpreted the *p*-value to be the probability that the null hypothesis is true and has thus reported what he thought to be the probability that the alternative hypothesis was true. Hopefully, you realize that this is not a valid conclusion. The *p*-value can only tell us the probability of observing these results *if* the election was *not* fraudulent. It cannot tell us the probability in the other direction—namely, the probability that the election was fraudulent based on observed results. This is akin to the "confusion of the inverse" discussed in Chapter 17.

There we saw that physicians sometimes confuse the (unknown) probability that the patient has a disease, given a positive test, with the (known) probability of a positive test, given that the patient has the disease.

You should also realize that the implication that the results of past elections would hold in this special election may not be correct. This point was raised by another statistician involved with the case. The *New York Times* report notes:

> *Paul Shaman, a professor of statistics at the Wharton School at University of Pennsylvania . . . exploits the limits in Professor Ashenfelter's reasoning. Relationships between machine and absentee voting that held in the past, he argues, need not hold in the present. Could not the difference, he asks, be explained by Mr. Stinson's "engaging in aggressive efforts to obtain absentee votes?"* (Passell, 11 April 1994, p. A15)

The case went to court two more times, but the original decision made by Judge Newcomer was upheld each time. The Republican Bruce Marks held the seat until December 1994. As a footnote, in the regular election in November 1994, Bruce Marks lost by 393 votes to Christina Tartaglione, the daughter of the chair of the board of elections, one of the people allegedly involved in the suspected fraud. This time, both candidates agreed that the election had been conducted fairly (Shaman, 28 November 1994). ■

FOR THOSE WHO LIKE FORMULAS

Some Notation for Hypothesis Tests

The null hypothesis is denoted by H_0, and the alternative hypothesis is denoted by H_1 or H_a.

"alpha" = α = desired probability of making a type 1 error when H_0 is true; we reject H_0 if p-value $\leq \alpha$.

"beta" = β = probability of making a type 2 error when H_1 is true; power = $1 - \beta$

Steps for Testing the Mean of a Single Population

Denote the population mean by μ and the sample mean and standard deviation by $\overline{X}$ and s, respectively.

Step 1. H_0: $\mu = \mu_0$, where μ_0 is the *chance* or *status quo* value.

H_1: $\mu \neq \mu_0$ for a two-sided test; H_1: $\mu < \mu_0$ or H_1: $\mu > \mu_0$ for a one-sided test, with the direction determined by the research hypothesis of interest.

Step 2. This test statistic applies only if the sample is large. The test statistic is

$$z = \frac{\overline{X} - \mu_0}{s / \sqrt{n}}$$

Step 3. The p-value depends on the form of H_1. In each case, we refer to the proportion of the standard normal curve above (or below) a value as the "area" above (or below) that value. Then we list the p-values as follows:

Alternative Hypothesis	p-Value
$H_1: \mu \neq \mu_0$	$2 \times$ area above $|z|$
$H_1: \mu > \mu_0$	area above z
$H_1: \mu < \mu_0$	area below z

Step 4. You must specify the desired α; it is commonly 0.05. Reject H_0 if p-value $\leq \alpha$.

Steps for Testing a Proportion for a Single Population
Steps 1, 3, and 4 are the same, except replace μ with the population proportion p and μ_0 with the hypothesized proportion p_0. The test statistic (step 2) is:

$$z = \frac{\hat{p} - p_0}{\sqrt{\dfrac{p_0(1 - p_0)}{n}}}$$

Steps for Testing for Equality of Two Population Means
Using Large Independent Samples
Steps 1, 3, and 4 are the same, except replace μ with $(\mu_1 - \mu_2)$ and μ_0 with 0.
Use previous notation for sample sizes, means, and standard deviations; the test statistic (step 2) is:

$$z = \frac{\overline{X}_1 - \overline{X}_2}{\sqrt{\dfrac{s_1^2}{n_1} + \dfrac{s_2^2}{n_2}}}$$

EXERCISES

1. In Exercise 12 in Chapter 19, we learned that in a survey of 507 adult American Catholics, 59% answered yes to the question, "Do you favor allowing women to be priests?"

   a. Set up the null and alternative hypotheses for deciding whether a majority of American Catholics favor allowing women to be priests.

   b. Using Example 3 in this chapter as a guide, compute the test statistic for this situation.

   c. If you have done everything correctly, the p-value for the test is about 0.00005. Based on this, make a conclusion for this situation. Write it in both statistical language and in words that someone with no training in statistics would understand.

2. Refer to Exercise 1. Is the test described there a one-sided or a two-sided test?

3. Suppose a one-sided test for a proportion resulted in a p-value of 0.03. What would the p-value be if the test were two-sided instead?

4. Suppose a two-sided test for a difference in two means resulted in a p-value of 0.08.

   a. Using the usual criterion for hypothesis testing, would we conclude that there was a difference in the population means? Explain.

   b. Suppose the test had been constructed as a one-sided test instead, and the evidence in the sample means was in the direction to support the alternative hypothesis. Using the usual criterion for hypothesis testing, would we be able to conclude that there was a difference in the population means? Explain.

5. Suppose you were given a hypothesized population mean, a sample mean, a sample standard deviation, and a sample size for a study involving a random sample from one population. What would you use as the test statistic?

6. In Example 3, we showed that the *Excel* command NORMSDIST(z) gives the area below the standardized score z. Use *Excel* or another computer or calculator function to find the p-value for each of the following examples and case studies, taking into account whether the test is one-sided or two-sided:

   a. Chapter 22, Example 2, $z = 2.17$, two-sided test

   b. Chapter 21, Example 1, $z = 2.00$, one-sided test

   c. Case Study 21.1, $z = 4.09$, one-sided test

7. Suppose you wanted to see whether a training program helped raise students' scores on a standardized test. You administer the test to a random sample of students, give them the training program, then readminister the test. For each student, you record the increase (or decrease) in the test score from one time to the next.

   a. What would the null and alternative hypotheses be for this situation?

   b. Suppose the mean change for the sample was 10 points and the accompanying standard error was 4 points. What would be the standardized score that corresponded to the sample mean of 10 points?

   c. Based on the information in part b, what would you conclude about this situation?

   d. What explanation might be given for the increased scores, other than the fact that the training program had an impact?

   e. What would have been a better way to design the study in order to rule out the explanation you gave in part d?

8. Specify what a type 1 and a type 2 error would be for Example 4, in which we tested whether there was a relationship between gender and driving after drinking alcohol. Remember that the Supreme Court used the data to determine whether a law was justified. The law differentiated between the ages at which young males and young females could purchase 3.2% beer. Explain what the consequences of the two types of error would be in that context.

9. On July 1, 1994, *The Press* of Atlantic City, NJ, had a headline reading, "Study: Female hormone makes mind keener" (p. A2). Here is part of the report:

> *Halbreich said he tested 36 post-menopausal women before and after they started the estrogen therapy. He gave each one a battery of tests that measured such things as memory, hand-eye coordination, reflexes and the ability to learn new information and apply it to a problem. After estrogen therapy started, he said, there was a subtle but statistically significant increase in the mental scores of the patients.*

Explain what you learned about the study, in the context of the material in this chapter, by reading this quote. Be sure to specify the hypotheses that were being tested and what you know about the statistical results.

10. Siegel (1993) reported a study in which she measured the effect of pet ownership on the use of medical care by the elderly. She interviewed 938 elderly adults. One of her results was reported as: "After demographics and health status were controlled for, subjects with pets had fewer total doctor contacts during the one-year period than those without pets (beta = −.07, $p < .05$)" (p. 164).

    a. State the null and alternative hypotheses Siegel was testing. Be careful to distinguish between a population and a sample.

    b. State the conclusion you would make. Be explicit about the wording.

11. Refer to Exercise 10. Here is another of the results reported by Siegel: "For subjects without a pet, having an above-average number of stressful life events resulted in about two more doctor contacts during the study year (10.37 vs. 8.38, p < .005). In contrast, the number of stressful life events was not significantly related to doctor visits among subjects with a pet" (1993, p. 164).

    a. State the null and alternative hypotheses Siegel is testing in this passage. Notice that two tests are being performed; be sure to cover both.

    b. Pretend you are a newspaper reporter, and write a short story describing the results reported in this exercise. Be sure you do not convey any misleading information. You are writing for a general audience, so do not use statistical jargon that would be unfamiliar to them.

12. In Example 2 in this chapter, we tested whether the average fat lost from 1 year of dieting versus 1 year of exercise was equivalent. The study also measured lean body weight (muscle) lost or gained. The average for the 47 men who exercised was a *gain* of 0.1 kg, which can be thought of as a loss of −0.1 kg. The standard deviation was 2.2 kg. For the 42 men in the dieting group, there was an average loss of 1.3 kg, with a standard deviation of 2.6 kg. Test to see whether the average lean body mass lost (or gained) would be different for the population. Specify all four steps of your hypothesis test.

13. Professors and other researchers use scholarly journals to publish the results of their research. However, only a small minority of the submitted papers are accepted for publication by the most prestigious journals. In many academic fields, there is a debate as to whether submitted papers written by women are

....

TABLE 22.2

	Pregnancy Occurred After		
	First Cycle	Two or More Cycles	Total
Smoker	29	71	100
Nonsmoker	198	288	486
Total	227	359	586

treated as well as those submitted by men. In the January 1994 issue of *European Science Editing* (Maisonneuve, January 1994), there was a report on a study that examined this question. Here is part of that report:

Similarly, no bias was found to exist at JAMA [Journal of the American Medical Association] *in acceptance rates based on the gender of the corresponding author and the assigned editor. In the sample of 1,851 articles considered in this study female editors used female reviewers more often than did male editors (P < 0.001).*

That quote actually contains the results of two separate hypothesis tests. Explain what the two sets of hypotheses tested are and what you can conclude about the *p*-value for each set.

14. For the data in Chapter 12, Example 4, we found that the risk of developing breast cancer was 1.33 times greater for women who had their first child at age 25 or older, compared with women who had their first child before age 25.

   a. Specify the null and alternative hypotheses for this study.

   b. The chi-squared statistic for this study is 1.75. Find the *p*-value and make a conclusion. Rewrite the conclusion in words that someone with no training in statistics would understand.

15. In Case Study 12.1, we explored data from an extrasensory perception study that used both static photos and videos as target material. We found that there was a statistically significant relationship between type of target and whether the trial was a success. The chi-squared statistic was 6.675. Go through the four steps of hypothesis testing for this example and fill in as many details as possible. In other words, state the hypotheses, the test statistic, the *p*-value, and the conclusion.

16. In Chapter 12, we computed the chi-squared statistic for the data shown in Table 22.2 to be 4.82.

   a. Specify the hypotheses that were being examined in this study.

   b. Verify that the *p*-value accompanying this test is about 0.03.

c. Suppose a friend were to try to explain to you that a *p*-value of 0.03 also tells us that there is a 97% chance that there is a relationship between smoking and cycles to pregnancy (first versus two or more). Is your friend correct? Explain why or why not.

d. Consider a smoker who wants to get pregnant, who is trying to decide whether to reject the null hypothesis for this study. Explain the consequences of making a type 1 or type 2 error for this situation.

17. On January 30, 1995, *Time* magazine reported the results of a poll of adult Americans, in which they were asked, "Have you ever driven a car when you probably had too much alcohol to drive safely?" The exact results were not given, but from the information provided we can guess at what they were. Of the 300 men who answered, 189 (63%) said yes and 108 (36%) said no. The remaining 3 weren't sure. Of the 300 women, 87 (29%) said yes while 210 (70%) said no, and the remaining 3 weren't sure.

a. Organize the data into a 2×2 contingency table, ignoring the "not sure" answers.

b. The chi-squared statistic for the table in part a is 70.43. Go through the four steps of hypothesis testing for this example. In other words, state the appropriate hypotheses, the test statistic, the *p*-value (at least to an order of magnitude), and the conclusion you would make.

18. In Example 2 in this chapter, we tested to see whether dieters and exercisers had different average fat loss. We concluded that they did because the difference for the samples under consideration was 1.8 kg, with a measure of uncertainty of 0.83 kg and a standardized score of 2.17. Fat loss was higher for the dieters.

a. Construct a 95% confidence interval for the population difference in mean fat loss. Consider the two different methods for presenting results: (1) the *p*-value and conclusion from the hypothesis test or (2) the confidence interval. Which do you think is more informative? Explain.

b. Suppose the alternative hypothesis had been that men who exercised lost more fat on average than men who dieted. Would the null hypothesis have been rejected? Explain why or why not. If yes, give the *p*-value that would have been used.

c. Suppose the alternative hypothesis had been that men who dieted lost more fat on average than men who exercised. Would the null hypothesis have been rejected? Explain why or why not. If yes, give the *p*-value that would have been used.

MINI-PROJECTS

1. Find three separate journal articles that report the results of hypothesis tests. For each one, do or answer the following:

a. State the null and alternative hypotheses.

b. Based on the *p*-value reported, what conclusion would you make?

c. What would a type 1 and a type 2 error be for the hypotheses being tested?

2. Conduct a test for extrasensory perception. You can either create target pictures or use a deck of cards and ask people to guess suits or colors. Whatever you use, be sure to randomize properly. For example, with a deck of cards you should always replace the previous target and shuffle very well.

a. Explain how you conducted the experiment.

b. State the null and alternative hypotheses for your experiment.

c. Report your results.

d. If you do not find a statistically significant result, can you conclude that extrasensory perception does not exist? Explain.

REFERENCES

Avorn, J., M. Monane, J. H. Gurwitz, R. J. Glynn, I. Choodnovskiy, and L. A. Lipsitz. (1994). Reduction of bacteriuria and pyuria after ingestion of cranberry juice. *Journal of the American Medical Association* 271, no. 10, pp. 751–754.

Gastwirth, J. L. (1988). *Statistical reasoning in law and public policy.* Vol. 2. *Tort law, evidence, and health.* New York: Academic Press.

Hurt, R., L. Dale, P. Fredrickson, C. Caldwell, G. Lee, K. Offord, G. Lauger, Z. Marusic, L. Neese, and T. Lundberg. (1994). Nicotine patch therapy for smoking cessation combined with physician advice and nurse follow-up. *Journal of the American Medical Association* 271, no. 8, pp. 595–600.

Maisonneuve, H. (January 1994). Peer review congress. *European Science Editing* no. 51, pp. 7–8.

Olds, D. L., C. R. Henderson, Jr., and R. Tatelbaum. (1994). Intellectual impairment in children of women who smoke cigarettes during pregnancy. *Pediatrics* 93, no. 2, pp. 221–227.

Passell, Peter. (11 April 1994). Probability experts may decide Pennsylvania vote. *New York Times,* p. A15.

Rauscher, F. H., G. L. Shaw, and K. N. Ky. (14 October 1993). Music and spatial task performance. *Nature* 365, p. 611.

Shaman, Paul. (28 November 1994). Personal communication.

Siegel, J. M. (1993). Companion animals: In sickness and in health. *Journal of Social Issues* 49, no. 1, pp. 157–167.

CHAPTER TWENTY-THREE

Significance, Importance, and Undetected Differences

THOUGHT QUESTIONS

1. Which do you think is more informative when you are given the results of a study: a confidence interval or a *p*-value? Explain.

2. Suppose you were to read that a new study based on 100 men had found that there was *no difference* in heart attack rates for men who exercised regularly and men who did not. What would you suspect was the reason for that finding? Do you think the study found *exactly* the same rate of heart attacks for the two groups of men?

3. An example in Chapter 22 used the results of a public opinion poll to conclude that a majority of Americans did not think Bill Clinton had the honesty and integrity they expected in a president. Would it be fair reporting to claim that "significantly fewer than 50% of American adults think Bill Clinton has the honesty and integrity they expect in a president"? Explain.

4. When reporting the results of a study, explain why a distinction should be made between "statistical significance" and "significance," as the term is used in ordinary language.

5. Remember that a type 2 error is made when a study fails to find a relationship or difference when one actually exists in the population. Is this kind of error more likely to occur in studies with large samples or with small samples? Use your answer to explain why it is important to learn the size of a study that finds no relationship or difference.

23.1   REAL IMPORTANCE VERSUS STATISTICAL SIGNIFICANCE

By now, you should realize that a statistically significant relationship or difference does not necessarily mean an important one. Further, a result that is "significant" in the statistical meaning of the word may *not* be "significant" in the common meaning of the word. Whether the results of a test are statistically significant or not, it is helpful to examine a confidence interval so that you can determine the *magnitude* of the effect. From the width of the confidence interval, you will also learn how much uncertainty there was in the sample results. For instance, from a confidence interval for a proportion, we can learn the "margin of error."

EXAMPLE 1   **IS THE PRESIDENT THAT BAD?**

In the previous chapter, we examined the results of a *Newsweek* poll that asked the question: "From everything you know about Bill Clinton, does he have the honesty and integrity you expect in a president?" (16 May 1994, p. 23). The poll surveyed 518 adults and 233, or 45% of them, said yes. There were 238 no answers (46%), and the rest were not sure. Using a hypothesis test, we determined that the proportion of the population who would answer yes to that question was statistically significantly less than half. From this result, would it be fair to report that "significantly less than 50% of all American adults think Bill Clinton has the honesty and integrity they expect in a president"?

What the Word *Significant* Implies The use of the word *significant* in that context implies that the proportion who feel that way is *much less* than 50%. However, using our computations from Chapter 22 and methods from Chapter 19, let's compute a 95% confidence interval for the true proportion who feel that way:

95% confidence interval = sample value ± 2 (standard deviation)
= 0.45 ± 2(0.022) = 0.45 ± 0.044 = 0.406 to 0.494

Therefore, it could be that the true proportion is as high as 49.4%! Although that value is less than 50%, it is certainly not "significantly less" in the usual, nonstatistical meaning of the word.

In addition, if we were to repeat this exercise for the proportion who answered no to the question, we would reach the opposite conclusion. Of the 518 respondents, 238, or 46%, answered no when asked, "From everything you know about Bill Clinton, does he have the honesty and integrity you expect in a president?" If we construct the test as follows:

Null hypothesis: The population proportion who would answer no is 0.50.

Alternative hypothesis: The population proportion who would answer no is less than 0.50.

the test statistic and p-value would be -1.82 and 0.034, respectively. Therefore, we would also accept the hypothesis that less than a majority would answer no to the question. In other words, we have now found that less than a majority would answer yes and less than a majority would answer no to the question.

The Importance of Learning the Exact Results The problem is that only 91% of the respondents gave a definitive answer. The rest of them had no opinion. Therefore, it would be misleading to focus only on the yes answers or only on the no answers without also reporting the percentage who were undecided. This example illustrates the importance of learning *exactly* what was measured and what results were used in a confidence interval or hypothesis test. ∎

EXAMPLE 2

IS ASPIRIN WORTH THE EFFORT?

In Case Study 1.2, we examined the relationship between taking an aspirin every other day and incidence of heart attack. In testing the null hypothesis (there is no relationship between the two actions) versus the alternative (there is a relationship), the chi-squared (test) statistic is over 25. Recall that a chi-squared statistic over 3.84 would be enough to reject the null hypothesis. In fact, the p-value for the test is less than 0.00001.

The Magnitude of the Effect These results leave little doubt that there is a strongly statistically significant relationship between taking aspirin and incidence of heart attack. However, the test statistic and p-value do not provide information about the *magnitude* of the effect. And remember, the p-value does not indicate the probability that there is a relationship between the two variables. It indicates the probability of observing a sample with such a strong relationship, *if* there is *no* relationship between the two variables in the population. Therefore, it's important to know the *extent* to which aspirin is related to heart attack outcome.

Representing the Size of the Effect The data show that the rates of heart attack were 9.4 per 1000 for the group who took aspirin and 17.1 per 1000 for those who took the placebo. Thus, there is a difference of slightly less than 8 people per 1000, or about one less heart attack for every 125 individuals who took aspirin. Therefore, if all men who are similar to those in the study were to take an aspirin every other day, the results indicate that out of every 125 men, 1 less would have a heart attack than would otherwise have been the case. Another way to represent the size of the effect is to note that the aspirin group had just over half as many heart attacks as the placebo group, indicating that aspirin could cut someone's risk almost in half. The original report in which these results were reported gave the relative risk as 0.53, with a 95% confidence interval extending from 0.42 to 0.67.

Whether that difference is large enough to convince a given individual to start taking that much aspirin is a personal choice. However, being told only the fact that there is very strong statistical evidence for a relationship between taking

TABLE 23.1

		Developed Breast Cancer?		
		Yes	No	Total
First Child Before Age 25?	Yes	65	4475	4540
	No	31	1597	1628
	Total	96	6072	6168

SOURCE: Carter et al., 1989, cited in Pagano and Gauvreau, 1993.

aspirin and incidence of heart attack does not, by itself, provide the information needed to make that decision. ■

23.2   THE ROLE OF SAMPLE SIZE IN STATISTICAL SIGNIFICANCE

If the sample size is large enough, almost any null hypothesis can be rejected. This is because there is almost always a slight relationship between two variables, or a difference between two groups, and if you collect enough data, you will find it.

EXAMPLE 3   **HOW THE SAME RELATIVE RISK CAN PRODUCE DIFFERENT CONCLUSIONS**

Consider an example discussed in Chapter 12: determining the relationship between breast cancer and age at which a woman had her first child. The results are shown in Table 23.1. The chi-squared statistic for the results is 1.746 with p-value of 0.19. Therefore, we would not reject the null hypothesis. In other words, we have not found a statistically significant relationship between age at which a woman had her first child and the subsequent development of breast cancer. This is despite the fact that the relative risk calculated from the data is 1.33.

Now suppose a larger sample size had been used and the same pattern of results had been found. In fact, suppose three times as many women had been sampled, but the relative risk was still found to be 1.33. In that case, the chi-squared statistic would also be increased by a factor of three. Thus, we would find a chi-squared statistic of 5.24 with p-value of 0.02! For the same pattern of results, we would now declare that there *is* a relationship between age at which a woman had her first child and the subsequent development of breast cancer. Yet, the reported relative risk would still be 1.33, just as in the earlier result. ■

TABLE 23.2

	Heart Attack	No Heart Attack	Total	Percent Heart Attacks	Rate per 1000
Aspirin	14	1486	1500	0.93	9.3
Placebo	26	1474	1500	1.73	17.3
Total	40	2960	3000		

23.3 NO DIFFERENCE VERSUS NO STATISTICALLY SIGNIFICANT DIFFERENCE

As we have seen, whether the results of a study are statistically significant can depend on the sample size. In the previous section, we found that the results in Example 3 were not statistically significant. We showed, however, that a larger sample size with the same pattern of results would have yielded a statistically significant finding.

There is a flip side to that problem. If the sample size is too small, an important relationship or difference can go undetected. In that case, we would say that the *power* of the test is too low. Remember that the power of a test is the probability of making the correct decision when the alternative hypothesis is true. But the null hypothesis is the status quo and is assumed to be true unless the sample values deviate from it enough to convince us that chance alone cannot reasonably explain the deviation. If we don't collect enough data (even if the alternative hypothesis is true), we may not have enough evidence to convincingly rule out the null hypothesis. In that case, a relationship that really does exist in a population may go undetected in the sample.

EXAMPLE 4   **ALL THAT ASPIRIN PAID OFF**

The relationship between aspirin and incidence of heart attacks was discovered in a long-term project called the Physician's Health Study (Physicians' Health Study Research Group, 1988). Fortunately, the study included a large number of participants (22,071)—or the relationship may never have been found.

The chi-squared statistic for the study was 25.01, highly statistically significant. But suppose the study had only been based on 3000 participants, still a large number. Further, suppose that approximately the same pattern of results had emerged, with about 9 heart attacks per 1000 in the aspirin condition and about 17 per 1000 in the placebo condition. The result using the smaller sample would be as shown in Table 23.2. For this table, we find the chi-squared statistic is 3.65, with a *p*-value of 0.06. This result is *not* statistically significant using the usual criterion of requiring a *p*-value of 0.05 or less. Thus, we would have to conclude that there is not a statistically significant relationship between taking aspirin and incidence of heart attacks. ■

<hr>

EXAMPLE 5   **IMPORTANT, BUT NOT SIGNIFICANT, DIFFERENCES IN SALARIES**

A number of universities have tried to determine whether male and female faculty members with equivalent seniority earn equivalent salaries. A common method for determining this is to use the salary and seniority data for men to find a regression equation to predict expected salary when seniority is known. The equation is then used to predict what each woman's salary should be, given her seniority; this is then compared with her actual salary. The differences between actual and predicted salaries are next averaged over all women faculty members to see if, on average, they are higher or lower than they would be if the equation based on the men's salaries worked for them.

Tomlinson-Keasey and colleagues (1994) used this method to study salary differences between male and female faculty members at the University of California at Davis. They divided faculty into 11 separate groups, by subject matter, to make comparisons more useful.

In each of the 11 groups, the researchers found that the women's actual pay was lower than what would be predicted from the regression equation, and they concluded that the situation should be investigated further. However, for some of the subject matter groups, the difference found was not statistically significant. For this reason, the researchers' conclusion generated some criticism.

Let's look at how large a difference would have had to exist for the study to be statistically significant. We use the data from the humanities group as an example. There were 92 men and 51 women included in that analysis. The mean difference between men's and women's salaries, after accounting for seniority and years since Ph.D., was $3612.

If we were to assume that the data came from some larger population of faculty members and test the null hypothesis that men and women were paid equally, then the *p*-value for the test would be 0.08. Thus, a statistically naive reader might conclude that no problem exists because the study found no statistically significant difference between average salaries for men and for women, adjusted for seniority.

Because of the natural variability in salaries, even after adjusting for seniority, the sample means would have to differ by over $4000 per year for samples of this size to be able to declare the difference to be *statistically significant*.

The conclusion that there is not a statistically significant difference between men's and women's salaries does not imply that there is not an important difference. It simply means that the natural variability in salaries is so large that a very large difference in means would be required to achieve statistical significance.

As one student suggested, the male faculty who complained about the study's conclusions because the differences were not statistically significant should donate the "nonsignificant" amount of $3612 to help a student pay fees for the following year. ■

CASE STUDY 23.1

Seen a UFO?
You May Be Healthier Than Your Friends

A survey of 5947 adult Americans taken in the summer of 1991 (Roper Organization, 1992) found that 431, or about 7%, reported having seen a UFO. Some authors have suggested that people who make such claims are probably psychologically disturbed and prone to fantasy. Nicholas Spanos and his colleagues (1993) at Carleton University in Ottawa, Canada, decided to test that theory. They recruited 49 volunteers who claimed to have seen or encountered a UFO. They found the volunteers by placing a newspaper ad that read, "Carleton University researcher seeks adults who have seen UFOs. Confidential" (Spanos et al., 1993, p. 625).

Eighteen of these volunteers, who reported seeing only "lights or shapes in the sky that they interpreted as UFOs" (p. 626), were placed in the "UFO nonintense" group. The remaining 31, who reported more complex experiences, were placed in the "UFO intense" group.

For comparison, 74 students and 53 community members were included in the study. The community members were recruited in a manner similar to the UFO groups, except that the ad read "for personality study" in place of "who have seen UFOs." The students received credit in a psychology course for participating.

All subjects were given a series of tests and questionnaires. Attributes measured were "UFO beliefs, Esoteric beliefs, Psychological health, Intelligence, Temporal lobe lability [to see if epileptic type episodes could account for the experiences], Imaginal propensities [such as Fantasy proneness], and Hypnotizability" (p. 626).

The *New York Times* reported the results of this work (Sullivan, 29 November 1993) with the headline "Study finds no abnormality in those reporting UFOs." The article described the results as follows:

> A study of 49 people who have reported encounters with unidentified flying objects, or UFOs, has found no tendency toward abnormality, apart from a previous belief that such visitations from beyond the earth do occur. . . . The tests [given to the participants] included standard psychological tests used to identify subjects with various mental disorders and assess their intelligence. The UFO group proved slightly more intelligent than the others. (p. 37)

Reading the *Times* report would leave the impression that there were no statistically significant differences found in the psychological health of the groups, although perhaps there was a significant difference in intelligence. In fact, that is *not* the case. On many of the psychological measures, the UFO groups scored statistically significantly *better*, meaning they were healthier than the student and community groups. The null hypothesis for this study was that there are no population differences in mean psychological scores for those who have encountered UFOs compared with those who have not. But the alternative hypothesis of interest to Spanos and his colleagues was *one-sided:* the speculation that UFO observers are *less* healthy. The data, indicating that they might be healthier, was not consistent with that alternative hypothesis, so the null hypothesis could not be rejected. Here is how Spanos and colleagues (1993) discussed the results in their report:

> *The most important findings indicate that neither of the UFO groups scored lower on any measures of psychological health than either of the comparison groups. Moreover, both UFO groups attained higher psychological health scores than either one or both of the comparison groups on five of the psychological health variables. In short, these findings provide no support whatsoever for the hypothesis that UFO reporters are psychologically disturbed. (p. 628)*

In case you are curious, the UFO nonintense group scored statistically significantly higher than any of the others on the IQ test, and there were no significant differences in fantasy proneness, other paranormal experiences, or temporal lobe lability.

23.4   A SUMMARY OF WARNINGS

From this discussion, you should realize that you can't simply rely on news reports to determine what to conclude from the results of studies. In particular, you should heed the following warnings:

1. If the word *significant* is used to try to convince you that there is an important effect or relationship, determine if the word is being used in the usual sense or in the statistical sense only.

2. If a study is based on a very large sample size, relationships found to be statistically significant may not have much practical importance.

3. If you read that "no difference" or "no relationship" has been found in a study, try to determine the sample size used. Unless the sample size was large, remember that an important relationship may well exist in the population but that not enough data were collected to detect it. In other words, the test could have had very low power.

4. If possible, learn what confidence interval accompanies the hypothesis test, if any. Even then you can be misled into concluding that there is no effect when there really is, but at least you will have more information about the magnitude of the possible difference or relationship.

5. Try to determine whether the test was one-sided or two-sided. If a test is one-sided, as in Case Study 23.1, and details aren't reported, you could be misled into thinking there is no difference, when in fact there was one in the direction opposite to that hypothesized.

6. Sometimes researchers perform a multitude of tests, and the reports focus on those that achieved statistical significance. Remember that if nothing interesting is happening and all of the null hypotheses tested are true, then 1 in 20 tests should achieve statistical significance just by chance. Beware of reports where

it is evident that many tests were conducted, but where results of only one or two are presented as "significant." ∎

Finding Loneliness on the Internet

It was big news. Researchers at Carnegie Mellon University (Kraut et al., 1998) found that "greater use of the Internet was associated with declines in participants' communication with family members in the household, declines in size of their social circle, and increases in their depression and loneliness" (p. 1017). An article in the *New York Times* reporting on this study was entitled, "Sad, lonely world discovered in cyberspace" (Harmon, 30 August 1998). The study included 169 individuals in 73 households in Pittsburgh, Pennsylvania, who were given free computers and Internet service in 1995. The participants answered a series of questions at the beginning of the study and again either 1 or 2 years later, measuring social contacts, stress, loneliness, and depression. The *New York Times* reported:

> *In the first concentrated study of the social and psychological effects of Internet use at home, researchers at Carnegie Mellon University have found that people who spend even a few hours a week online have higher levels of depression and loneliness than they would if they used the computer network less frequently . . . it raises troubling questions about the nature of "virtual" communication and the disembodied relationships that are often formed in cyberspace. (Harmon, 30 August 1998, p. A3)*

Given these dire reports, one would think that using the Internet for a few hours a week is devastating to your mental health. But a closer look at the findings reveals that the changes were actually quite small, although statistically significant.

Internet use averaged 2.43 hours per week for participants. The statistical analysis used in the study was more complicated than that in this book, but some simple statistics illustrate the magnitude of the results. The number of people in the participants' "local social network" decreased from an average of 23.94 people to an average of 22.90 people, hardly a noticeable loss. On a scale from 1 to 5, self-reported loneliness decreased from an average of 1.99 to 1.89; lower scores indicate greater loneliness. And on a scale from 0 to 3, self-reported depression dropped from an average of 0.73 to an average of 0.62; lower scores indicate higher depression. Further, more measures were taken than those reported, but only those that were statistically significant received attention, a problem discussed in Warning 6. For instance, the study measured social support, defined as "the number of people with whom they can exchange social resources," and reported that "although the association between Internet use and subsequent social support is negative, the effect does not approach statistical significance (p > 0.40)" (Kraut et al., 1998, p. 1023). The news report did not mention that "social support" was measured but not found to be significant.

The *New York Times* did report the magnitude of some of the changes near the end of the article, noting for instance that "one hour a week on the Internet was associated, on average, with an increase of 0.03, or 1 percent on the depression scale."

But the attention the research received masked the fact that the impact of Internet use on depression, loneliness, and social contact was actually quite small. ■

....

EXERCISES

1. An advertisement for Claritin, a drug for seasonal nasal allergies, made this claim: "Clear relief without drowsiness. In studies, the incidence of drowsiness was similar to placebo" (*Time,* 6 February 1995, p. 43). The advertisement also reported that 8% of the 1926 Claritin takers and 6% of the 2545 placebo takers reported drowsiness as a side effect. A one-sided test of whether a higher proportion of Claritin takers than placebo takers would experience drowsiness in the population results in a *p*-value of about 0.005.

    a. From this information, would you conclude that the incidence of drowsiness for the Claritin takers is statistically significantly higher than that for the placebo takers?

    b. Does the answer to part a contradict the statement in the advertisement that the "incidence of drowsiness was similar to placebo"? Explain.

    c. Use this example to discuss the importance of making the distinction between the common use and the statistical use of the word *significant.*

2. Refer to Case Study 23.2, in which a report stated that Internet use was associated with a statistically significant increase in depression.

    a. Would it have been more appropriate to use one-sided or two-sided hypothesis tests for that research? Explain.

    b. Explain what would have constituted a type 1 and type 2 error when testing whether Internet use is associated with greater loneliness. Which type of error could have been committed in this study?

3. An article in *Time* magazine (Gorman, 6 February 1995) reported that an advisory panel recommended that the Food and Drug Administration (FDA) allow an experimental AIDS vaccine to go forward for testing on 5000 volunteers. The vaccine was developed by Jonas Salk, who developed the first effective polio vaccine. The AIDS vaccine was designed to boost the immune system of HIV-infected individuals and had already been tested on a small number of patients, with mixed results but no apparent side effects.

    a. In making its recommendation to the FDA, the advisory panel was faced with a choice similar to that in hypothesis testing. The null hypothesis was that the vaccine was not effective and therefore should not be tested further, whereas the alternative hypothesis was that it might have some benefit. Explain the consequences of a type 1 and a type 2 error for the decision the panel was required to make.

    b. The chairman of the panel, Dr. Stanley Lemon, was quoted as saying, "I'm not all that excited about the data I've seen . . . [but] the only way the concept is going to be laid to rest . . . is really to try [the vaccine] in a large population" (p. 53). Explain why the vaccine should be tested on a larger group, when it had not proven effective in the initial tests on a small group.

4. In Example 4 of Chapter 22, we revisited data from Case Study 6.3, regarding testing to see if there was a relationship between gender and driving after drinking. We found that we could not rule out chance as an explanation for the sample data; the chi-squared statistic was 1.637 and the p-value was 0.20. Now suppose that the sample contained three times as many drivers, but the proportions of males and females who drank before driving were still 16% and 11.6%, respectively.

   a. What would be the value of the chi-squared statistic for this hypothetical larger sample? (*Hint:* See Example 3 in this chapter.)

   b. The p-value for the test based on the larger sample would be about 0.03. Restate the hypotheses being tested and explain which one you would choose on the basis of this larger hypothetical sample.

   c. In both the original test and the hypothetical one based on the larger sample, the probability of making a type 1 error if there really is no relationship is 5%. Assuming there is a relationship in the population, would the power of the test—that is, the probability of correctly finding the relationship—be higher, lower, or the same for the sample three times larger compared with the original sample size?

5. *New Scientist* (Mestel, 12 November 1994) reported a study in which psychiatrist Donald Black used the drug fluvoxamine to treat compulsive shoppers:

   In Black's study, patients take the drug for eight weeks, and the effect on their shopping urges is monitored. Then the patients are taken off the drug and watched for another month. In the seven patients examined so far, the results are clear and dramatic, says Black: the urge to shop and the time spent shopping decrease markedly. When the patient stops taking the drug, however, the symptoms slowly return. (p. 7)

   a. Explain why it would have been preferable to have a double-blind study, in which shoppers were randomly assigned to take fluvoxamine or a placebo.

   b. What are the null and alternative hypotheses for this research?

   c. Can you make a conclusion about the hypotheses in part b on the basis of this report? Explain.

6. In reporting the results of a study to compare two population means, explain why researchers should report each of the following:

   a. A confidence interval for the difference in the means

   b. A p-value for the results of the test, as well as whether it was one- or two-sided

   c. The sample sizes used

   d. The number of separate tests they conducted during the course of the research

7. Explain why it is important to learn what sample size was used in a study for which "no difference" was found.

8. The top story in *USA Today* on December 2, 1993, reported that "two research teams, one at Harvard and one in Germany, found that the risk of a heart attack during heavy physical exertion . . . was two to six times greater than during less strenuous activity or inactivity" (Snider, 2 December 1993, p. 1A). It was implicit, but not stated, that these results were for those who do not exercise regularly.

   a. There is a confidence interval reported in this statement. Explain what it is and what characteristic of the population it is measuring.

   b. The report also stated that "frequent exercisers had slight or no increased risk." (In other words, frequent exercisers had slight or no increased risk of heart attack when they took on a strenuous activity compared with when they didn't.) Do you think that means the relative risk was actually 1.0? If not, discuss what the statement really does mean in the context of what you have learned in this chapter.

9. We have learned that the probability of making a type 1 error when the null hypothesis is true is usually set at 5%. The probability of making a type 2 error when the alternative hypothesis is true is harder to find. Do you think that probability depends on the size of the sample? Explain your answer.

10. Refer to Case Study 22.1, concerning the ganzfeld procedure for testing ESP. In earlier studies using the ganzfeld procedure, the results were mixed in terms of whether they were statistically significant. In other words, some of the experiments were statistically significant and others were not. Critics used this as evidence that there was really nothing going on, even though more than 5% of the experiments were successful. Typical sample sizes were from 10 to 100 participants. Give an explanation for why the results would have been mixed.

11. When the Steering Committee of the Physicians' Health Study Research Group (1988) reported the results of the effect of aspirin on heart attacks, committee members also reported the results of the same aspirin consumption, for the same sample, on strokes. There were 80 strokes in the aspirin group and only 70 in the placebo group. The relative risk was 1.15, with a 95% confidence interval ranging from 0.84 to 1.58.

   a. What value for relative risk would indicate that there is no relationship between taking aspirin and having a stroke? Is that value contained in the confidence interval?

   b. Set up the appropriate null and alternative hypotheses for this part of the study. The original report gave a *p*-value of 0.41 for this test. What conclusion would be made on the basis of that value?

   c. Compare your results from parts a and b. Explain how they are related.

   d. There was actually a higher percentage of strokes in the group that took aspirin than in the group that took a placebo. Why do you think this result did not get much press coverage, whereas the result indicating that aspirin reduces the risk of heart attacks did get substantial coverage?

12. The authors of the report in Case Study 23.1, comparing the psychological health of UFO observers and nonobservers, presented a table in which they compared the four groups of volunteers on each of 20 psychological measures. For each of the measures, the null hypothesis was that there were no differences in population means for the various types of people on that measure. If there truly were no population differences, and if all of the measures were independent of each other, for how many of the 20 measures would you expect the null hypothesis to be rejected, using the 0.05 criterion for rejection? Explain how you got your answer.

13. In the study by Lee Salk, reported in Case Study 1.1, he found that infants who listened to the sound of a heartbeat in the first few days of life gained more weight than those who did not. In searching for potential explanations, Salk wrote the following. Discuss Salk's conclusion and the evidence he provided for it. Do you think he effectively ruled out food intake as being important?

    In terms of actual weight gain the heartbeat group showed a median gain of 40 grams; the control group showed a median loss of 20 grams. There was no significant difference in food intake between the two groups. . . . There was crying 38 percent of the time in the heartbeat group of infants; in the control group one infant or more cried 60 percent of the time. . . . Since there was no difference in food intake between the two groups, it is likely that the weight gain for the heartbeat group was due to a decrease in crying. (Salk, May 1973, p. 29)

14. An advertisement for NordicTrack, an exercise machine that simulates cross-country skiing, claimed that "in just 12 weeks, research shows that *people who used a NordicTrack lost an average of 18 pounds."* Forgetting about the questions surrounding how such a study might have been conducted, what additional information would you want to know about the results before you could come to a reasonable conclusion?

15. Explain why it is not wise to *accept* a null hypothesis.

16. Now that you understand the reasoning behind making inferences about populations based on samples (confidence intervals and hypothesis tests), explain why these methods require the use of random, or at least representative, samples instead of convenience samples.

17. Would it be easier to reject hypotheses about populations that had a lot of natural variability in the measurements or a little variability in the measurements? Explain.

MINI-PROJECTS

1. Find a newspaper article that you think presents the results of a hypothesis test in a misleading way. Explain why you think it is misleading. Rewrite the appropriate part of the article in a way that you consider to be not misleading. If necessary, find the original journal report of the study so you can determine what details are missing from the newspaper account.

2. Find two journal articles, one that reports on a statistically significant result and one that reports on a nonsignificant result. Discuss the role of the sample size in

the determination of statistical significance, or lack thereof, in each case. Discuss whether you think the researchers would have reached the same conclusion if they had used a smaller or larger sample size.

REFERENCES

Carter, C. L., D. Y. Jones, A. Schatzkin, and L. A. Brinton. (1989). A prospective study of reproductive, familial, and socioeconomic risk factors for breast cancer using NHANES I data, *Public Health Reports* 104, January–February, pp. 45–49.

Gorman, Christine. (6 February 1995). Salk vaccine for AIDS. *Time,* p. 53.

Harmon, Amy. (30 August 1998). Sad, lonely world discovered in cyberspace. *New York Times,* p. A3.

Kraut, R., V. Lundmark, M. Patterson, S. Kiesler, T. Mukopadhyay, and W. Scherlis. (1998). Internet paradox: A social technology that reduces social involvement and psychological well-being? *American Psychologist* 53, no. 9, 1017–1031.

Mestel, Rosie. (12 November 1994). Drug brings relief to big spenders. *New Scientist,* no. 1951, p. 7.

Pagano, M. and K. Gauvreau. (1993). *Principles of biostatistics.* Belmont, CA: Duxbury Press.

Physicians' Health Study Research Group. Steering Committee. (28 January 1988). Preliminary report: Findings from the aspirin component of the ongoing Physicians' Health Study. *New England Journal of Medicine* 318, no. 4, pp. 262–264.

Roper Organization. (1992). *Unusual personal experiences: An analysis of the data from three national surveys.* Las Vegas: Bigelow Holding Corp.

Salk, Lee. (May 1973). The role of the heartbeat in the relations between mother and infant. *Scientific American,* pp. 26–29.

Snider, M. (2 December 1993). "Weekend warriors" at higher heart risk. *USA Today,* p. 1A.

Spanos, N. P., P. A. Cross, K. Dickson, and S. C. DuBreuil. (1993). Close encounters: An examination of UFO experiences. *Journal of Abnormal Psychology* 102, no. 4, pp. 624–632.

Sullivan, Walter. (29 November 1993). Study finds no abnormality in those reporting U.F.O.s. *New York Times,* p. B7.

Tomlinson-Keasey, C., J. Utts, and J. Strand. (1994). Salary equity study at UC Davis, 1994. Technical Report, Office of the Provost, University of California at Davis.

CHAPTER TWENTY-FOUR

Meta-Analysis: Resolving Inconsistencies across Studies

THOUGHT QUESTIONS

1. Suppose a new study involving 14,000 participants has found a relationship between a particular food and a certain type of cancer. The report on the study notes that "past studies of this relationship have been inconsistent. Some of them have found a relationship, but others have not." What might be the explanation for the inconsistent results from the different studies?

2. Suppose ten similar studies, all on the same kind of population, have been conducted to determine the relative risk of heart attack for those who take aspirin and those who don't. To get an overall picture of the relative risk we could compute a separate confidence interval for each study or combine all the data to create one confidence interval. Which method do you think would be preferable? Explain.

3. Suppose two studies have been done to compare surgery versus relaxation for sufferers of chronic back pain. One study was done at a back specialty clinic and the other at a suburban medical center. The result of interest in each case was the relative risk of further back problems following surgery versus following relaxation training. To get an overall picture of the relative risk, we could compute a separate confidence interval for each study or combine the data to create one confidence interval. Which method do you think would be preferable? Explain.

4. Refer to Thought Questions 2 and 3. If two or more studies have been done to measure the same relative risk, give one reason why it would be better to combine the results and one reason why it would be better to look at the results separately for each study.

24.1   THE NEED FOR META-ANALYSIS

The obvious questions were answered long ago. Most research today tends to focus on relationships or differences that are not readily apparent. For example, plants that are obviously poisonous have been absent from our diets for centuries. But we may still consume plants (or animals) that contribute to cancer or other diseases over the long term. The only way to discover these kinds of relationships is through statistical studies.

Because most of the relationships we now study are moderate in size, researchers often fail to find a statistically significant result. As we learned in the last chapter, the number of participants in a study is a crucial factor in determining whether the study finds a "significant" relationship or difference, and many studies are simply too small to do so. As a consequence, reports are often published that appear to conflict with earlier results, confusing the public and researchers alike.

The Vote-Counting Method

One way to address this problem is to gather all the studies that have been done on a topic and try to assimilate the results. Until recently, if researchers wanted to examine the accumulated body of evidence for a particular relationship, they would find all studies conducted on the topic and simply count how many of these had found a statistically significant result. They would often discount entirely all studies that had not, and then attempt to explain any remaining differences in study results by subjective assessments.

As you should realize from Chapter 23, this **vote-counting method** is seriously flawed unless the number of participants in each study is taken into account. For example, if ten studies of the effect of aspirin on heart attacks had each been conducted on 2200 men, rather than one study on 22,000 men, none of the ten studies would be likely to show a relationship. In contrast, as we saw in Chapter 12, the one study on 22,000 men showed a very significant relationship. In other words, the same data, had it been collected by ten separate researchers, would have resulted in exactly the opposite conclusion to what was found in the one large study.

What Is Meta-Analysis?

Since the early 1980s researchers have become increasingly aware of the problems with traditional research synthesis methods. They are now more likely to conduct a *meta-analysis* of the studies. **Meta-analysis** is a collection of statistical techniques for combining studies. These techniques focus on the *magnitude of the effect* in each study, rather than on either a vote-count or a subjective evaluation of the available evidence.

Quantitative methods for combining results have been available since the early 1900s, but it wasn't until 1976 that the name "meta-analysis" was coined. The seminal paper was called, "Primary, secondary and meta-analysis of research." In that paper, Gene Glass (1976) showed how these methods could be used to synthesize the results of studies comparing treatments in psychotherapy. It was an idea whose time had come. Researchers and methodologists set about to create new techniques for combining and comparing studies, and thousands of meta-analyses were undertaken.

What Meta-Analysis Can Do

Meta-analysis has made it possible to find definitive answers to questions about small and moderate relationships by taking into account data from a number of different studies. It is not without its critics, however, and as with most statistical methods, difficulties and disasters can befall the naive user. In the remainder of this chapter, we examine some decisions that can affect the results of a meta-analysis, discuss some benefits and criticisms of meta-analysis, and look at two case studies.

24.2

TWO IMPORTANT DECISIONS FOR THE ANALYST

You have already learned how to do a proper survey, experiment, or observational study. Conducting a meta-analysis is an entirely different enterprise because it does not involve using new participants at all. It is basically a study of the studies that have already been done on a particular topic.

A variety of decisions must be made when conducting a meta-analysis, many of which are beyond the scope of our discussion. However, there are two important decisions of which you should be informed when you read the results of a meta-analysis. The answers will help you determine whether to believe the results.

When reading the results of meta-analysis, you should know

1. Which studies were included

2. Whether the results were compared or combined

Question 1: Which Studies Should Be Included?

Types of Studies

As you learned early in this book, studies are sometimes conducted by special interest groups. A study may be also conducted by a student to satisfy requirements for a Ph.D. degree but then never published in a scholarly journal. Sometimes studies are reported at conferences and published in the proceedings. These studies are not carefully reviewed and criticized in the same way they would be for publication in a scholarly journal. Thus, one decision that must be made before conducting a meta-analysis is whether to include all studies on a particular topic or only those that have been published in properly reviewed journals.

Timing of the Studies

Another consideration is the timing of the studies. For example, in Case Study 24.2, we review a recent meta-analysis that attempted to answer the question of whether mammograms should be used for the detection of breast cancer in women in their 40s. Should the analysis have included studies started many years ago, when neither the equipment nor the technicians may have been as sophisticated as they are today?

Quality Control

Sometimes studies are investigated for quality before they are included. For instance, Eisenberg and colleagues (15 June 1993) conducted a meta-analysis on whether behavioral techniques such as biofeedback and relaxation were effective in lowering blood pressure in people whose blood pressure was too high. They identified a total of 857 articles for possible inclusion but used only 26 of them in the meta-analysis. In order to be included, the studies had to meet stringent criteria, including the use of one or more control conditions, random assignment into the experimental and control groups, detailed descriptions of the intervention techniques for treatment and control, and so on. They also excluded studies that used children or that used only pregnant women.

Thus, they decided to exclude studies that had some of the problems you learned about earlier in this book, such as nonrandomized treatments or no control group. Using the remaining 26 studies, they found that both the behavioral techniques and the placebo techniques (such as "sham meditation") produced effects, but that there was not a significant difference between the two. Their conclusion was that "cognitive interventions for essential hypertension are superior to no therapy but not superior to credible sham techniques or self-monitoring" (p. 964). If they had included a multitude of studies that did not use placebo techniques, they may not have realized that those techniques could be as effective as the real thing.

Accounting for Differences in Quality

Some researchers believe all possible studies should be included in the meta-analysis. Otherwise, they say, the analyst could be accused of including studies that support the desired outcome and finding excuses to exclude those that don't. One solution to this problem is to include all studies but account for differences in quality in the process of the analysis. A meta-analysis of programs designed to prevent adolescents from smoking used a compromise approach:

> Evaluations of 94 separate interventions were included in the meta-analysis. Studies were screened for methodological rigor and those with weaker methodology were segregated from those with more defensible methodology; major analyses focused on the latter. (Bruvold, 1993, p. 872)

Assessing Quality

Chalmers and his co-authors (1981) constructed a generic set of criteria for deciding which papers to include in a meta-analysis and for assessing the quality of studies.

Their basic idea is to have two researchers, who are blind as to who conducted the studies and what the results were, independently make decisions about the quality of each study. In other words, there should be two independent assessments about whether to include each study, and those assessments should be made without knowing how the study came out. This technique should eliminate inclusion biases based on whether the results were in a desired direction.

When you read the results of a meta-analysis, you should be told how the authors decided which studies to include. If there was no attempt to discount flawed studies in some way, then you should realize that the combined results may also be flawed.

Question 2: Should Results Be Compared or Combined?

In Chapters 11 and 12, you learned about the perils of combining data from separate studies. Recall that Simpson's Paradox can occur when you combine the results of two separate contingency tables into one. The problem occurs when a relationship is different for one population than it is for another, but the results for samples from the two are combined.

Meta-Analysis and Simpson's Paradox

Meta-analysis is particularly prone to the problem of Simpson's Paradox. For example, consider two studies comparing surgery to relaxation as treatments for chronic back pain. Suppose one is conducted at a back-care specialty clinic, whereas the other is conducted at a suburban medical center. It is likely that the patients with the most severe problems will seek out the specialty clinic. Therefore, relaxation training may be sufficient and preferable for the suburban medical center, but surgery may be preferable for the specialty clinic.

Populations Must Be the Same and Methods Similar

A meta-analysis should never attempt to combine the data from all studies into one hypothesis test or confidence interval unless it is clear that the same populations were sampled and similar methods were used. Otherwise, the analysis should first attempt to see if there are readily explainable differences among the results.

For example, in Case Study 24.1, we will investigate a meta-analysis of the effects of cigarette smoking on sperm density. Some of the studies included used infertility clinics as the source of participants; others used the general population. The researchers discovered that the relationship between smoking and sperm density was more variable across the studies for which the participants were from infertility clinics. They speculated that "it is possible that smoking has a greater effect on normal men since infertility clinic populations may have other reasons for the lowered sperm density" (Vine et al., January 1994, p. 41).

Because of the difference, they reported results separately for the two sources of participants. If they had not, the relationship between smoking and sperm density would have been underestimated for the general population.

When you read the results of a meta-analysis, you need to ascertain whether something like Simpson's Paradox could have been a problem. If dozens of studies have been combined into one result, without any mention of whether the potential for this problem was investigated, you should be suspicious.

Smoking and Reduced Fertility

Vine and colleagues (January 1994) identified 20 studies, published between January 1966 and October 1, 1992, that examined whether men who smoked cigarettes had lower sperm density than men who did not smoke. They found that most of the studies reported reduced sperm density for men who smoked, but the difference was statistically significant for only a few of the studies. Thus, in the traditional scientific review, the accumulated results would seem to be inconsistent and inconclusive.

Studies were excluded from the meta-analysis for only two reasons. One was if the study was a subset of a larger study that was already included. The other was if it was obvious that the smokers and nonsmokers differed with respect to fertility status. One study was excluded for that reason. The nonsmokers all had children, whereas the smokers were attending a fertility clinic.

A variety of factors differed among the studies. For example, only ten of the studies reported that the person measuring the sperm density was blind to the smoking status of the participants. In six of the studies, a "smoker" was defined as someone who smoked at least ten cigarettes per day. Thirteen of the studies used infertility clinics as a source of participants, seven did not. All of these factors were checked to see if they resulted in a difference in outcome.

None of these factors resulted in a difference. However, the authors were suspicious of 2 studies conducted on infertility clinic patients because those 2 studies showed a much larger relationship between smoking and sperm density than the remaining 11 studies on that same population. When those two studies were omitted, the remaining studies using infertility clinic patients showed a smaller average effect than the studies using normal men. The authors conducted an analysis using all of the data, as well as separate analyses on the two types of populations: those attending infertility clinics and those not attending. They omitted the two studies with larger effects when studying only the infertility clinics.

The authors found that there was indeed lower average sperm density for men who smoked. Using all of the data combined, but giving more weight to studies with larger sample sizes, they found that the reduction in sperm density for smokers compared with nonsmokers was 12.6%, with a 95% confidence interval extending from 8.0% to 17.1%. A test of the null hypothesis that the reduction in sperm density for smokers in the population is actually zero resulted in a p-value less than 0.0001. The estimate of reduction in sperm density for the normal (not infertile) men only was even higher, at 23.3%; a confidence interval was not provided.

The authors also conducted their own study of this question using 88 volunteers recruited through the newspaper. In summarizing the findings of past reviews, their meta-analysis, and their own study of 88 men, Vine and colleagues (January 1994) illustrate the importance of meta-analysis:

The results of this meta-analysis indicate that smokers' sperm density is on average 13% [when studies are weighted by sample size] to 17% [when studies are given equal weight] lower than that of nonsmokers. . . . The reason for the inconsistencies in published findings with regard to the association between smoking and sperm density appears to be the result of random error and small sample sizes in most studies. Consequently, the power is low and the chance of a false negative finding is high. Results of the authors' study support these findings. The authors noted a 22.8% lower sperm density among smokers compared with nonsmokers. However, because of the inherent variability in sperm density among individuals, the study sample size (n = 88) was insufficient to produce statistically significant results. (p. 40)

As with individual studies, results from a meta-analysis like this one should not be interpreted to imply a causal relationship. As the authors note, there are potential confounding factors. They mention studies that have shown smokers are more likely to consume drugs, alcohol, and caffeine and are more likely to experience sex early and to be divorced. Those are all factors that could contribute to reductions in sperm density, according to the authors. ■

24.3   SOME BENEFITS OF META-ANALYSIS

We have just seen one of the major benefits of meta-analysis. When a relationship is small or moderate, it is difficult to detect with small studies. A meta-analysis allows researchers to rule out chance for the combined results, when they could not do so for the individual studies. It also allows researchers to get a much more precise estimate of a relationship—that is, a narrower confidence interval—than they could get with individual small studies.

> *Some benefits of meta-analysis are*
> 1. Detecting small or moderate relationships
> 2. Obtaining a more precise estimate of a relationship
> 3. Determining future research
> 4. Finding patterns across studies

Determining Future Research

Summarizing what has been learned so far in a research area can lead to insights about what to ask next. In addition, identifying and enumerating flaws from past studies can help illustrate how to better conduct future studies.

EXAMPLE 1   **DESIGNING BETTER EXPERIMENTS**

In Case Study 21.1, we examined the ganzfeld experiments used to study extra-sensory perception. The experiments in that case study were conducted using an automated procedure that was developed in response to an earlier meta-analysis. The earlier meta-analysis had found highly significant differences when compared with chance (Honorton, 1985). However, a critic who was involved with the analysis insisted that the results were, in fact, due to flaws in the procedure (Hyman, 1985). He identified flaws such as improper randomization of target pictures and potentially sloppy data recording. He also identified more subtle flaws. For example, some of the early studies used photographs that a sender was supposed to transmit mentally to a receiver. If the sender was holding a photograph of the true target and the receiver was later asked to pick which of four pictures had been the true target, the receiver could have seen the sender's fingerprints, and these—and not some psychic information—could have provided the answer. The new experiments, reported in Case Study 21.1, were designed to be free of all the flaws that had been identified in the first meta-analysis. ■

EXAMPLE 2   **CHANGES IN FOCUS OF THE RESEARCH**

Another example is provided by a meta-analysis on the impact of sexual abuse on children (Kendall-Tackett et al., 1993). In this case, the authors did not suggest corrections of flaws but rather changes in focus for future research. For example, they noted that many studies of the effects of sexual abuse combine children across all age groups into one search for symptoms. In their analysis across studies, they were able to look at different age groups to see if the consequences of abuse differed. They found:

> For preschoolers, the most common symptoms were anxiety, nightmares, general PTSD [post-traumatic stress disorder], internalizing, externalizing, and inappropriate sexual behavior. For school-age children, the most common symptoms included fear, neurotic and general mental illness, aggression, nightmares, school problems, hyperactivity, and regressive behavior. For adolescents, the most common behaviors were depression; withdrawn, suicidal, or self-injurious behaviors; somatic complaints; illegal acts; running away; and substance abuse. (p. 167)

In essence, the authors are warning researchers that they could encounter Simpson's Paradox if they do not recognize the differences among various age groups. They provide advice for anyone conducting future studies in this domain:

> Many researchers have studied subjects from very broad age ranges (e.g., 3–18 years) and grouped boys and girls together . . . this grouping together of all ages can mask particular developmental patterns of the occurrence of some symptoms. At a minimum, future researchers should divide children into

preschool, school, and adolescent age ranges when reporting the percentages of victims with symptoms. (p. 176) ∎

Finding Patterns across Studies

Sometimes patterns that are not apparent in single studies become apparent in a meta-analysis. This could be due to small sample sizes in the original studies or to the fact that each of the original studies considered only one part of a question.

In Case Study 24.1, in which smokers were found to have lower sperm density, the authors were able to investigate whether the decrease was more pronounced for heavier smokers. In the 12 studies for which relevant information was provided, 8 showed evidence that the magnitude of decrease in sperm density was related to increasing numbers of cigarettes smoked. None of these studies alone provided sufficient evidence to detect this pattern.

EXAMPLE 3

GROUPING STUDIES ACCORDING TO ORIENTATION

Bruvold (1993) compared adolescent smoking prevention programs. He characterized programs as having one of four orientations. The "rational" orientation focused on lectures and displays of substances; the "developmental" orientation used lectures with discussion, problem solving, and some role playing; the "social norms" orientation used participation in community and recreational activities; and the "social reinforcement" orientation used discussion, role playing, and public commitment not to smoke.

Individual studies are likely to focus on one orientation only. But a meta-analysis can group studies according to which orientation they used and do a comparison. Bruvold (1993) found that

the rational orientation had very little impact on behavior, that the social norms and developmental orientations had approximately the same intermediate impact on behavior, and that the social reinforcement orientation had the greatest impact on behavior. (pp. 877–878)

Of course, caution must be applied when comparing studies on one feature only. As with most research, confounding factors are a possibility. For example, if the programs using the social reinforcement orientation had also been the more recent studies, that would have confounded the results because there has been much more societal pressure against smoking in recent years. ∎

24.4   CRITICISMS OF META-ANALYSIS

Meta-analysis is a relatively new endeavor and not all of the surrounding issues have been resolved. Some valid criticisms are still being debated by methodologists. You have already encountered one of them—namely, that Simpson's Paradox or its equivalent could apply.

Some possible problems with meta-analysis are

1. Simpson's Paradox
2. Confounding variables
3. Subtle differences in treatments of the same name
4. The file drawer problem
5. Biased or flawed original studies
6. Statistical significance versus practical importance
7. False findings of "no difference"

The Possibility of Confounding Variables

Because meta-analysis is essentially observational in nature, the various treatments cannot be randomly assigned across studies. Therefore, there may be differences across studies that are confounded with the treatments used. For example, it is common for the studies considered in meta-analysis to have been carried out in a variety of countries. It could be that cultural differences are confounded with treatment differences. Thus, a meta-analysis of studies relating dietary fat to breast cancer may find a strong link, but it may be because the same countries that have high-fat diets are also unhealthful in other ways.

Subtle Differences in Treatments of the Same Name

Another problem of meta-analysis is that different studies may involve subtle differences in treatments but call the differences by the same name. For instance, chemotherapy may be applied weekly in one study but biweekly in another. The higher frequency may be too toxic, whereas the lower frequency is beneficial. When researchers are combining hundreds of studies, not uncommon in meta-analysis, they may not take the time to discover these subtle differences, which may result in major differences in the outcomes under study.

The File Drawer Problem

Related to the question of which studies to include in a meta-analysis is the possibility that numerous studies may not be discovered by the meta-analyst. These are likely to be studies that did not achieve statistical significance and thus were never published. This is called the **file drawer problem** because the assumption is that these studies are filed away somewhere but not publicly accessible. If the statistically significant studies of a relationship are the ones that are more likely to be available, then the meta-analysis may overestimate the size of the relationship. This is akin to selecting only those with strong opinions in sample surveys.

One way researchers deal with the file drawer problem is to contact all persons known to be working on a particular topic and ask them if they have done studies that were never published. In smaller fields of research, this is probably an effective method of retrieving studies. Another way to deal with the problem, suggested by psychologist Robert Rosenthal (1991, p. 104), is to estimate how many such studies would be needed to reduce the relationship to nonsignificance. If the answer is an absurdly large number, then the relationship is probably a real one.

Biased or Flawed Original Studies

If the original studies were flawed or biased, so is the meta-analysis. In *Tainted Truth*, Cynthia Crossen (1994, pp. 43–53) discusses the controversy surrounding whether oat bran lowers cholesterol. She notes that the final word on the controversy came in the form of a meta-analysis published in June 1992 (Ripsin et al., 1992) that was funded by Quaker Oats. The study concluded that "this analysis supports the hypothesis that incorporating oat products into the diet causes a modest reduction in blood cholesterol level" (Ripsin et al., 1992, p. 3317).

Crossen raises questions about the studies that were included in the meta-analysis. First, she comments that "of the entire published literature of scientific research on oat bran in the U.S., the lion's share has been as least partly financed by Quaker Oats" (Crossen, 1994, p. 52). In response to Quaker Oats's defense that no strings were attached to the money and that scientists are not going to risk their reputations for a small research grant, Crossen replies:

> *In most cases, that is true. Nothing is more damaging to scientists' reputations —and their economic survival—than suspicions of fraud, corruption or dishonesty. But scientists are only human, and in the course of any research project they make choices. Who were the subjects of the study? Young or old, male or female, high cholesterol or low? How were they chosen? How long did the study go on? . . . Since cholesterol levels vary season to season, when did the study begin and when did it end? (p. 49)*

Statistical Significance versus Practical Importance

We have already learned that a statistically significant result is not necessarily of any practical importance. Meta-analysis is particularly likely to find statistically significant results because typically the combined studies provide very large sample sizes. Thus, it is important to learn the magnitude of an effect found in a meta-analysis and not just that it is statistically significant.

For example, the oat bran meta-analysis did indeed show a statistically significant reduction in cholesterol, a conclusion that led to headlines across the country. However, the magnitude of the reduction was quite small. A 95% confidence interval extended from 3.3 mg/dL to 8.4 mg/dL. The average American has a cholesterol level of about 210 mg/dL.

False Findings of "No Difference"

It is also possible that a meta-analysis will erroneously reach the conclusion that there is no difference or no relationship—when in fact there was simply not enough data to find one that was statistically significant. Because a meta-analysis is supposed to be so comprehensive, there is even greater danger than with single studies that the conclusion will be taken to mean that there really is no difference. We will see an example of this problem in Case Study 24.2.

Most meta-analyses include sufficient data so that important differences will be detected, and that can lead to a false sense of security. The main danger is when not many studies have been done on a particular subset or question. It may be true that there are hundreds of thousands of participants across all studies, but if only a small fraction of them are in a certain age group, received a certain treatment, and so on, then a statistically significant difference may still not be found for that subset. When you read about a meta-analysis, be careful not to confuse the overall sample size with the one used for any particular subgroup.

CASE STUDY 24.2 ## Controversy over Mammograms

In the fall of 1993, a debate raged between the National Cancer Institute and the American Cancer Society. Both organizations had been recommending that at age 40 women should begin to have mammograms, or breast X rays, as a breast cancer screening device.

In February 1993, the National Cancer Institute convened an international conference to bring together experts from around the world to help conduct a meta-analysis of studies on the effectiveness of mammography as a screening device. Their conclusion about women aged 40–49 years was: "For this age group it is clear that in the first 5–7 years after study entry, there is no reduction in mortality from breast cancer that can be attributed to screening" (Fletcher et al., 20 October 1993, p. 1653).

The results of the meta-analysis amounted to a withdrawal of the National Cancer Institute's support for mammograms for women under 50. The American Cancer Society refuted the study and announced that it would not change its recommendation. As noted in a front page story in the *San Jose Mercury News* on November 24, 1993:

> *A spokeswoman for the American Cancer Society's national office said Tuesday that the . . . study would not change the group's recommendation because it was not big enough to draw definite conclusions. The study would have to screen 1 million women to get a certain answer because breast cancer is so uncommon in young women.*

In fact, the entire meta-analysis considered eight randomized experiments conducted over 30 years, including nearly 500,000 women. That sounds impressive. But not all of the studies included women under 50. As noted by Sickles and Kopans (20 October 1993):

> *Even pooling the data from all eight randomized controlled trials produces insufficient statistical power to indicate presence or absence of benefit from screening. In the eight trials, there were only 167,000 women (30% of the participants) aged 40–49, a number too small to provide a statistically significant result. (p. 1622)*

There were other complaints about the studies as well. Even the participants in the conference recognized that there were methodological flaws in some of the studies. About one of the studies on women aged 40–49, they commented:

> *It is worrisome that more patients in the screening group had advanced tumors, and this fact may be responsible for the results reported to date. . . . The technical quality of mammography early in this trial is of concern, but it is not clear to what extent mammography quality affected the study outcome. (Fletcher et al., 20 October 1993, p. 1653)*

There is thus sufficient concern about the studies themselves to make the results inconclusive. But they were obviously statistically inconclusive as well. Here is the full set of results for women aged 40–49:

> *A meta-analysis of six trials found a relative risk of 1.08 (95% confidence interval = 0.85 to 1.39) after 7 years' follow-up. After 10–12 years of follow-up, none of four trials have found a statistically significant benefit in mortality; a combined analysis of Swedish studies showed a statistically insignificant 13% decrease in mortality at 12 years. One trial (Health Insurance Plan) has data beyond 12 years of follow-up, and results show a 25% decrease in mortality at 10–18 years. Statistical significance of this result is disputed, however. (Fletcher et al., 20 October 1993, p. 1644)*

A debate on the *MacNeil-Lehrer News Hour* between American Cancer Society spokesman Dr. Harmon Eyre and one of the authors of the meta-analysis, Dr. Barbara Rimer, illustrated the difficulty of interpreting results like these for the public. Dr. Eyre argued that "30% of those who will die could be saved if you look at the data as we do; the 95% confidence limits in the meta-analysis could include a 30% decrease in those who will die."

Dr. Rimer, on the other hand, simply kept reiterating that mammography for women under age 50 simply "hasn't been shown to save lives." In response, Dr. Eyre accused the National Cancer Institute of having a political agenda. If women under 50 do not have regular mammograms, it could save a national health-care program large amounts of money.

By now, you should recognize the real nature of the debate here. The results are inconclusive because of small sample sizes and possible methodological flaws. The question is not a statistical one—it is a public policy one. Given that we do not know for certain that mammograms save lives for women under 50, should health insurers be required to pay for them? ■

EXERCISES

1. Explain why the vote-counting method is not a good way to summarize the results in a research area.

2. Explain why the person deciding which studies to include in a meta-analysis should be told how each study was conducted but not the results of the study.

3. According to a report in *New Scientist* (Vine, 21 January 1995), the UK Cochrane Centre in Oxford is launching the *Cochrane Database of Systematic Reviews,* which will "focus on a number of diseases, drawing conclusions about which treatments work and which do not from all the available randomized controlled trials" (p. 14). The *Reviews* will be made available electronically to practicing physicians and will contain meta-analyses of various areas of medical research.

   a. Why do you think they are only planning to include "randomized controlled trials" (that is, randomized experiments) and not observational studies?

   b. Do you think they should include studies that did *not* find statistically significant results when they do their meta-analyses? Why or why not?

4. Refer to Exercise 3. Pick one of the benefits of meta-analysis listed in Section 24.3 and explain how that benefit applies to the reviews in the *Cochrane Database.*

5. Refer to Exercise 3. Pick three of the possible problems with meta-analysis listed in Section 24.4 and discuss how they might apply to the reviews in the *Cochrane Database.*

6. An article in the *Sacramento Bee* (15 April 1998, pp. A1, A12), titled "Drug reactions a top killer, research finds," reported a study estimating that between 76,000 and 137,000 deaths a year in the United States occur due to adverse reactions to medications. Here are two quotes from the article:

   The scientists reached their conclusion not in one large study but by combining the results of 39 smaller studies. This technique, called meta-analysis, can enable researchers to draw statistically significant conclusions from studies that individually are too small. (p. A12)

   [A critic of the research] said the estimates in Pomeranz's study might be high, because the figures came from large teaching hospitals with the sickest patients, where more drug use and higher rates of drug reactions would be expected than in smaller hospitals. (p. A12)

   a. Pick one of the benefits in Section 24.3 and one of the criticisms in Section 24.4 and apply them to this study.

   b. The study estimated that between 76,000 and 137,000 deaths a year occur due to adverse reactions to medications, with a mean estimate of 100,000. If this is true, then adverse drug reaction is the fourth-leading cause of death in the United States. Based on these results, which one of the criticisms given in Section 24.4 is definitely not a problem? Explain.

7. In a meta-analysis, researchers can choose either to combine results across studies to produce a single confidence interval or to report separate confidence intervals for each study and compare them. Give one advantage and one disadvantage of combining the results into one confidence interval.

8. Would the file drawer problem be more likely to present a substantial difficulty in a research field with 50 researchers or in one with 1000 researchers? Explain.

9. Give two reasons why researchers might not want to include all possible studies in a meta-analysis of a certain topic.

10. Suppose a meta-analysis on the risks of drinking coffee included 100,000 participants of all ages across 80 studies. One of the conclusions was that a confidence interval for the relative risk of heart attack for women over 70 who drank coffee, compared with those who didn't, was 0.90 to 1.30. The researchers concluded that coffee does not present a problem for women over 70. Explain why their conclusion is not justified.

11. Eisenberg and colleagues (15 June 1993) used meta-analysis to examine the effect of "cognitive therapies" such as biofeedback, meditation, and relaxation methods on blood pressure. Their abstract states that "cognitive interventions for essential hypertension are superior to no therapy but not superior to credible sham techniques" (p. 964). Here are some of the details from the text of the article:

Mean blood pressure reductions were smallest when comparing persons experimentally treated with those randomly assigned to a credible placebo or sham intervention and for whom baseline blood pressure assessments were made during a period of more than 1 day. Under these conditions, individuals treated with cognitive behavioral therapy experienced a mean reduction in systolic blood pressure of 2.8 mm Hg (CI, –0.8 to 6.4 mm Hg) . . . relative to controls. (p. 967)

The comparison included 12 studies involving 368 subjects. Do you agree with the conclusion in the abstract that cognitive interventions are not superior to credible sham techniques? What would be a better way to word the conclusion?

12. *Science News* (25 January 1995) reported a study on the relationship between levels of radon gas in homes and lung cancer. It was a case-control study, with 538 women with lung cancer and 1183 without lung cancer. The article noted that the average level of radon concentration in the homes of the two groups over a 1-year period was exactly the same. The article also contained the following quote, reporting on an editorial by Jonathan Samet that accompanied the original study:

For the statistical significance needed to assess accurately whether low residential exposures constitute no risk . . . the study would have required many more participants than the number used in this or any other residential-radon study to date. . . . But those numbers may soon become available, he adds, as researchers complete a spate of new studies whose data can be pooled for reanalysis. (p. 26)

Write a short report interpreting the quote for someone with no training in statistics.

13. An article in *Science* (Mann, 11 November 1994) describes two approaches used to try to determine how well programs to improve public schools have worked. The first approach was taken by an economist named Eric Hanushek. Here is part of the description:

Hanushek reviewed 38 studies and found the "startlingly consistent" result that "there is no strong or systematic relationship between school expenditures and student performance." . . . Hanushek's review used a technique called "vote-counting." (p. 961)

The other approach was a meta-analysis and the results are reported as follows:

[The researchers] found systematic positive effects. . . . Indeed, decreased class size, increased teacher experience, increased teacher salaries, and increased per-pupil spending were all positively related to academic performance. (p. 962)

Explain why the two approaches yielded different answers and which one you think is more credible.

14. Suppose ten studies were done to assess the relationship between watching violence on television and subsequent violent behavior in children. Suppose that none of the ten studies detected a statistically significant relationship. Is it possible for a vote-counting procedure to detect a relationship? Is it possible for a meta-analysis to detect a relationship? Explain.

....

MINI-PROJECTS

1. Find a journal article that presents a meta-analysis. (They are commonly found in medical and psychology journals.)

   a. Give an overview of the article and its conclusions.

   b. Explain what the researchers used as the criteria for deciding which studies to include.

   c. Explain which of the benefits of meta-analysis were incorporated in the article.

   d. Discuss each of the potential criticisms of meta-analysis and how they were handled in the article.

   e. Summarize your conclusions about the topic based on the article and your answers to parts b–d.

2. Find a newspaper report of a meta-analysis. Discuss whether important information is missing from the report. Critically evaluate the conclusions of the meta-analysis, taking into consideration the criticisms listed in Section 24.4.

REFERENCES

Bruvold, W. H. (1993). A meta-analysis of adolescent smoking prevention programs. *American Journal of Public Health* 83, no. 6, pp. 872–880.

Chalmers, T. C., H. Smith, Jr., B. Blackburn, B. Silverman, B. Schroeder, D. Reitman, and A. Ambroz. (1981). A method for assessing the quality of a randomized control trial. *Controlled Clinical Trials* 2, pp. 31–49.

Crossen, C. (1994). *Tainted truth: The manipulation of fact in America.* New York: Simon and Schuster.

Eisenberg, D. M., T. L. Delbance, C. S. Berkey, T. J. Kaptchuk, B. Kupelnick, J. Kuhl, and T. C. Chalmers. (15 June 1993). Cognitive behavioral techniques for hypertension: Are they effective? *Annals of Internal Medicine* 118, no. 12, pp. 964–972.

Fletcher, S. W., B. Black, R. Harris, B. K. Rimer, and S. Shapiro. (20 October 1993). Report of the International Workshop on Screening for Breast Cancer. *Journal of the National Cancer Institute* 85, no. 20, pp. 1644–1656.

Glass, G. V. (1976). Primary, secondary and meta-analysis of research. *Educational Researcher* 5, pp. 3–8.

Honorton, C. (1985). Meta-analysis of psi ganzfeld research: A response to Hyman. *Journal of Parapsychology* 49, pp. 51–91.

Hyman, R. (1985). The psi ganzfeld experiment: A critical appraisal. *Journal of Parapsychology* 49, pp. 3–49.

Kendall-Tackett, K. A., L. M. Williams, and D. Finkelhor. (1993). Impact of sexual abuse on children: A review and synthesis of recent empirical studies. *Psychological Bulletin* 113, no. 1, pp. 164–180.

Mann, C. C. (11 November 1994). Can meta-analysis make policy? *Science* 266, pp. 960–962.

Ripsin, C. M., J. M. Keenan, D. R. Jacobs, Jr., P. J. Elmer, R. R. Welch, L. Van Horn, K. Liu, W. H. Turnbull, F. W. Thye, and M. Kestin et al. (1992). Oat products and lipid lowering: A meta-analysis. *Journal of the American Medical Association* 267, no. 24, pp. 3317–3325.

Rosenthal, R. (1991). *Meta analytic procedures for social research.* Rev. ed. Newbury Park, CA: Sage Publications.

Sickles, E. A., and D. B. Kopans. (20 October 1993). Deficiencies in the analysis of breast cancer screening data. *Journal of the National Cancer Institute* 85, no. 20, pp. 1621–1624.

Vine, Gail (21 January 1995). Is there a database in the house? *New Scientist,* no. 1961, pp. 14–15.

Vine, M. F., B. H. Margolin, H. I. Morrison, and B. S. Hulka. (January 1994). Cigarette smoking and sperm density: A meta-analysis. *Fertility and Sterility* 61, no. 1, pp. 35–43.

CHAPTER TWENTY-FIVE

Putting What You Have Learned to the Test

This chapter consists solely of case studies. Each case study begins with a full or partial article from a newspaper or journal. Read each article and think about what might be misleading or missing, or simply misinformation. A discussion following each article summarizes some of the points I thought should be raised. You may be able to think of others. I hope that as you read these case studies you will realize that you have indeed become an educated consumer of statistical information.

Cranberry Juice and Bladder Infections

SOURCE: Juice does prevent infection, 9 March 1994, *Davis (CA) Enterprise*, p. A9.

CHICAGO (AP)—A scientific study has proven what many women have long suspected: Cranberry juice helps protect against bladder infections.

Researchers found that elderly women who drank 10 ounces of a juice drink containing cranberry juice each day had less than half as many urinary tract infections as those who consumed a look-alike drink without cranberry juice.

The study, which appeared today in the *Journal of the American Medical Association*, was funded by Ocean Spray Cranberries, Inc., but the company had no role in the study's design, analysis or interpretation, *JAMA* said.

"This is the first demonstration that cranberry juice can reduce the presence of bacteria in the urine in humans," said lead researcher Dr. Jerry Avorn, a specialist in medication for the elderly at Harvard Medical School. ■

Discussion

This study was well conducted, and the newspaper report is very good, including information such as the source of the funding and the fact that there was a placebo ("a look-alike drink without cranberry juice"). A few details are missing from the news account, however, that you may consider to be important. They are contained in the original report (Avorn, et al., 1994):

1. There were 153 subjects.

2. The participants were randomly assigned to the cranberry or placebo group.

3. The placebo was fully described as "a specially prepared synthetic placebo drink that was indistinguishable in taste, appearance, and vitamin C content, but lacked cranberry content."

4. The study was conducted over a 6-month period, with urine samples taken monthly.

5. The measurements taken were actually bacteria counts from urine, rather than a more subjective assessment of whether an infection was present. The news story claims "less than half as many urinary tract infections," but the original report gave the odds of bacteria levels exceeding a certain threshold. The odds in the cranberry juice group were only 42% of what they were in the control group. Unreported in the news story is that the odds of remaining over the threshold from one month to the next for the cranberry juice group were only 27% of what they were for the control group.

6. The participants were elderly women volunteers with a mean age of 78.5 years and high levels of bacteria in their urine at the start of the study.

7. The original theory was that if cranberry juice was effective, it would work by increasing urine acidity. That was not the case, however. The juice inhibited bacterial growth in some other way.

CASE STUDY 25.2   ## Children on the Go

SOURCE: Brenda C. Coleman, Children who move often are problem prone, 15 September 1993, *West Hawaii Today.*

CHICAGO—Children who move often are 35 percent more likely to fail a grade and 77 percent more likely to have behavioral problems than children whose families move rarely, researchers say.

A nationwide study of 9915 youngsters ages 6 to 17 measured the harmful effects of moving. The findings were published in today's issue of the *Journal of the American Medical Association.*

About 19 percent of Americans move every year, said the authors, led by Dr David Wood of Cedars-Sinai Medical Center in Los Angeles. The authors cited a 1986–87 Census Bureau study.

The authors said that our culture glorifies the idea of moving through maxims such as "Go West, young man."

Yet moving has its "shadow side" in the United States, where poor and minority families have been driven from place to place by economic deprivation, eviction and racism, the researchers wrote.

Poor families move 50 percent to 100 percent more often than wealthier families, they said, citing the Census Bureau data.

The authors used the 1988 National Health Interview Survey and found that about one-quarter of children had never moved, about half had moved fewer than three times and about three-quarters fewer than four times.

Ten percent had moved at least six times, and the researchers designated them "high movers."

Compared with the others, the high movers were 1.35 times more likely to have failed a grade and 1.77 times more likely to have developed at least four frequent behavioral problems, the researchers said. Behavioral problems ranged from depression to impulsiveness to destructiveness.

Frequent moving had no apparent effect on development and didn't appear to cause learning disabilities, they found.

The researchers said they believe their study is the first to measure the effects of frequent relocation on children independent of other factors that can affect school failure and behavioral problems.

Those factors include poverty, single parenting, belonging to a racial minority and having parents with less than a high school education.

Children in families with some or all of those traits who moved often were much more likely to have failed a grade—1.8 to 6 times more likely—than children of families with none of those traits who seldom or never moved.

The frequently relocated children in the rougher family situations also were 1.8 to 3.6 times more likely to have behavioral problems than youngsters who stayed put and lived in more favorable family situations.

"A family move disrupts the routines, relationships and attachments that define the child's world," researchers said. "Almost everything outside the family that is familiar is lost and changes."

Dr. Michael Jellinek, chief of child psychiatry at Massachusetts General Hospital, said he couldn't evaluate whether the study accurately singled out the effect of moving. ■

Discussion

There are some problems with this study and with the reporting of it in the newspaper. First, it is not until well into the news article that we finally learn that the reported figures have already been adjusted for confounding factors, and even then the news article is not clear about this. In fact, the 1.35 relative risk of failing a grade and the 1.77 relative risk of at least four frequent behavioral problems apply *after* adjusting for poverty, single parenting, belonging to a racial minority, and having parents with less than a high school education.

The news report is also missing information about baseline rates. From the original report (Wood et al., 1993), we learn that 23% of children who move frequently have repeated a grade, whereas only 12% of those who never or infrequently move have repeated a grade.

The news report also fails to mention that the data were based on surveys with parents. The results could therefore be biased by the fact that some parents may not have been willing to admit that their children were having problems.

Although the news report implies that moving is the cause of the increase in problems, this is an observational study and causation cannot be established. Numerous other confounding factors that were not controlled for could account for the results. Examples are number of days missed at school, age of the parents, and quality of the schools attended.

CASE STUDY 25.3   ## It Really Is True about Aspirin

SOURCE: Peter Aldhous, A hearty endorsement for aspirin, 7 January 1994, *Science* 263, p. 24.

Aspirin is one of the world's most widely used drugs, but a major study being published this week suggests it is not used widely enough. If everybody known to be at high risk of vascular disease were to take half an aspirin a day, about 100,000 deaths and 200,000 nonfatal heart attacks and strokes could be avoided worldwide each year, the study indicates. "This is one of the most cost-effective drug interventions one could have in developed countries," says Oxford University epidemiologist Richard Peto, one of the coordinators of the study, a statistical overview, or meta-analysis, of clinical trials in which aspirin was used to prevent blood clots.

The new meta-analysis, which covers both aspirin and more expensive antiplatelet drugs, combined the results of 300 trials involving 140,000 patients. Its recommendation: A regime of half a tablet of aspirin a day is valuable for all victims of heart attack and stroke, and other at-risk patients such as angina sufferers and recipients of coronary bypass grafts. . . The full analysis will be published in three consecutive issues of the *British Medical Journal* beginning on January 8.

| TABLE 25.1 | Meta-Analysis of Aspirin Studies |

	Summary of Results of Aspirin Trials		Proportion Who Suffered Nonfatal Stroke, Nonfatal Heart Attack, or Death During Trial	
Type of Patient	Length of Treatment	Number of Patients	Treatment Group	Control Group
Suspected acute heart attack	1 month	20,000	10%	14%
Previous history of heart attack	2 years	20,000	13%	17%
Previous history of stroke	3 years	10,000	18%	22%
Other vascular diseases	1 year	20,000	7%	9%

SOURCE: Aldhous, 7 January 1994, p. 24.

While the results are likely to bring consensus to the field, there are still a few areas of uncertainty. One major disagreement surrounds the study's finding that aspirin can reduce thrombosis in patients immobilized by surgery, where some of the studies were not double-blinded, and therefore could be biased. Worried about this possibility . . . [researchers] have reanalyzed a smaller set of data. They included only results of trials that had faultless methodology and looked separately at patients in general surgery and those who had had procedures such as hip replacements, where the surgery itself can damage veins in the leg and further increase the risk of thrombosis. Their conclusion: Aspirin is beneficial only following orthopedic surgery.

A table accompanying the article summarized the studies (see Table 25.1). ■

Discussion

This article, which is only partially quoted here, is a good synopsis of a major meta-analysis. It does not deal with the influence of aspirin on healthy people, but it does seem to answer the question of whether aspirin is beneficial for those who have had cardiovascular problems. Table 25.1 provides information on the magnitude of the effect. Although no confidence limits are provided, you can get a rough estimate of them yourself because you are told the sample sizes.

It does seem odd that the numbers in the table only sum to 70,000, although the article reports that there were 140,000 patients. No explanation is given for the discrepancy.

The main problem you should have recognized occurs at the end of the reported text. It is indicated that aspirin is not beneficial following surgery except for orthopedic surgery. This should have alerted you to the problem of reduced sample size when only a subset of the data is considered. In fact, the article continued as follows:

> Oxford cardiologist Rory Collins, another coordinator of the study, counters that splitting the data into such small chunks defeats the object of conducting a meta-analysis of many trials. And he notes that if the data for general and orthopedic surgery are considered together, there is still a significant benefit from aspirin, even if methodologically suspect trials are excluded.

In other words, analyzing the studies with faultless methodology only, the researchers concluded that there is still benefit to using aspirin after surgery. If the studies examined are further reduced to consider only those following general surgery, the added benefit of taking aspirin is not statistically significant. Remember, that *does not* mean that there is no effect.

CASE STUDY 25.4

You Can Work and Get Your Exercise at the Same Time

SOURCE: White collar commute, 11 March 1994, *Davis* (CA) *Enterprise,* p. C5.

One in five clerical workers walks about a quarter mile a day just to complete routine functions like faxing, copying and filing, a national survey on office efficiency reports. The survey also shows that the average office worker spends close to 15 percent of the day just walking around the office. Not surprisingly, the survey was commissioned by Canon U.S.A., maker of—you guessed it—office copiers and printers that claim to cut the time and money "spent running from one machine to the next." ∎

Discussion

We are not given any information that would allow us to evaluate the results of this survey. Further, what does it mean to say that one in five workers walk that far? Do the others walk more or less? Is the figure based on an average of the top 20% (one in five) of a set of numbers, in which case outliers for office delivery personnel and others are likely to distort the mean?

CASE STUDY 25.5

Sex, Alcohol, and the First Date

SOURCE: Teen alcohol-sex survey finds the unexpected, 11 May 1994, *Sacramento Bee,* p. A11.

WASHINGTON—Young couples are much more likely to have sex on their first date if the male partner drinks alcohol and the woman doesn't, new research shows.

The research, to be presented today at an American Psychological Association conference on women's health, contradicts the popular male notion that plying a woman with alcohol is the quickest path to sexual intercourse.

In fact, interviews with 2052 teenagers found that they reported having sex on a first date only 6 percent of the time if the female drank alcohol while the male did not. That was lower than the 8 percent who reported having sex when neither partner drank.

Nineteen percent of the teens reported having sex when both partners drank, but the highest frequency of sex on the first date—24 percent—was reported when only the male drank.

The lead researcher in the study, Dr. M. Lynne Cooper of the State University of New York at Buffalo, said that drinking may increase a man's willingness "to self-disclose things about himself, be more likely to communicate feelings, be more romantic—and the female responds to that."

Rather than impairing a woman's judgment, alcohol apparently makes many women more cautious, Cooper said. "Women may sort of smell danger in alcohol, and it may trigger some warning signs," she said. "It makes a lot of women more anxious." ■

Discussion

This is an excellent example of a misinterpreted observational study. The authors of both the original study and the news report are making the assumption that drinking behavior influences sexual behavior. Because the drinking behavior was clearly not randomly assigned, there is simply no justification for such an assumption.

Perhaps the causal connection is in the reverse direction. The drinking scenario that was most frequently associated with sex on the first date was when the male drank but the female did not. If a couple suspected that the date would lead to sex, perhaps they would plan an activity in which they both had access to alcohol, but the woman decided to keep her wits.

The simplest explanation is that the teenagers did not tell the truth about their own behavior. It would be less socially acceptable for a female to admit that she drank alcohol and then had sex on a first date than it would be for a male. Therefore, it could be that males and females both exaggerated their behavior, but in different directions.

CASE STUDY 25.6 ## Unpalatable Pâté

SOURCE: S. Plous, Psychological mechanisms in the human use of animals, 1993, *Journal of Social Issues* 49, no. 1, pp. 11–52.

This article is about human psychological perceptions of the use of animals. Just one paragraph will be used for this example:

With the exception of several recent public opinion polls on animal rights and a few studies on vegetarianism and hunting, a computer-assisted literature review yielded only six published research programs specifically investigating attitudes toward the use of animals. The first was an exploratory study in which roughly 300 Australian students were asked whether they approved or disapproved of certain uses of animals (Braithwaite and Braithwaite, 1982). As it turned out, students frequently condemned consumptive practices while endorsing consumption itself. For example, nearly three fourths of the students disapproved of "force-feeding geese to make their livers swell up to produce pâté for restaurants," but the majority did not disapprove of "eating pâté produced by the force-feeding of geese." The authors interpreted these findings as evidence of an inconsistency between attitudes and behaviors. (p. 13) ■

Discussion

This is a good example of how a slight variation of the wording of a question can produce very different results. I don't know about you, but to me the first version of the question is bound to produce negative results. It would have been more appropriate to ask the students how they felt about "force-feeding geese to produce pâté for restaurants," without mention of the livers swelling. I think the authors have overinterpreted the results by claiming that this illustrates a difference between attitudes and behaviors.

CASE STUDY 25.7   # Nursing Moms Can Exercise, Too

SOURCE: Aerobics OK for breast-feeding moms, 17 February 1994, *Davis* (CA) *Enterprise*, p. A7.

Moderate aerobic exercise has no adverse effects on the quantity or quality of breast milk produced by nursing mothers, and can significantly improve the mothers' cardiovascular fitness, according to UC Davis researchers.

For 12 weeks the study monitored 33 women, beginning six to eight weeks after the births of their children. All were exclusively breast-feeding their infants, with no formula supplements, and had not previously been exercising. Eighteen women were randomly assigned to an exercise group and 15 to a non-exercising group. The exercise group participated in individual exercise programs, including rapid walking, jogging or bicycling, for 45 minutes each day, five days per week. . . . At the end of the 12-week study, [the researchers] found:

Women in both groups experienced weight loss. The rate of weight loss and the decline in the percentage of body fat after childbirth did not differ between the exercise and control groups, because women in the exercise group compensated for their increased energy expenditure by eating more.

There was an important improvement in the aerobic fitness of the exercising women, as measured by the maximal oxygen consumption.

There was no significant difference between the two groups in terms of infant breast-milk intake, energy output in the milk or infant weight gain.

Prolactin levels in the breast milk did not differ between the two groups, suggesting that previously observed short-term increases in the level of that hormone among non-lactating women following exercise do not influence the basal level of prolactin. ■

Discussion

This is an excellent study and excellent reporting. The study was an experiment and not an observational study, so we can rule out confounding factors resulting from the fact that mothers who choose to exercise differ from those who do not. The mothers were randomly assigned to either the exercise group or a nonexercising control group. To increase ecological validity and generalizability, they were allowed to choose their own forms of exercise. All mothers were exclusively breast feeding, ruling out possible interactions with other food intake on the part of the infants. The study could obviously not be performed blind because the women knew whether they were exercising. It presumably could have been single blind, but was not.

The article reports that there was an important improvement in fitness; it does not give the actual magnitude. The reporter has done the work of determining that the improvement is not only statistically significant but also of practical importance. From the original research report, we learn that "maximal oxygen uptake increased by 25 percent in the exercising women but by only 5 percent in the control women ($P < .001$)" (Dewey et al., 1994, p. 449).

What about the differences that were reported as nonsignificant or nonexistent, such as infant weight gain? There were no obvious cases of misreporting important but not significant differences. Most of the variables for which no significant differences were found were very close for the two groups. In the original report, confidence intervals are given for each group, along with a p-value for testing whether there is a statistically significant difference. For instance, 95 percent confidence intervals for infant weight gain are 1871 to 2279 grams for the exercise group and 1733 to 2355 grams for the control group ($p = 0.86$). The p-value indicates that if aerobic exercise has no impact on infant weight gain in the population, we would expect to see sample results differ this much or more very often. Therefore, we cannot rule out chance as an explanation for the very small differences observed. The same is true of the other variables for which no differences were reportedly found.

CASE STUDY 25.8   # So You Thought Spinach Was Good for You?

SOURCE: R. Nowak, Beta-carotene: Helpful or harmful? 22 April 1994, *Science* 264, pp. 500–501.

Over the past decade, it's become a tenet of cancer prevention theory that taking high doses of antioxidant vitamins—like vitamin E or A—will likely protect against cancer. So in light of that popular hypothesis, cancer prevention experts are having to struggle to make sense of the startling finding, published in the 14 April *New*

England Journal of Medicine, that supplements of the antioxidant beta-carotene markedly increased the incidence of lung cancer among heavy smokers in Finland.

The result is particularly worrying because it comes from a large, randomized clinical trial—the gold standard test of a medical intervention. And as well as dumbfounding the experts, the Finnish study has triggered calls for a moratorium on health claims about antioxidant vitamins (beta-carotene is converted into vitamin A in the body), and prompted close scrutiny of several other large beta-carotene trials that are currently under way.

What mystifies the experts is that the Finnish trial goes against all the previously available evidence. Beta-carotene's biological activity suggests that it should protect against cancer. It's an antioxidant that can sop up chemicals called free radicals that may trigger cancer. And over a hundred epidemiologic surveys indicate that people who have high levels of beta-carotene in their diet and in their blood have lower risks of cancer, particularly lung cancer. Finally, the idea that beta-carotene would have only beneficial effects on cancer is buttressed by the results of the only other large-scale clinical trial completed thus far. It found that a combination of beta-carotene, vitamin E, and selenium reduced the number of deaths from stomach cancer by 21% among 15,000 people living in Linxian County in China, compared with trial participants who did not take the supplements. . . .

. . . When all the data were in and analyzed at the end of the trial, it became apparent that the incidence of lung cancer was 18% higher among the 14,500 smokers who took beta-carotene than among the 14,500 who didn't. The probability that the increase was due to chance is less than one in one hundred. In clinical trials, a difference is taken seriously when there is less than a one-in-twenty probability that it happened by chance.

The trial organizers were so baffled by the results that they even wondered whether the beta-carotene pills used in the study had become contaminated with some known carcinogen during the manufacturing process. Tests have ruled out that possibility. ■

Discussion

The researchers seem to be getting very upset about what could simply be the luck of the draw. Notice that the article mentions that "over a hundred epidemiologic surveys indicate that people who have high levels of beta-carotene in their diet and in their blood have lower risks of cancer, particularly lung cancer." Further, the significance level associated with this test is "less than one in one hundred." It is true by the nature of chance that occasionally a sample will result in a statistically significant finding that is the reverse of what is true in the population.

It may sound like a sample of size 29,000 is large enough to rule out that explanation. But from the original report, we find that there were only 876 new cases of lung cancer among all the groups. Further, although "the incidence of lung cancer was 18% higher among the 14,500 smokers who took beta-carotene than among the 14,500 who didn't," a 95% confidence interval for the true percentage in the population ranged from 3% to 36% higher. Although the entire confidence interval still lies in the direction of higher rates for those who took beta-carotene, the lower endpoint of 3% is close to zero.

It is also the case that the researchers performed a multitude of tests, any one of which may have been surprising if it had come out in the direction opposite to what was predicted. For example, when they looked at total mortality, they found that it was only 8% higher for the beta-carotene group (mainly due to the increased lung-cancer deaths), and a 95% confidence interval ranged from 1% to 16%.

It may turn out that beta-carotene is indeed harmful for some people. But the furor caused by this study appears to be overblown, given the number of studies that have been done on this topic and the probability that eventually a study will show a relationship opposite to that in the population just by chance.

CASE STUDY 25.9   # Chill Out—Move to Honolulu

SOURCE: Tim Friend, Cities of discord: Home may be where the heart disease is, 13 May 1994, *USA Today*, p. 5D.

Do you count the number of items that the selfish person in front of you is holding in the express lane at the grocery store?

Been fantasizing about slamming your brakes the next time some jerk tailgates you in rush hour?

Yeah? Then you may have a "hostile" personality, on the road to an early death.

"Anger kills," says Dr. Redford Williams, Duke University Medical Center, whose book by the same name is out today in paperback.

Williams also releases today his study revealing the USA's five most hostile and five mellowest cities. It's based on a poll that measured citizens' hostility levels and compared them with cities' death rates.

"The study was stimulated by research of people with hostile personalities and high levels of cynical mistrust of others—people who have frequent anger and who express that anger overtly," Williams says.

More than two decades of research involving anger and heart disease show that people with higher hostility levels have higher rates of heart disease deaths and over-all deaths.

"The hypothesis for the new study was since we know that hostility is a risk factor for high death rates, are cities characterized by high hostility levels also characterized by high death rates? Is what's bad for an individual, bad for a population?" Williams asks.

He and the Gallup Organization found that cities with higher hostility scores consistently had higher death rates. Those with lower hostility scores had lower death rates. The new finding suggests hostile cities may want to chill out.

SOURCE: Tim Friend, Philly hostile? "Who asked ya?" 13 May 1994. *USA Today*, p. 1D.

Philadelphia—the City of Brotherly Love—appears to be the USA's most hostile town, and may be paying for it with higher death rates.

Dr. Redford Williams, Duke University Medical Center, asked the Gallup Organization to measure the levels of hostility and mistrust in 10 cities from states with the highest and lowest heart disease death rates.

Studies have shown that people with hostile personalities have an increased risk of dying from heart disease, he says.

He wanted to find out if that applied to cities, too.

Hostility levels were measured, ranked and then paired with cities' death rates:

- Philadelphia had the highest hostility score and highest death rate. New York was second, Cleveland third.

- Honolulu had the lowest hostility score and the lowest death rate. Seattle was second, Minneapolis was third lowest.

- Death rates in the five cities with the highest hostility were 40% higher than in the cities with the lowest score.

Statistically, the probability of the correlation occurring by chance is less than 1 in 10,000. ■

Discussion

These two stories appeared on the same day in *USA Today*. They discuss a study that Dr. Redmond Williams commissioned the Gallup Organization to conduct and which he released on the same day the paperback edition of his book was to be released. That sounds like a good advertising gimmick.

The study sounds interesting, but we are not provided with sufficient details to ascertain whether there are major confounding effects. For example, was the interviewer also from the city that was being interviewed? Remember that the interviewer can bias responses to questions. Did the interviewers know the purpose of the study? Also, one piece of information is available only by consulting a graph accompanying the articles. The graph, titled "Hostility leads to death," shows a scatterplot of death rate versus hostility for the ten cities. At the bottom is a footnote: "[Hostility index] adjusted for race, education, age, income, gender." What does that mean?

The main problem with these articles is the implication that if the residents of cities like Philadelphia would lower their hostility index, they would also lower their death rates. But this is an observational study, and there are numerous potential confounding factors. For instance, do you think the weather and pollution levels are better for you in Honolulu or in Philadelphia? Do you think you are more likely to be murdered in New York City or in Seattle? Are the populations equally likely to be elderly in all of the cities?

The implication that these results would occur by chance less than once in 10,000 is also misleading. As admitted in the article, the study purposely used cities that were outliers in both directions on death rates. The graph shows that the death rate (per 1000) in Honolulu is less than half of what it is in Philadelphia. Remember from Chapter 11 that outliers can have a large impact on correlations.

So You Thought Hot Dogs Were Bad for You?

SOURCE: Constance Holden, Hot dog hazards, 27 May 1994, *Science* 264, p. 1255.

Parents worried about whether electromagnetic fields (EMFs) cause cancer now have a more all-American concern: Hot dogs, warns a study from the University of Southern California (USC), are more than 4 times as likely as EMFs to be linked with childhood leukemia. But the researchers caution that the data may not cut the mustard.

In the May issue of *Cancer Causes and Control,* three research groups report a link between cured-meat consumption and cancer. The most striking evidence comes from a group led by USC epidemiologist John Peters, which earlier found that EMF exposure is associated with a doubling of the risk for childhood leukemia. Among the 232 cases in the study, children who ate 12 or more hot dogs a month were nine times as likely as hot dog–free controls to develop leukemia. Peters also found an increased risk for kids whose fathers ate a lot of hot dogs.

Not that Mom can eat with impunity either: The other two studies linked maternal intake of hot dogs and cured meats during pregnancy with childhood brain tumors. Looking at 234 cases of various childhood cancers, University of North Carolina epidemiologist David Savitz found that children whose mothers downed hot dogs at least once a week were more than twice as likely as controls to develop brain tumors. And in a study at Children's Hospital of Philadelphia, epidemiologist Greta Bunin found a weak link between maternal hot dog intake during pregnancy and increased risk of a brain tumor, astrocytic glioma, in their children.

The epidemiologists say that the effects they observe could be from the N-nitroso compounds in cured meats, such as nitrites, which cause cancer in lab animals. And that makes sense, says Savitz, because vitamins—with their carcinogen-fighting antioxidant properties—appear to have a protective effect in juvenile hot dog eaters.

Savitz warns that the studies are far from conclusive. They all suffer from a lack of data on subjects' exposures to other N-nitroso compounds. And, says Bunin, "the cured-meat association could be an indicator of a diet poor in other ways." What's needed now, Savitz says, is research looking more closely at diet. "Who knows, maybe it's the condiments," he says. ■

Discussion

From the information available in this story, it appears that the three studies were all of the case-control type. In other words, researchers located children who already had leukemia or other cancers and similar "control" children, and then interviewed the families about their habits.

One of the problems with trying to link diet to cancer in a case-control study is that people have poor memories about what they ate years ago. Parents who are searching for an answer to why their child has cancer are probably more likely to "remember" feeding the child an unhealthful diet than parents whose children do not have cancer.

Another obvious problem, partially addressed in the story, is that a diet that includes 12 or more hot dogs a month is likely to be low in consumption of more healthful foods such as fruits and vegetables. So, the fact that the children who consumed more hot dogs were more likely to have developed cancer could be confounded with the fact that they consumed fewer vitamin-rich foods, had higher fat in their diets, or a variety of other factors. The article does mention that "vitamins . . . appear to have a protective effect in juvenile hot dog eaters," indicating that those who consumed vitamin-rich foods, or vitamin supplements, did not have the same increased risk as those who did not.

One additional problem with the report of these studies is that no baseline information about leukemia rates in children is provided. Because hot dogs are a convenient food that children like to eat, parents should be provided with the information needed to weigh the risks against the benefits. According to an article in the Atlantic City *Press* (1 July 1994, p. C10), leukemia rates for children in New Jersey counties range from about 3.1 to 7.9 per 100,000. Remember that those figures include the whole population, so they do not mean that the rate for hot dog consumers would be nine times those rates. Therefore, although a relative risk of 9.0 for developing leukemia is frighteningly high, the overall rate of leukemia in children is still extremely low.

REFERENCES

Avorn, J., M. Monane, J. H. Gurwitz, R. J. Glynn, L. Choodnovskiy, and L. A. Lipsitz. (1994). Reduction of bacteriuria and pyuria after ingestion of cranberry juice. *Journal of the American Medical Association* 271, no. 10, pp. 751–754.

Braithwaite, J., and V. Braithwaite. (1982). Attitudes toward animal suffering: An exploratory study. *International Journal for the Study of Animal Problems* 3, pp. 42–49.

Dewey, K. G., C. A. Lovelady, L. A. Nommsen-Rivers, M. A. McCrory, and B. Lönnerdal. (1994). A randomized study of the effects of aerobic exercise by lactating women on breast-milk volume and composition. *New England Journal of Medicine* 330, no. 7, pp. 449–453.

Wood, D., N. Halfon, D. Scarlata, R. Newacheck, and S. Nessim. (1993). Impact of family relocation on children's growth, development, school function, and behavior. *Journal of the American Medical Association* 270, no. 11, pp. 1334–1338.

Solutions to Selected Exercises

Chapter 1

3. If the measurements of interest are extremely variable, then a large sample will be required to detect real differences between groups.

8. Visit both numerous times because waiting times vary, then compare the averages.

9. a. No. It would be unethical to randomly assign people to smoke cigars or not.
   b. No. People who smoke cigars may also drink more alcohol, eat differently, or have some other characteristic that is the actual cause of the higher rate of cancer.

13. a. Experiment
   b. Yes, because of the randomization

15. The sample is the few thousand asked; the population is all adults in the nation.

17. a. Observational study
   b. No. Meditators may already be different.

Chapter 2

3. a. Biased, the guards; unbiased, trained independent interviewers

5. Major and grade-point average

7. No. For example, college major

10. a. Version 2
   b. Version 1

12. A statistical difference may not have any practical importance.

15. Use Component 3; volunteer respondents versus nationwide random sample

Chapter 3

1. a. Gender (male/female)
   b. Time measured on a clock that is 5 minutes fast
   c. Weight of packages on a postal scale that weighs items 1 ounce high half the time and 1 ounce low half the time

4. a. Deliberate bias and unnecessary complexity

5. a. Do you support or not support banning prayers in schools?

6. a. Measurement
   b. Categorical

7. a. Discrete
   b. Continuous

9. It would be easier with only a little natural variability.

13. Although only one-fifth favored forbidding public speeches (version A), almost one-half did not want to allow them (version B).

16. Anonymous

19. a. No, there is too much overlap.
   b. Route 1 is always 14 minutes, and Route 2 is always 16 minutes.

20. Beer consumption measured in bottles (discrete) or ounces (continuous)

Chapter 4

1. a. Cluster sampling
   b. Systematic sampling
   c. Stratified sampling
   d. Convenience or haphazard sampling

4. a. 7%
   b. 73% to 87%

5. c. Cluster sampling

9. a. Survey
   b. Experiment

15. b. Yes

17. a. Case study

18. No

22. Not reaching the individuals selected

Chapter 5

2. a. Single-blind; neither
   b. Single-blind; block design
   c. Double-blind; block design

5. a. The color of the cloth is a confounding variable.

8. The placebo effect

11. Explanatory is form of exercise; response is weight loss.

16. Yes. Randomly assign treatments to the volunteers. Generalizability may be a problem if volunteers don't represent the population.

22. a. Ecological validity; experimenter effect
   b. Hawthorne effect

Chapter 7

4. Sales prices of cars

9. a. 65
   b. 54, 62.5, 65, 70, 78

13. a. Skewed to the right
   b. Median

15. Incomes for musicians

21. Range

25. Temperatures for the entire year

Chapter 8

2. a. 10%
   b. 60%
   c. 1%

4. a. 2.05
   c. −0.67

8. a. 34th percentile

10. 208.8

12. a. Upper quartile is $z = 0.67$

15. 68% in 84 to 116, 95% in 68 to 132, 99.7% in 52 to 148

18. 17.56 minutes

20. 0.75 or 75%

Chapter 9

1. a. Pie chart
   b. Histogram
   c. Bar graph
   d. Scatterplot

5. The horizontal axis does not maintain a constant scale.

8. b. Because population has steadily increased, numbers of violent crime would increase even if the rate stayed constant.

Chapter 10

1. Yes, about 5 of them

3. a. $y = 0.96 (x - 0.5) = -0.48 + 0.96x$ (in pints)

7. A correlation of −.6 implies a stronger relationship.

11. a. Predicted ideal weight for a 150-pound woman is 133.9 pounds.

12. a. Negative, because the slope is negative.
   b. Predicted time is 36.86, just 0.53 second too high.
   c. They decrease by 0.1094 second per year, or about 0.44 second per 4-year Olympics.

15. Yes. The strength of the relationship is the same as for the golfers and the sample is much larger.

Chapter 11

2. No. People who prefer to exercise (walk) may also be people who keep their weight under control.

5. Winning time in an Olympic running event and the cost of running shoes are likely to be negatively correlated.

7. **d.** Combining groups inappropriately

10. Salary and years of education in a company where the president makes $1 million a year but has an eighth-grade education

14. No. There are confounding factors, such as more sports cars are red.

Chapter 12

4. Relative risk = 10

6. **a.** Retrospective observational study, probably case control

9. **a.** 13.9%, 0.106, 212/1525 or 0.139, 465 to 3912

10. **c.** Proportion

17. **a.** 0.28 or 28% for males; 0.23 or 23% for females
    b. 8.64; statistically significant
    c. No. The chi-squared statistic would be about 0.86.

Chapter 13

2. **a.** [partial answer] There was a 72% increase from 1940 to 1950.

3. 233.33 for 1981 and 466 for 1995

7. $56.07

17. "Average duration of unemployment, in weeks" is lagging.

Chapter 14

1. **a.** Positive
   b. Nonexistent (ignoring possible global warming)

8. You would expect a positive correlation.

10. **a.** Seasonal and cycles
    b. Trend and seasonal
    c. Possible seasonal and cycles

15. **a.** 1994 salary would be $119,646.
    b. 1994 salary would be $487,356

Chapter 15

2. **a.** 9% or .09

6. The probability that the earth will be uninhabitable by the year 2050

9. **a.** .10 or 10%
   b. (.10)(.10) or .01; assume payments (or not) are independent for the two customers.

14. **a.** Observing the relative frequency

17. **a.** Not likely to be independent; they are likely to share political preferences.

24. 3.3; you would not expect this each time, but you would expect it in the long run.

26. Discrete data, where the expected value is also the mode; for instance, suppose a small box of raisins contains 19 pieces with probability .1, 20 with probability .8, and 21 with probability 1.

Chapter 16

2. Anchoring

9. Statement A has a higher probability, but people would think statement B did.

14. The rate of the disease for people similar to you

16. Optimism

Chapter 17

2. See Exercise 17 for an example.

6. There are numerous meanings to the word *pattern,* and it is easy to find something that fits one.

9. Yes, there are limited choices and people know their friends' calling behaviors.

12. **a.** 9,900/10,700 = 0.925
    b. 9,900/99,000 = 0.10
    c. 10,700/100,000 = 0.107

20. **a.** Choice A, because for choice B the expected value is $9.00.

Chapter 18

4. **b.** Virtually impossible; standardized score over 10

11. **a.** Bell-shaped, mean = 0.20, standard deviation = 0.008
    b. No, the standardized score for 0.17 is −3.75.

13. **b.** No, the standardized score for a mean of 7.1 is about 7.24.

14. **a.** No, the expected number not meeting standards is only 3.
    b. Yes, conditions are met.

17. Bell-shaped, with a mean of 3.1 and standard deviation of 0.05.

Chapter 19

2. a. 0.16 to 0.28

5. a. 68%
  b. 90%

7. No, the volunteer sample is not representative.

15. a. 20% to 24%
  b. 0.02 ± 0.002 or 1.8% to 2.2%

16. 25.6% to 48.4%; it almost covers chance (25%).

Chapter 20

3. a. SEM is 1/12 hour, or 5 minutes.

6. a. The population means are probably different.
  b. The population means could be the same.

8. a. $29 \pm (1.645)(16.3)$ or 2.2 to 55.8
  b. Yes; the interval does not cover zero.

16. 12% to 23%; there is a change because 0 is not in the interval.

17. Method 1; the difference within couples is desired, not the difference across sexes.

Chapter 21

3. b. Type 1 error

5. a. Null: Psychotherapy and desipramine are equally effective in treating cocaine use. Alternative: One method is more effective than the other.
  c. Type 1; concluding there is a difference when there isn't

12. c. There is a link between vertex baldness and heart attacks, but the evidence in the sample wasn't strong enough to conclusively detect it.

16. a. Minor disease with serious treatment—for example, tonsillitis and the treatment is surgery
  b. Being infected with HIV

17. They should use a higher p-value to reduce the probability of a type 2 error.

Chapter 22

2. One-sided

4. a. No
  b. Yes, the p-value would be .04.

7. a. Null: The training program has no effect on test scores.
  b. 2.5
  c. Alternative; scores would be higher after the program.

16. a. Null: There is no relationship between smoking behavior and time to pregnancy.
  c. No, the p-value is not the probability that the null is true.

Chapter 23

1. a. Yes
  b. No, the magnitude wasn't much different, but the large sample sizes led to a statistically significant difference.

4. a. 4.911

8. a. A confidence interval for the relative risk of heart attack during heavy versus minimal physical exertion is from 2.0 to 6.0.

10. The small sample sizes would result in low power, or high probability of a type 2 error even if ESP were present.

12. One

Chapter 24

3. a. They would like to make causal conclusions.
  b. Yes

6. b. Statistical significance versus practical importance

8. 1000 researchers; 50 could easily be contacted.

9. One reason is that some studies used outdated technology.

14. Vote counting could not detect a relationship, but meta-analysis could, by increasing the overall power.

Index

A

Abuses of statistics, 7–9
Accumulated probability, 265–267
Accuracy of sample surveys, 53, 54
Adhikari, A., 69
Adolescent sexuality, 37–38
Advertising questions, 39–40
Aging
  and meditation case study, 95–98
  risk statistics and, 212
Aldhous, Peter, 441, 442
Alper, Joe, 143
Alternative hypothesis, 374–375
  for categorical variables, 393–394
  in ESP case study, 382
  notation for, 400
American Institute of Public Opinion, 63
Anastasi, Anne, 39
Anchoring, 283–284
Anonymity of participants, 34
Arraf, Jane, 261
Ashenfelter, Orley, 399
Aspirin and heart attacks, 6, 196–197, 409–410, 411–412, 441–443
Attitudes, measuring, 39
Availability heuristic, 282–286
Average. *See* Mean
Avorn, Jerry, 390, 439
Axes
  changing labeling of, 147, 148, 149
  failure to label, 146
  not starting at zero, 146–147, 148

B

Baird, D. D., 198
Baldness and heart attacks, 81–82, 85, 362
Bar graphs, 142–143
Barking dogs, 22–23
Barnett, A., 293
Baseline risk, 211
Base rates, 286, 303
Base year for CPI, 228–229
Basketball, streak shooting in, 304–305
Bayes' Rule, 303, 309
Belief in the law of small numbers, 301
Bell-shaped curve, 129
  Rule for Sample Proportions and, 321–322
Bell-shaped data sets, 113
Bem, D., 206, 381
Beta-carotene case study, 446–448
Betting
  gambler's fallacy and, 301
  on sports events, 307–309
Bias
  in measurements, 42–43
  meta-analysis and, 431
  in questions, 33
  in sample surveys, 63
Bickel, P. J., 221
Bimodal shape, 113
Birthdays
  death days and, 270–271
  probability of sharing, 299–300
Bishop, Jerry E., 338
Black, B., 212
Black, Donald, 417
Blackenhorn, David, 272
Bladder infection/cranberry juice case study, 389–390, 439
Blinding procedures, 75–76
Block designs, 76
Blocks, 76
Boland, Philip J., 122
Bower, B., 384
Boxplots, 116–119
  creating, 117–118
  interpreting, 118
Braithwaite, J., 445
Braithwaite, V., 445
Breast cancer risk, 208–209, 210, 302–303, 432–433
Brinton, L. A., 208
British Medical Journal, 441
British Open, 307–309
Brooks Shoe Manufacturing Company case study, 26–27
Bruvold, W. H., 424, 429
Bryson, M. C., 63
Bunin, Greta, 450
Bureau of Economic Analysis (BEA), 234
Bureau of Labor Statistics (BLS), 52, 227, 232
Bush, George, 7, 236
Bushyhead, J. B., 290

Credits

Page 8: Figure 1.1, "Study: Smoking May Lower Kids' IQs." Associated Press, February 11, 1994. Reprinted with permission.

Page 95: Case Study 6.2, "The Effects of Meditation on Aging." Science Notes, *Noetic Sciences Review,* Summer 1993, p. 28. Reprinted with permission. **Page 101:** Case Study 6.4, "Study: Smoking May Lower Kids' IQs." Associated Press, February 11, 1994. Reprinted with permission. **Page 103:** Case Study 6.5, "Gun Ownership as a Risk Factor for Homicide in the Home." *The Washington Post,* Weekly Edition, October 17–23, 1993. Reprinted with permission.

Page 149: Figure 9.9, "Incomes of Doctors vs. Other Professionals." *The Washington Post,* January 11, 1979; from "Pay Practices of Doctors on the Examining Table" by Victor Cohn and Peter Milius; graph credited to Alice Kresse. Reprinted with permission. **Page 150:** Figure 9.10, "Rising Postal Rates." Copyright *USA Today.* Reprinted with permission.

Page 358: Table 20.1, "Study: Smoking May Lower Kids' IQs." Associated Press, February 11, 1994. Reprinted with permission.

Page 439: Case Study 25.1, Article on Cranberry Juice and Bladder Infections. Appearing in *The Davis Enter-prise,* March 9, 1994, p. A9. Reprinted with permission. **Page 440:** Case Study 25.2, "Study: Children Who Move Often Are Problem Prone." *West Hawaii Today,* September 15, 1993. Reprinted with permission. **Page 441:** Case Study 25.3, "A Hearty Endorsement for Aspirin." Reprinted with permission from *Science,* Vol. 263, January 7, 1994, p. 24. Copyright 1994 American Association for the Advancement of Science. **Page 443:** Case Study 25.4, "White Collar Commute." *The Davis Enterprise,* March 11, 1994, p. C5. Reprinted with permission. **Page 443:** Case Study 25.5, "Teen Alcohol-Sex Survey Finds the Unexpected." *The Sacramento Bee/Cox News Service,* May 11, 1994. Reprinted with permission. **Page 444:** Case Study 25.6, S. Plous, "Psychological Mechanisms in the Human Use of Animals." *Journal of Social Issues,* Vol. 49, No. 1, pp. 11–52. Reprinted with permission. **Page 445:** Case Study 25.7, "Aerobics OK for Breast-Feeding Moms." *The Davis Enterprise,* February 17, 1994, p. A7. Reprinted with permission. **Page 446:** Case Study 25.8, "Beta-Carotene: Helpful or harmful?" Reprinted with permission from *Science,* Vol. 264, April 22, 1994, pp. 500–501. Copyright 1994 American Association for the Advancement of Science. **Page 448:** Case Study 25.9, "Cities of Discord: Home May Be Where the Heart Disease Is." Copyright 1994, *USA Today.* Reprinted with permission. **Page 448:** Excerpt, 'Philly Hostile? "Who Asked Ya?"' Copyright 1994, *USA Today.* Reprinted with permission. **Page 450:** Case Study 25.10, "Hot Dog Hazards." Reprinted with permission from *Science,* Vol. 264, May 27, 1994, p. 255. Copyright 1994 American Association for the Advancement of Science.